D1474769

ENVIRONMENTAL ACCOUNTING

ENVIRONMENTAL ACCOUNTING

EMERGY and Environmental Decision Making

HOWARD T. ODUM
Center for Environmental Policy
Environmental Engineering Sciences
University of Florida, Gainesville

JOHN WILEY & SONS, INC.
New York / Chichester / Brisbane / Toronto / Singapore

This text is printed on acid-free paper.

This publication is designed to provide accurate and authoritative information in regard to the subject matter covered. It is sold with the understanding that the publisher is not engaged in rendering legal, accounting, or other professional services. If legal advice or other expert assistance is required, the services of a competent professional person should be sought.

Library of Congress Cataloging in Publication Data:

Odum, Howard T., 1924–
Environmental accounting : EMERGY and environmental decision making / Howard T. Odum.
p. cm.
Includes bibliographical references (p.).
ISBN0-471-11442-1 (acid-free paper)
1. Natural resources—Accounting. 2. Environmental impact analysis. I. Title.
HF5686.N3038 1996
657′.73–dc20 95-11683

Printed in the United States of America

10 9 8 7 6 5 4 3 2 1

CONTENTS

PREFACE

Starting with studies of energy in ecosystems in the 1950s, ways were found to evaluate the work of systems of many scales. In the 1960s it became apparent that environmental and economic systems could be evaluated on a common basis. Preliminary efforts were published in *Environment, Power, and Society* in 1971 and *Energy Basis for Man and Nature* in 1976. Then, with the definition of transformity and EMERGY, it became easier to make calculations and explain that these concepts measure real wealth. For a decade we have used longer versions of this book as a text for a course in energy analysis, and the methods have been widely applied.

This short book introduces EMERGY accounting for evaluation of environmental and economic use. EMERGY, a measure of real wealth, is the work previously required to generate a product or service. For over a century theorists have sought ways of relating resource limitations to economic-environmental systems, often using energy as a common metric. These had limited success because different kinds of available energy are not equivalent. Now commodities, services, and environmental work of different types are put on a common basis as EMERGY. Transformity, the EMERGY per unit energy, identifies the scale of energy phenomena. Tables of transformity facilitate EMERGY evaluations.

Expressing EMERGY in emdollars (Em$) indicates the part of the gross economic product based on that real wealth. The emdollar value of something helps people visualize its public policy importance. Calculating the EMERGY of storages and processes provides a new scale for evaluating environment, resources, human service, information, and alternatives for development. This book recommends achieving public and planetary welfare by maximizing empower, the rate of production and use of EMERGY.

The energy systems method starts with a network diagram to identify sources and pathways to be evaluated. Macroscopic minimodels to represent the systems of humanity and nature were summarized in 1983 with the book *Systems Ecology,* reprinted in 1994 by the University Press of Colorado as *Ecological and General Systems.* The appropriate systems diagram includes the issue under discussion and the causal factors from the larger surroundings. EMERGY values are written on the flows and storages of energy systems diagrams for quantitative overview. EMERGY indices are calculated and used to select among environmental alternatives. Computer simulation of simple overview models can provide a time-and-sensitivity perspective.

If EMERGY evaluation can show which environmental management maximizes economic vitality with less trial and error, society may improve efficiencies, innovate with fewer failures, and adapt to change more rapidly. By applying EMERGY concepts to nature and society, this book suggests a better way to arrive at public policies on resources and environment.

Chapter 1 introduces the window of systems overview and its use to evaluate EMERGY. Chapter 2 contains the scientific basis of the EMERGY value concept in the natural energy hierarchy of the universe. Chapter 3 estimates the EMERGY budget of the earth. Chapter 4 relates EMERGY and money. Chapter 5 summarizes the procedure for making an EMERGY evaluation table. Then several chapters show how to use EMERGY to evaluate environments, minerals, waters, primary energy sources, economic developments, nations, and international trade. Chapter 13 on the time dimension considers EMERGY in oscillations according to scales of size and time. Chapter 14 contains comparisons with other approaches and responses to criticisms. Although there is not room for many results in this book, the last chapter suggests areas for fruitful applications to policy.

HOWARD T. ODUM

Gainesville, Florida
September 1995

ACKNOWLEDGMENTS

This synthesis was made possible by the open academic policies of my universities, the excitement of students, the dedication of associates, and research projects of government agencies and foundations. Mark T. Brown has been for two decades my esteemed partner in energy systems projects, coteaching, and graduate instruction. I also acknowledge collaboration with and manuscript critique by scholars elsewhere: John Alexander, Suzanne Bayley, Ben Fusaro, Shu-Li Huang, Bengt-Owe Jansson, Ann-Mari Jansson, George Knox, Shengfang Lan, M. J. Lavine Suk Mo Lee, Clay Montague, Per O. Nilsson, Gonzague Pillet, David Scienceman, Sergio Ulgiati, F. C. Wang, James Zucchetto, and many others. Development of EMERGY concepts and application in recent years was made possible by interest and support of the Cousteau Society, Captain Jacques Cousteau, Jean Michelle Cousteau, and Richard Murphy. An earlier draft (*EMERGY and Policy*) was written as Visiting Professor at the Lyndon Baines Johnson School of Public Affairs of the University of Texas. The quest to extend EMERGY value to education and the public is shared with Elisabeth C. Odum, Niki Meith, and Robert King. Joan Breeze was my editorial assistant.

ENVIRONMENTAL ACCOUNTING

CHAPTER 1

INTRODUCTION: EMERGY AND REAL WEALTH

The earth's environment provides necessary life support for society and its economy, fertile soils, clean waters, clear air, good climate, healthy ecological systems, and aesthetic surroundings. Based on reserves of these natural resources, the frenzied growth of free capitalism has been one of the wonders of the world. But now, after two centuries of expansion, the economic development of the diminishing resources of the earth has reached a new stage. Enterprises for private profit are consuming the environmental systems that are the basis of public welfare. The conflicts between those intent on protecting the environment and those intent on further economic development are becoming increasingly important in public policy discussions and political elections.

Whereas environmental issues are now characterized by adversarial decision making, rancor, and confusion, these conflicts may not be necessary in the future. A science-based evaluation system is now available to represent both the environmental values and the economic values with a common measure. EMERGY, spelled with an "m," measures both the work of nature and that of humans in generating products and services. By selecting choices that maximize EMERGY production and use, policies and judgments can favor those environmental alternatives that maximize real wealth, the whole economy, and the public benefit. In this book, environmental accounting with EMERGY is introduced with its theoretical basis, calculation procedures, and examples of its application.

ENERGY AND EMERGY

A real wealth product, such as a forest log, contains available energy (potential energy which can do work). By burning the wood, that energy can be released and degraded into heat. The available energy can thus be measured in calories. The calorie was originally defined as the heat necessary to raise one gram of water one degree centigrade (one degree Celsius, or 1°C). One kilocalorie (kilogram calorie, abbreviated kcal) is equal to 1000 calories. International commissions recommended that the joule (rather than the calorie) be used as the unit of energy. There are 4186 joules in a kilocalorie. Whatever the units used, the forest log contains potential energy.

In contrast, EMERGY is a measure of the available energy that has already been used up (degraded during transformations) to make the log. EMERGY is a record of previously used-up available energy that is a property of the smaller amount of available energy in the transformed product (the log). Sometimes we refer to EMERGY as "energy memory." See Scienceman (1987) for discussion of the historical roots of the prefix "em" and for linguistic background for the word EMERGY.

SCALES AND THE ENVIRONMENTAL WINDOW

It has been customary and enlightening in many fields to show graphs that correlate size with time. As in Figure 1.1a, phenomena on the lower left, which occupy small areas, turn over and are replaced rapidly, whereas those to the upper right have a larger territory and are replaced more slowly. As shown, the activities of nature and humanity operate on many scales, from the molecules to the stars. In the middle of the range is the environment, which includes the interface between the ecological systems and the human economic society.

Because humans operate at the environmental scale, they see so much detail that they sometimes think that this realm has no organization or principles. Many who concentrate on the science of small, fast things think of the larger scale of environment as whatever results from a disorganized struggle of organisms for existence. Their environmental view is one without many principles, an anarchistic viewpoint from the small scale. Many who study the larger scale of human economy and choice also see the environment as being without principle other than the human choice and what people are willing to pay money for. Their viewpoint is an anarchistic view from the larger scale.

Yet, when we step back and view what is important to the scale of time and space of the environment, we find some of the same patterns of organization and process that are found in the small-scale sciences of chemistry and biology and in the larger-scale sciences of the earth and the universe. Patterns and processes on any scale can be represented by diagrams that have principal

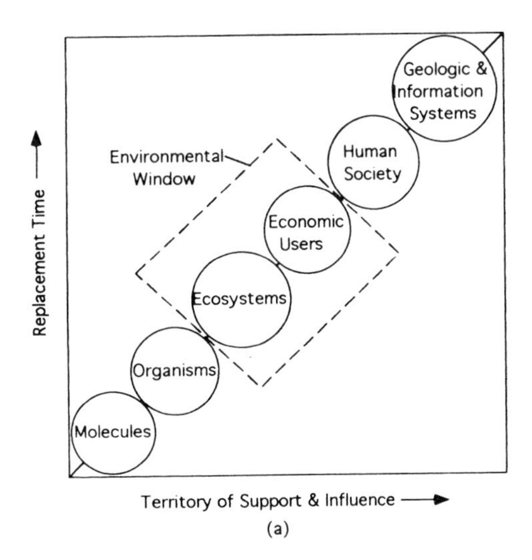

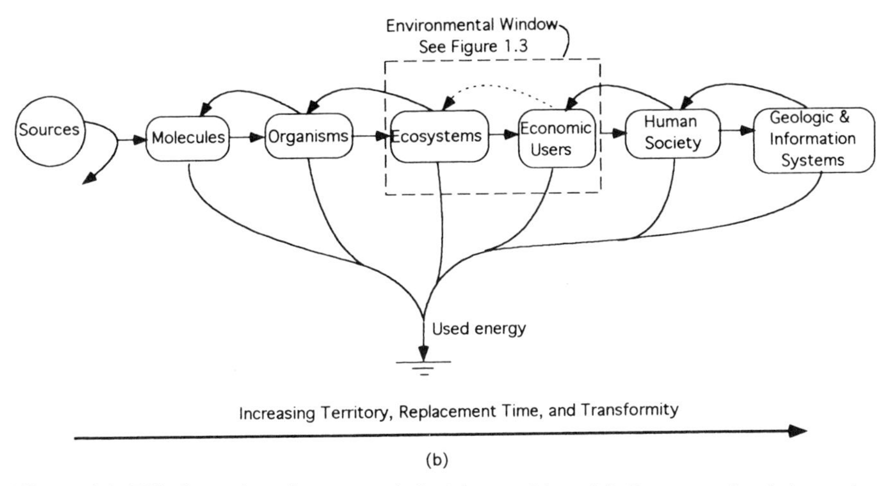

Figure 1.1. Window of environmental decision making. **(a)** On a graph of the scales of territory and time, **(b)** On a comparable energy systems diagram.

parts and the connections by which they interact with each other. These pictures of structure and function are *systems diagrams.*

Possibly everything in the universe is connected to everything else directly or indirectly. To diagram the entire realm would probably not be possible, even if simplified to items of major importance. But we can concentrate our attention on a "window" in the scale of time and space. The small environmental window in Figure 1.1 is appropriate for many questions of resource use and microeconomics. A somewhat large window is necessary for the larger scale that includes wildlife, chemical cycles, human society, macroeconomics, and networks of shared information.

In universities there are many fields of study that deal with parts of the environmental realm such as geology, oceanography, meteorology, forestry, fishery science, agricultural science, ecology, biogeochemistry, economics, sociology, political science, and engineering. Yet there is just one interconnected system, which has not been studied enough as a unified whole. Evaluating alternatives and designs of the environmental realm requires a systems overview. What is known about the parts and processes from all the piecemeal fields can be combined in a systems synthesis with simplified models, at first verbally, and then diagrammed to help the mind see parts and wholes together. The diagrams are a necessary step to evaluate EMERGY.

ENERGY SYSTEMS DIAGRAMS

Diagramming is done with energy systems symbols (Figure 1.2). Diagrams of our world are called energy systems because everything has some energy. Pathways may indicate causal interactions, show material cycles, or carry information, but always with some energy. The systems diagram also defines the equations that are used for systems simulation (see example in Table 13.1). The symbol language, including the mathematical relationships, has been given in great detail previously (H. T. Odum, 1983b, 1994a).

Figure 1.3 uses the symbols to show a highly aggregated systems view of the main components of an environmental window. Shown are ecological components, economic users, and circulation of money to people (final-demand consumers) outside of the window to the right. Energy systems diagrams, like the graphs of territory and time, have items in order of size and turnover time from left to right. Whether parts are aggregated or disaggregated, energy systems diagrams should have all the known inputs and outputs of the system crossing the selected window.

Energy Laws

The pathways of the energy systems diagram of the network within the window trace the energy flows coming in from outside sources through parts, and interactions leaving the window either as an energy output to the surrounding system or as used energy (degraded energy). The *first energy law* states that energy entering a system is neither created nor destroyed. All inflow energy is either stored in one of the tanks inside or flows out through pathways to the outside. Sometimes it is useful to write numbers of kilocalories on the pathways indicating the rates of energy flow and numbers in the storages to indicate the kilocalories stored here. Although energy is conserved (accounted for without change in quantity) in passing through a network, the forms of energy are very different, and kilocalories of different kinds of energy are not

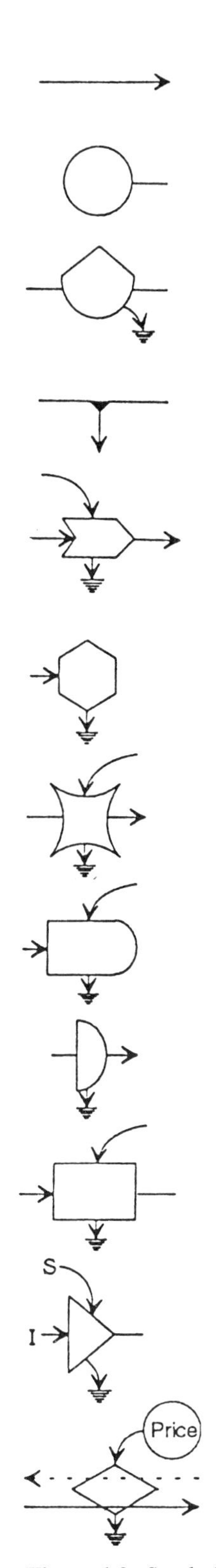

Energy circuit: A pathway whose flow is proportional to the quantity in the storage or source upstream.

Source: Outside source of energy delivering forces according to a program controlled from outside; a forcing function.

Tank: A compartment of energy storage within the system storing a quantity as the balance of inflows and outflows; a state variable.

Heat sink: Dispersion of potential energy into heat that accompanies all real transformation processes and storages; loss of potential energy from further use by the system.

Interaction: Interactive intersection of two pathways coupled to produce an outflow in proportion to a function of both; control action of one flow on another; limiting factor action; work gate.

Consumer: Unit that transforms energy quality, stores it, and feeds it back autocatalytically to improve inflow.

Switching action: A symbol that indicates one or more switching actions.

Producer: Unit that collects and transforms low-quality energy under control interactions of high-quality flows.

Self-limiting energy receiver: A unit that has a self-limiting output when input drives are high because there is a limiting constant quality of material reacting on a circular pathway within.

Box: Miscellaneous symbol to use for whatever unit or function is labeled.

Constant-gain amplifier: A unit that delivers an output in proportion to the input I but is changed by a constant factor as long as the energy source S is sufficient.

Transaction: A unit that indicates a sale of goods or services (solid line) in exchange for payment of money (dashed line). Price is shown as an external source.

Figure 1.2. Symbols of the energy systems language (H. T. Odum, 1971a,b, 1983b).

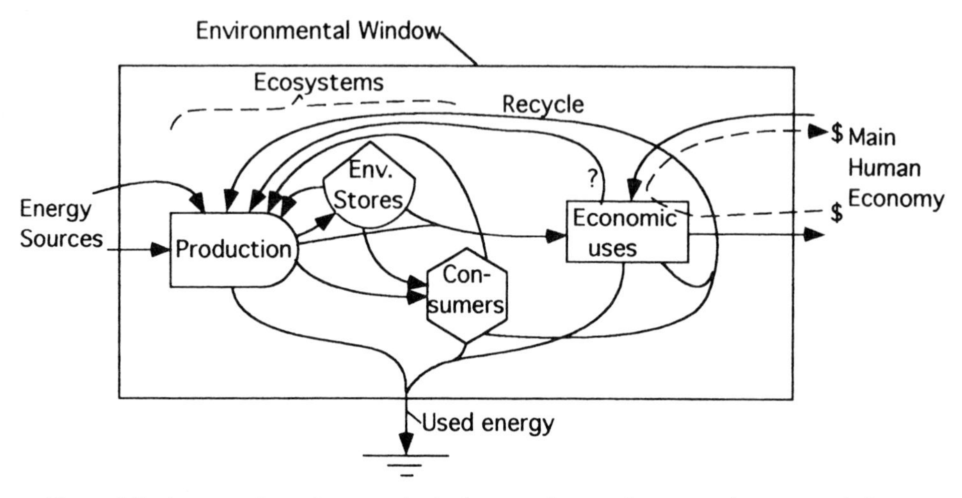

Figure 1.3. An overview of some principal parts of an environmental systems window.

equivalent in ability to do work. Used energy passing through the heat sink cannot do any work.

Potential energy capable of doing work comes in from the sides, and leaves through the bottom pathway to the "heat sink" symbol, representing the dispersal of used energy, no longer capable to doing work. Almost all processes are accompanied by such energy dispersal. The principle of universal depreciation is often called the *second energy law.* Everything that is recognizable in our biosphere has a natural tendency to depreciate and be dispersed. The things that depreciate are the real wealth of our lives. They require continual repair and replacement. As energy loses its concentration and ability to do work, it leaves the system in degraded form.

STORED WEALTH AND ITS EMERGY

Examples of real wealth are the products of work, such as clothes, books, food, minerals, fuels, information, art, technology, species, electricity, biodiversity, and so on. Everything which we regard as being of real value has to be produced and maintained by work processes from the environment, sometimes helped by people and sometimes not. Figure 1.4 shows the general pattern of environmental production of real wealth. Included is a feedback pathway (from right to left) representing the way the storage assists the production process.

The environmental production system shown in Figure 1.4 uses inflows of potential energy to make and store real wealth. Energy sources are indicated with a circular symbol. Energies are crossing into the system's window from the left. When energy is transformed into something new, work has been

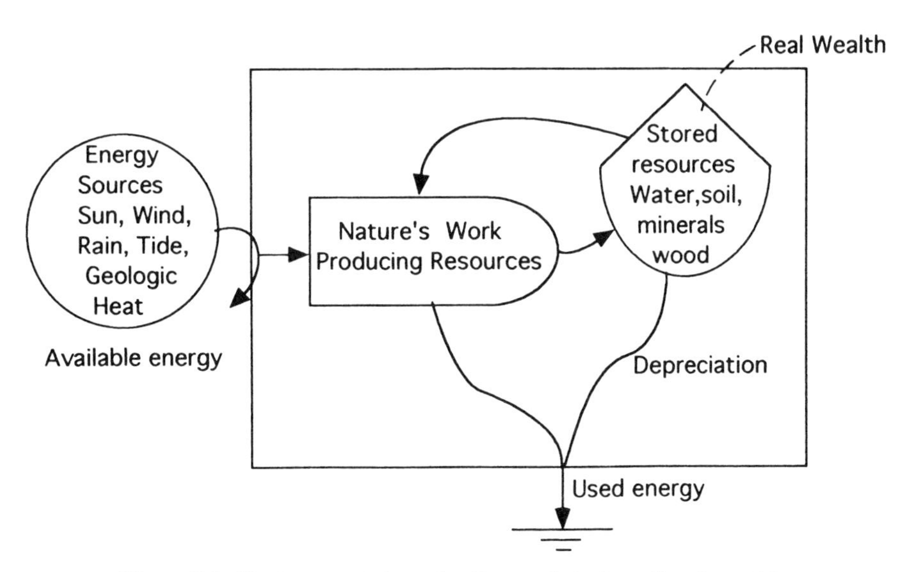

Figure 1.4. Environmental production and storing of real wealth.

accomplished, and the process is called production. In Figures 1.3 and 1.4, production is shown occurring within the "producer" symbol. The products are shown going into storage. The "storage" symbol resembles water tanks that were once common over the landscape in midcentury. Energies from sources were used in energy transformation processes to produce the quantities stored in the tank. However, most of the energy with potential to do work (called available energy) loses its ability to do work in the process of making new products. The new products have a new form of energy but in smaller quantity.

To build and maintain the storage of available resources, environmental work has to be done, requiring energy use and transformation. We can quantitatively evaluate the storage by the work done in its formation. Work of energy transformation can be measured by the availability of the energy that is used up. Thus, real wealth can be measured by the work previously done. EMERGY is a scientific measure of real wealth in terms of the energy required to do the work of production.

EMERGY is the available energy of one kind of previously used up directly and indirectly to make a service or product. Its unit is the emjoule (H. T. Odum, 1986a, 1988; Scienceman, 1987).

However, energy of one kind is not equivalent in its ability to do work to energy of another kind. Therefore, in adding up the available energies of different kinds contributing to a production process, all have to be expressed in units of one kind of energy. In this book, for convenience, all forms of energy contributing are expressed in units of solar energy that would be required to generate all the inputs. Thus, wealth is measured by the solar EMERGY required to accumulate it. Chapter 2 has much more on kinds of energy and their relation to EMERGY.

Solar EMERGY is the available solar energy used up directly and indirectly to make a service or product. Its unit is the solar emjoule (abbreviated sej).

EMERGY estimates the magnitude of work involved in the production. For example (Figure 1.5), the sum of environmental contributions to produce 7.8×10^{10} J of spruce forest wood in Sweden is a solar EMERGY flow of $30{,}000 \times 10^{10}$ sej/yr (Doherty et al., 1995).

Because EMERGY measures what comes into a systems window, it is a property of the larger network surrounding the window. It cannot be evaluated without some knowledge of the larger surroundings of the geobiosphere. Often it is desirable to use two systems windows to evaluate something, one for the local evaluation and a larger one to understand how the smaller window is being affected by the surroundings.

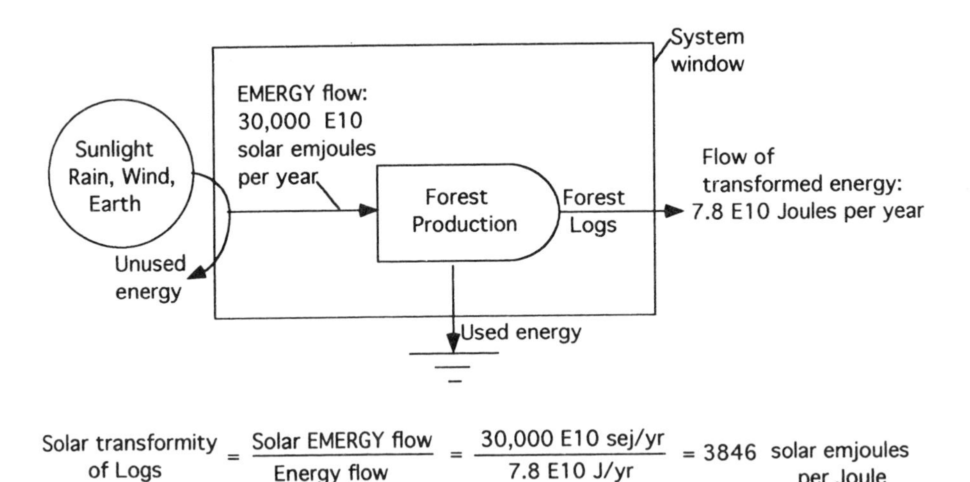

$$\text{Solar transformity of Logs} = \frac{\text{Solar EMERGY flow}}{\text{Energy flow}} = \frac{30{,}000 \text{ E10 sej/yr}}{7.8 \text{ E10 J/yr}} = 3846 \text{ solar emjoules per Joule}$$

Figure 1.5. Average wood production of one hectare (1 ha) of Swedish spruce forest (Doherty et al., 1995). (In this and later figures, large numbers or very small numbers are expressed in the E notation of computers where the number after E is the number of decimals. For example, 7.8 E5 is the abbreviation for 780,000. This is also equivalent to the scientific notation 7.8×10^5. An example of a number less than one is 2.0 E −3 which is equivalent to 0.002 and 2.0×10^{-3} in scientific notation).

EMERGY IN A STORAGE DURING ITS GROWTH, STEADY STATE, OR DECLINE

With the help of Figure 1.6, we can consider how energy and EMERGY are processed in a storing process. An example is the storage of standing forest logs as the trees grow. In this figure we consider the simplest case, when there is no outflow from storage to other users. Figure 1.6a shows the flow of energy going into storage, while at the same time some of the stored energy is being

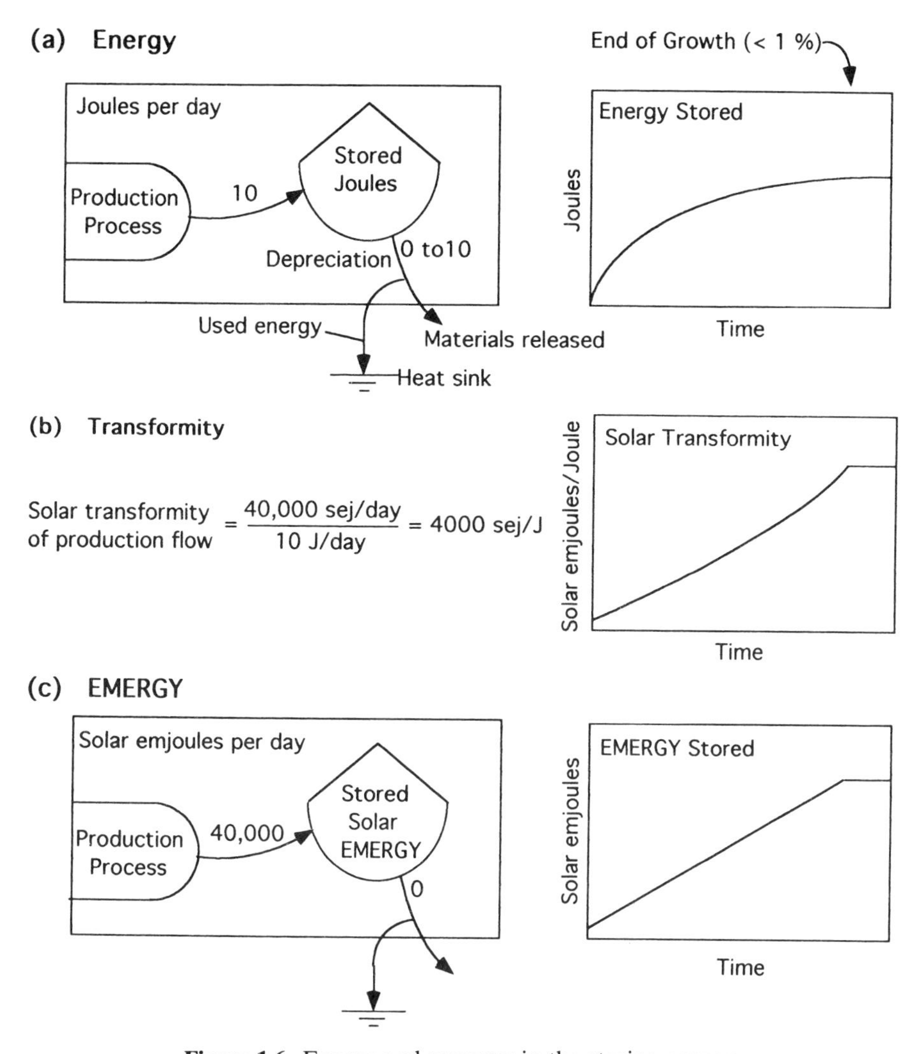

Figure 1.6. Energy and EMERGY in the storing process.

dispersed, degraded, and lost from the storage through the heat-sink pathway. The larger the storage becomes, the greater is the proportionate depreciation outflow. Eventually, as the graph shows, the energy inflow balances the sink outflow and the storage becomes constant. When inflows and outflows balance, a system is said to be in a *steady state.* When the system in Figure 1.6a reaches a steady state (storage constant), depreciation is 10 J/day, equal to the inflow of energy.

Shown in Figure 1.6c is the solar EMERGY flow. The storage receives and accumulates the solar EMERGY required to develop the wood storage. The solar EMERGY stored is that required to make the storage, in spite of the depreciation going on. Degraded energy going down the heat sink pathway is not available to do work, and thus has no EMERGY. The EMERGY accumulates as long as energy is increasing.

As soon as there is no further increase in storage, by definition EMERGY storing stops. Let us restate this important property of EMERGY:

When an energy storage is constant, its stored EMERGY is constant.

Next, consider the case of a declining storage, for example, a declining stand of forest trees. Using Figure 1.6 again, suppose the production process is stopped. Then the storage decreases as its depreciation processes continue (not shown). The energy storage decreases, and with it the stored EMERGY is lost. In the example, when half of the energy is gone, half of the EMERGY is gone too.

Next, still using Figure 1.6, suppose that the storage was removed from the window. For example, suppose logging harvested half the trees. Now half of the energy has been removed and with it half of the EMERGY. Thus, we can restate what happens with declining storage (assuming that the nature of the remaining storage is unchanged):

When an energy storage declines, its stored EMERGY declines in proportion.

THE TRANSFORMITY

The quotient of a product's EMERGY divided by its energy is defined as its *transformity* (H. T. Odum. 1976b, 1988). The units of the transformity are emjoules per joule. (Thus it is not a dimensionless ratio.) In this book where we are using only solar EMERGY, the units are solar emjoules per joule (abbreviated sej/J). For a Swedish spruce forest log (Figure 1.5) there are about 3846 sej/J.

Solar transformity is the solar EMERGY required to make one joule of a service or product. Its units are solar emjoule per joule (sej/J). A product's solar transformity is its solar EMERGY divided by its energy.

The more energy transformations there are contributing to a product, the higher is the transformity. This is because at each transformation, available energy is used up to produce a smaller amount of energy of another form. Thus, the EMERGY increases but the energy decreases, and therefore the EMERGY per unit energy increases sharply. The forest log has a larger transformity than the sun and wind that contributed to produce it.

Referring again to the forest production system in Figure 1.5, the energy in the outflow of one hectare (1 ha) of forest production of spruce logs is 7.8×10^{10} J/yr. The combined solar EMERGY contributed by natural processes in making the logs is $30{,}000 \times 10^{10}$ sej/yr. The solar transformity is the solar EMERGY flow divided by the energy flow ($30{,}000 \times 10^{10}$ sej/yr divided by 7.8×10^{10} J/yr), which is 3846 sej/yr.

Goods and services that have required the most work to make and have the least energy have the largest transformities. Examples are human services and information (the energy of information is that of the information carrier—paper, computer disk, human brain, and so forth). As we will discuss later, most energy transformations are controlled by inputs of high transformity, whose energy contribution is small but whose EMERGY contribution may be large. An example is the control of a forest by people, a pathway not shown in Figure 1.5 but included in later chapters.

During the growth of a storage (Figure 1.6a), some of the input energy is being lost from the storage *M* a necessary consequence of the second energy law. The EMERGY required to develop the growth is accumulating in the storage. Hence, the transformity of the storage (EMERGY per unit energy) increases as a storage grows. The solar transformity of the stored energy in Figure 1.6b is higher than that of the inflow from the production process.

EQUATIONS AND EMERGY SIMULATION

Definitions of EMERGY and transformity and their relation to storage may be given in equation form (Figure 1.7a). These represent the relationships already described in the preceding text. Figure 1.7 represents the systems already given in Figure 1.6 except that an outside use pathway has been included ($k_2 * Q$).

These differential equations for energy and EMERGY storage may also be adapted in a computer program EMTANK (Appendix F) to simulate growth. The graphs in Figure 1.7b are a typical result.

EMPOWER

As generally used in science and engineering, power is the flow of energy per unit time. For example, the power output of the forest wood production in Figure 1.4 is 7.8×10^{10} J/ha/pyr. Usually what is intended with the word "power" is the available energy (useful energy) flowing per unit time, although

$$\text{Energy } Q: \quad dQ/dt = J - k1*Q - k2*Q$$

$$\text{EMERGY } E: \quad \text{IF } dQ > 0 \text{ then } dE/dt = Tr_j*J - Trq*k_2*Q$$

$$\text{IF } dQ = 0 \text{ then } dE/dt = 0$$

$$\text{If } dQ < 0 \text{ then } dE/dt = Tr_q*dQ/dt$$

$$\text{Where Transformity of } J = Tr_j \text{ and}$$

$$\text{Transformity of } Q = Tr_q = E/Q$$

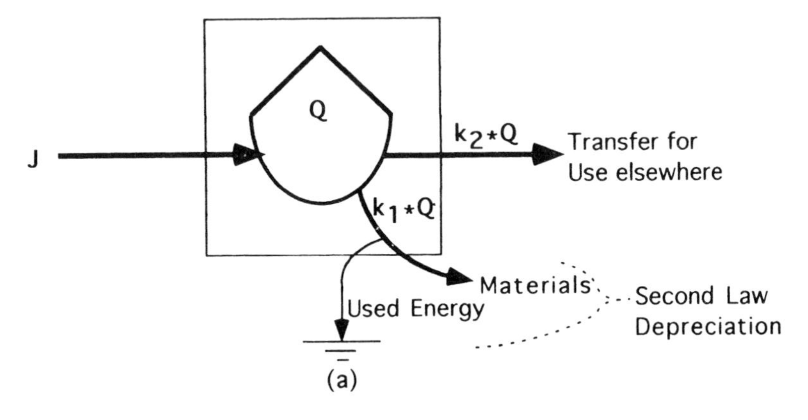

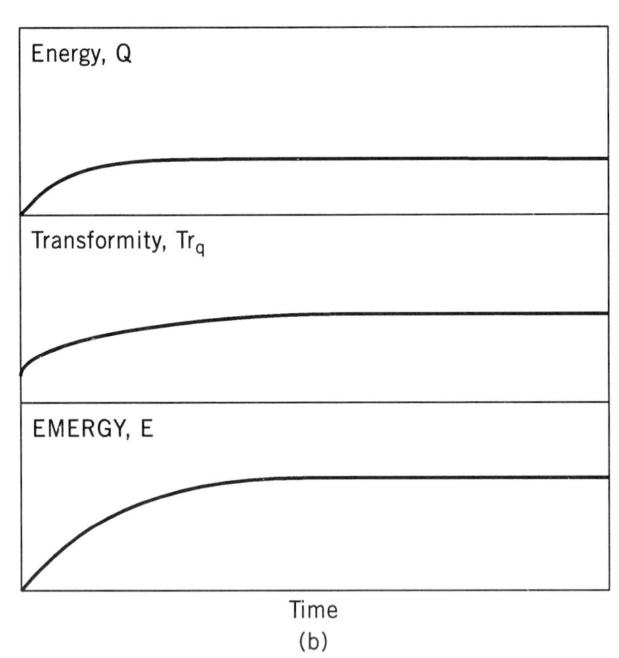

Figure 1.7. Equations for defining and simulating EMERGY production and storage. **(a)** Differential equations and energy diagram. **(b)** Results of microcomputer simulation with a BASIC language program EMTANK (Table F.1).

what is actually measured may be the outflow of degraded energy per time (used energy).

A flow of EMERGY is measured in emjoules per time. Another word defined for this is "empower." For example, the empower output of the Swedish forest production in Figure 1.5 is $30{,}000 \times 10^{10}$ sej/ha/yr. As later chapters will indicate, empower may be a measure that is closer to the general English meaning of "power" than it is to the scientific usage of "power."

EMERGY AND SOCIETY'S ORIGINAL CONCEPT OF ENERGY

A commonplace idea among people everywhere is that some things take more effort than others. Long before physical science made a narrow quantitative definition of energy, the word "energy" was used to refer to the work done. For example, it was said that a house took more energy to build than a chair. The universal idea among common folk was, and still is, that putting more energy into something generates more value. The concept EMERGY is scientifically defined to give a quantitative measure to this ancient principle.

Because of the way it was defined, the scientific use of "energy" does not coincide with the folk concept. The practical, operational scientific measure

TABLE 1.1. Summary of Definitions*

Available energy: Potential energy capable of doing work and being degraded in the process (exergy) (units: kilocalories, joules, etc.).
Useful energy: Available energy used to increase system production and efficiency.
Power: Useful energy flow per unit time.
EMERGY: Available energy of one kind previously required directly and indirectly to make a product or service (units: emjoules).
Empower: EMERGY flow per unit time (units: emjoules per unit time).
Transformity: EMERGY per unit available energy (units: emjoule per joule).
Solar EMERGY: Solar energy required directly and indirectly to make a product or service (units: solar emjoules).
Solar empower: Solar EMERGY flow per unit time (units: solar emjoules per unit time).
Solar transformity: Solar EMERGY per unit available energy (units: solar emjoules per joule).

* Another commonly used unit of energy is the kilocalorie (4186 Joules). If the kilocalorie, abbreviated kcal, is used, then the following are the appropriate units:

Power: kcal/time
EMERGY: emkcal.
Empower: emkcal/time.
Transformity: emkcal/kcal (numerically equal to emjoule/Joule).
Solar EMERGY: semkcal.
Solar Empower: semkcal/time.
Solar Transformity: semkcal/kcal (numerically equal to sej/J).

of energy is the heat generated when various forms of energy are converted. The scientific concept makes no allowance for different kinds of energy representing different levels of effort. The scientific concept rates a calorie of sunlight, electricity, nuclear fission, and human thinking as equal. In other words, the different levels of prior effort involved in generating different kinds of energy are ignored. One reason students have so much trouble learning physics is because they have to unlearn their prior concepts of energy.

By using the EMERGY concept, we retain the folk concept of work but apply scientific measures. By defining EMERGY in terms of the previous use of energy, vast data available on energy are readily available to compute EMERGY in nearly everything. Data on available energy assembled for exergy analysis may be converted to EMERGY values by multiplying each item by its transformity. See Chapter 5.

EMERGY is a measure that looks back upstream to record what energy went into the train of transformation processes. The computations recognize that each type of energy has a different upstream energy input which must be included in order to summarize all the energy of one type that is required to generate the output (see Chapter 2).

SUMMARY

By drawing a systems window frame and diagramming its contents in an aggregated form, we can put our mind's eye on a part of the environmental realm so as to evaluate what goes into a product or service. Using standard data on inflows of resources, we can evaluate the EMERGY that is being produced or stored, a measure of its real wealth. The definitions of EMERGY and related quantities are given again in Table 1.1.

CHAPTER 2

EMERGY AND THE ENERGY HIERARCHY[1]

With definitions from Chapter 1, we consider next the thermodynamic basis for EMERGY and its relationship to previous uses of energy in science and engineering. Here we explain energy flow networks, use transformities to quantify an *energy hierarchy,* and show the way EMERGY accounting is based on the fundamentals of energy distribution.

To evaluate the solar EMERGY of something means calculating the energy of each kind required for another. The new accounting concepts came from recognizing that the universe is organized in a hierarchy of energy transformations quantitatively described by transformities. By recognizing work as an energy transformation, works of different kinds are compared by using transformity of those energy conversions to calculate their EMERGY contributions.

ENERGY CONCEPTS

Whereas this is not the place for a textbook introduction to energy concepts, it is necessary to show which parts of the basic energetics used in science and engineering are used here in their normal context, which are further generalized, and which are derived from recognition of the energy transformation hierarchy. Starting with the operational definition of energy as something convertible to heat, Table 2.1 gives the definitions and laws as they are generally used, but here the definitions are related to energy systems networks. Table

[1] Readers primarily concerned with *uses* of EMERGY accounting may want to skip to Chapter 3.

Table 2.1. Energy Definitions and Concepts

Heat: The collective motions of molecules, whose average intensity is the temperature which may be measured by expansion of matter in a thermometer.
Energy: May be defined as anything that can be 100% converted into heat.
First law: Energy is neither created nor destroyed in circulation and transformations in systems. That energy is conserved is a consequence of its definition above.
Available energy (exergy): Potential energy capable of doing work and being degraded in the process (units: joules, kilocalories, etc).
Work: An energy transformation process that results in a change in concentration or form of energy (see Figure 2.1).
Power: Flow of energy per unit time.
Second law (time's arrow): Concentrations (storages) of available energy are continuously degraded (depreciating). Available energy is degraded in any energy transformation process.
Third law: As the heat content approaches zero, the temperature on the Kelvin scale approaches absolute zero (−273°C), molecules are in simple crystalline states, and the entropy of the state is defined as zero.
Molecular entropy: A measure of the complexity of molecular states due to heat energy. Starting at zero degrees Kelvin (0 K), molecular entropy is the integrated sum of heat added divided by the Kelvin temperature at which it is added (units: kilocalories per degree).

2.2 states principles of energy networks that may eventually be recognized as laws of energy systems.

Figure 2.1 shows a typical energy transformation process illustrating the first and second laws. According to the second law, the overall entropy change counting that in the window and that in the environment is always an increase. Figure 2.1a shows the inputs of energy availability, the energy dispersed, and the smaller yield of a new product. Included is the low energy but important controlling input feedback that is generally present (J_F) (often omitted in introductory energetics texts). Figure 2.1a also has the various energy flows as normally labeled in open system thermodynamics. Expressions for EMERGY

Table 2.2. Network Energy Concepts in EMERGY Accounting

Time's speed regulator: Power in an energy transformation depends on the workload. Maximum output power occurs with an optimum intermediate efficiency.*
Maximum Empower principle (fourth law?): In the competition among self-organizing processes, network designs that maximize empower will prevail.
Energy transformation hierarchy (fifth law?): Energy flows of the universe are organized in an energy transformation hierarchy. The position in the energy hierarchy is measured with transformities.

* For a quantitative treatment of the optimum efficiency for maximum power, see chap. 7 in *Systems Ecology* (H. T. Odum, 1983b, 1994a).

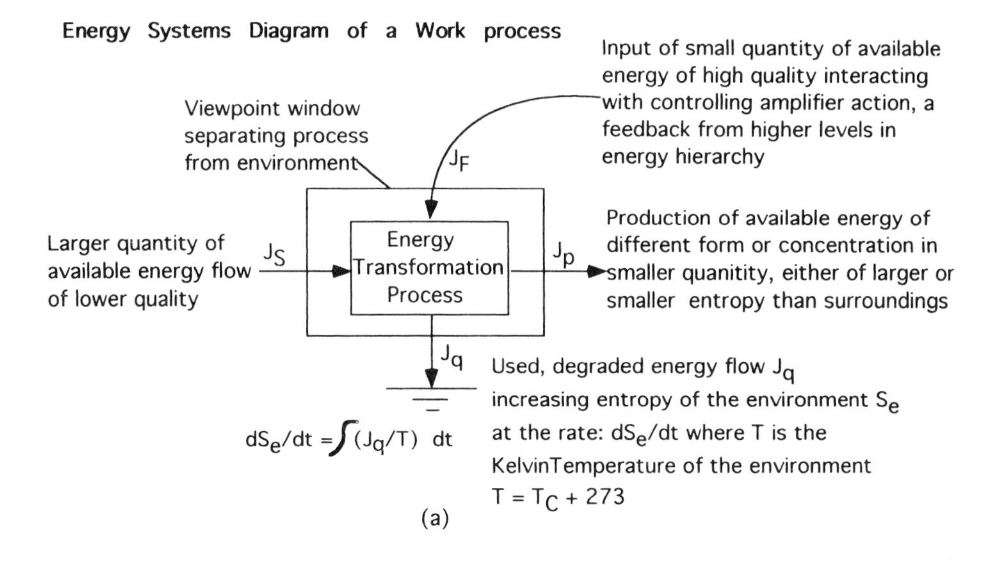

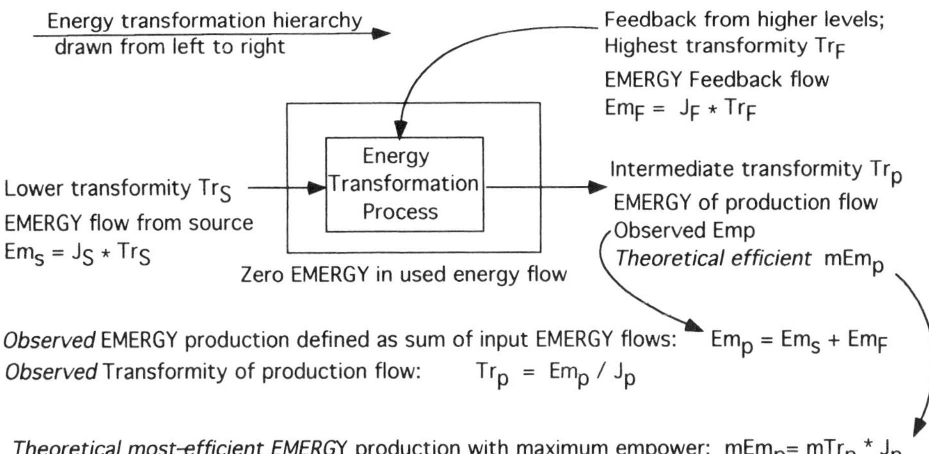

Figure 2.1. Systems diagram of an energy transformation process identifying the component energy flows. **(a)** Energetics. **(b)** EMERGY and transformities.

and transformity are given in Figure 2.1b, where we see, for example, that the EMERGY flow from a source is equal to the product of its energy flow times its transformity.

When transformations are arranged with different loadings, different combinations of speed and efficiency result. The transformity that accompanies the optimum efficiency for maximum power transfer has a theoretical lower limit that open systems may approach after a long period of self-organization. We can look for the empower transformations with the best efficiencies in systems that have been in environmental and economic competition for a long time.

Many systems, especially those newly formed, may require more EMERGY for the same output because they are not operating at maximum efficiency commensurate with maximum power. In this case, higher transformities may be observed than the best possible one. Where there are several transformities for the same product, the least may be used to measure the inefficiencies of the others.

Whereas mechanical work is often defined as the product of force times distance, a more general definition of work is required for most processes where many kinds of energy are interacting. Most energy transformations that make up the network of the biosphere and human economy involve four or more kinds of energy, as shown in Figure 2.1. Therefore, work is defined here as an energy transformation.

Using available energy to define work is correct as long as only one kind of energy is being compared. For example, using twice as much available energy to raise twice as much water against gravity means twice as much work gets done. However, judging work of two kinds of available energy by their energy contents, when those contents are in different levels of energy hierarchy is not correct, as considered next.

ENERGY TRANSFORMATION HIERARCHY

A hierarchy, such as a military organization, has many units of one kind (soldiers) that contribute to and are controlled by a unit at a higher level (corporal). Similarly, many corporals contribute to and are controlled by a unit at the next level (sergeant), and so on. In any energy transformation (such as Figure 2.1), many joules of available energy of one kind are required in a transformation process to produce a unit of energy of another kind. The energy thus generated by the work of transformation constitutes a higher level in the series of transformations. As already explained (Figure 2.1), higher levels of energy feed back to interact with, and thus control, the contributing energy flows. The output of one energy transformation contributes and converges energy to produce an even smaller output at the next higher level in an energy transformation chain. In other words, networks of energy transformations comprise an energy hierarchy (Figure 2.2).

The universe is hierarchically organized and represents a manifestation of energy. Transformity is a measure of the hierarchy of energy and is apparently applicable to all quantities of matter, energy, or information. Transformities have as many orders of magnitude as there are energy levels in the universe.

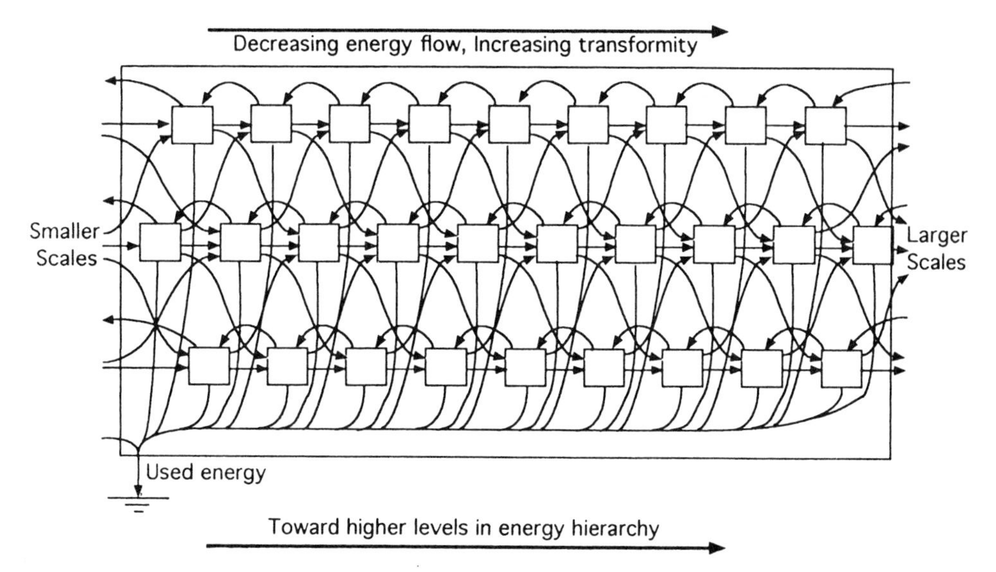

Figure 2.2. Hierarchical network of energy transformation processes that results from the loss of available energy at each transformation due to the second law.

For example, the solar transformity within the geobiosphere ranges from 1 for sunlight to 1×10^{32} sej/J for some categories of genetic information.

AUTOCATALYTIC STORAGE AND MAXIMUM EMPOWER

Energy systems networks contain energy storages autocatalytically coupled to energy transformation processes, as drawn in Figure 2.3. The term *autocatalytic* refers to the differential equation for the storage where the term for production is a product of the input source and the feedback from storage (Figure 2.3). When the "producers" (bullet-shaped) and "consumers" symbols (hexagon-shaped) are used (Figure 1.2), the autocatalytic design of Figure 2.3 is implied, unless diagrammed differently within.

As suggested by Lotka (1922a,b, 1925) as the fourth law of thermodynamics, autocatalytic feedback designs develop because they maximize power. Designs that process more useful energy, will prevail in competition with alternate designs because more available energy provides contingency needs and better adaptation to surrounding conditions. Historically there have been two ways of describing the same autocatalytic design in Figure 2.3, which reflect the two scales of view of different fields of science. The first is a corollary of the maximum power principle from Lotka in 1925 and elaborated by others later. This way of explaining design came from those thinking on a network scale.

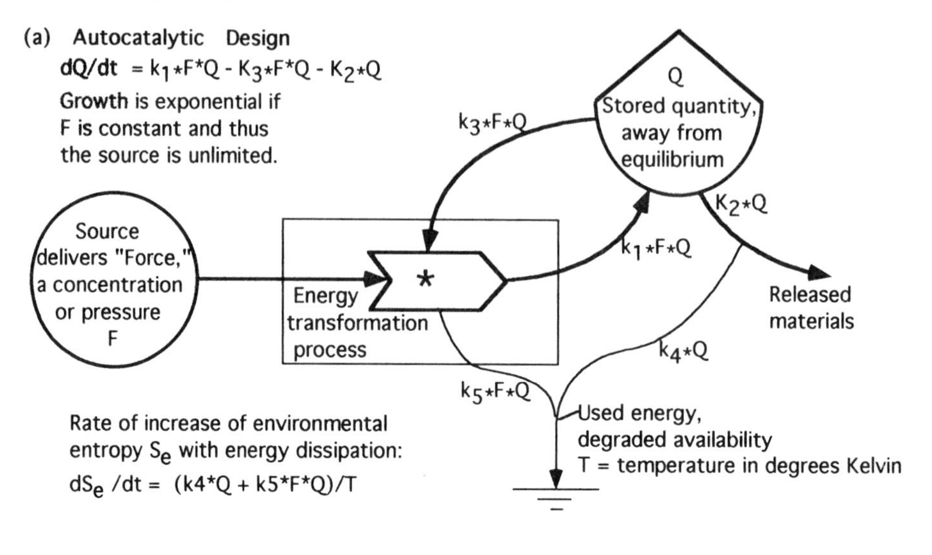

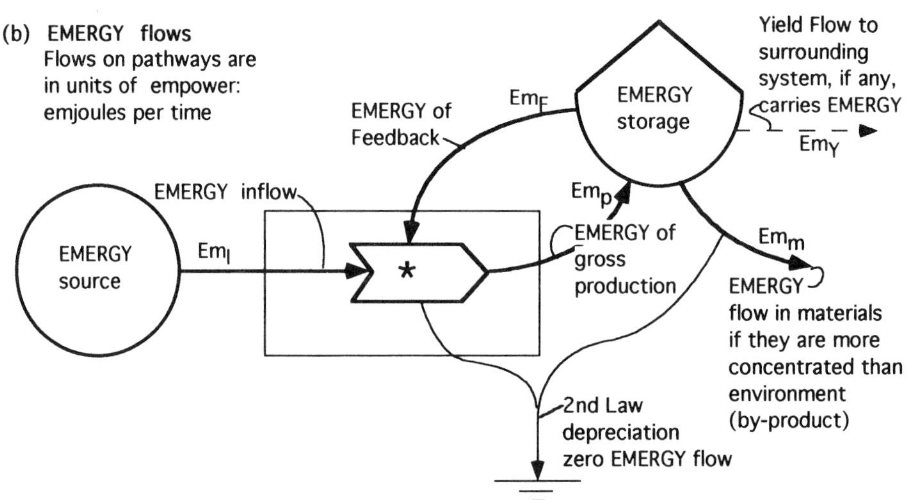

Figure 2.3. Systems diagram of the autocatalytic coupling of storage to energy transformation processes.

1. Self-organizing systems develop autocatalytic storages to maximize useful power transformations. (Maximizing power use also maximizes the rate of dissipation of available energy and the rate of entropy production.)

Much later the same result was discovered among molecular scientists but viewed in a different way without crediting the earlier systems descriptions, apparently without recognizing the similarity of concepts.

2. Self-organizing systems disperse energy faster, maximizing the rate of entropy production by developing autocatalytic dissipative structures. (Maximizing dissipation also maximizes useful power transformations.)

In version 1, which has evolutionary and deterministic biological roots, emphasis is on evolving new progress. In version 2 the emphasis is on statistical processes and random energy dispersal. Both are ways of looking at the same system (Figure 2.3).

TRANSFORMITIES IN AN AQUARIUM ECOSYSTEM

Feedback amplifiers also develop between units. Like the autocatalytic single unit, feedbacks between units reinforce productive flows and develop optimum efficiencies. Typically, ecological and economic systems have a closed-loop design connecting dispersed producers and more concentrated consumers. A simple ecosystem in a sealed glass aquarium is diagrammed in Figure 2.4, illustrating EMERGY and transformities in ecosystems. Although the aquarium is closed to matter, it absorbs solar energy and allows the used, degraded heat to diffuse out. Producing plant tissues manufacture organic matter that consuming plant tissues, animals, and microbes utilize, and in the process they regenerate raw materials (carbon dioxide, water, phosphates, nitrates, and so forth) for the plants to use again. Typical steady-state flow values are printed on the pathways. The nutrients have a small chemical potential energy relative to the concentrations in the water as they are released and recycled.

From left to right, items are concentrated, starting with the sun, which is collected by plants and further collected into the bodies of animals. From right to left items are dispersed; the raw materials released from the consumers are dispersed back into the waters available for plants to concentrate again. Georgescu-Roegen (1977) suggested that systems such as the industrial economy will run down because minerals all become dispersed. However, ecosystems show how materials can be concentrated from very low concentrations if there is abundant available energy to drive the accumulation process.

In the aquarium system (Figure 2.4) solar EMERGY flow is the same for all the pathways, but the energy flow decreases at each step. The EMERGY that entered from the sun ends as the last of the available energy that started at the source is used up. This happens as the feedbacks of recycling materials are utilized in the production processes (the intersection and interaction of nutrients with producers). Solar transformities of each pathway are calculated in the small table (in the caption) as the quotient of the input EMERGY and the energy flows. Transformities increase along the pathways, and are highest in the feedback of nutrients.

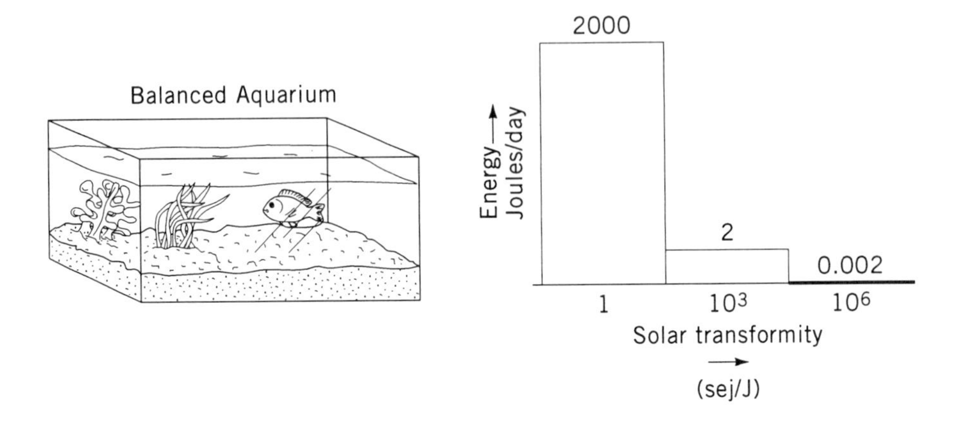

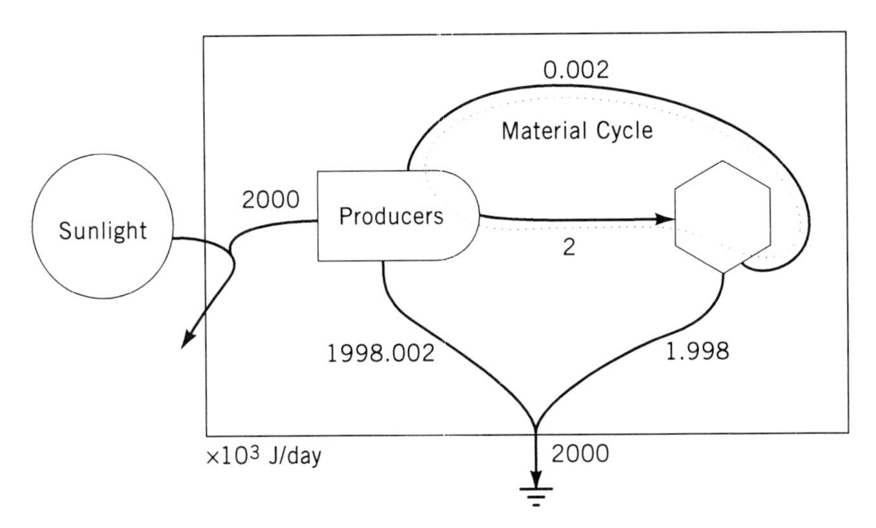

Figure 2.4. Diagram of the ecosystem in a sealed aquarium with daily flow of solar energy. The cycle and reuse of chemical materials is shown with a dotted line accompanying the energy flows. Solar EMERGY and transformities are included in the table.

Path	Energy (×10^3 J/day)	Solar EMERGY (×10^3 sej/day)	Solar Transformity (sej/j)
Sunlight used	2000	2000	1
Organic produce	2	2000	1,000
Material recycle	0.002	2000	1,000,000

ENERGY TRANSFORMATION NETWORKS

The web of connected energy transformation processes in Figure 2.2 is shown with more detail in Figure 2.5. When energy is transformed to a smaller fraction and then divided again and again, a fractal hierarchy develops. The

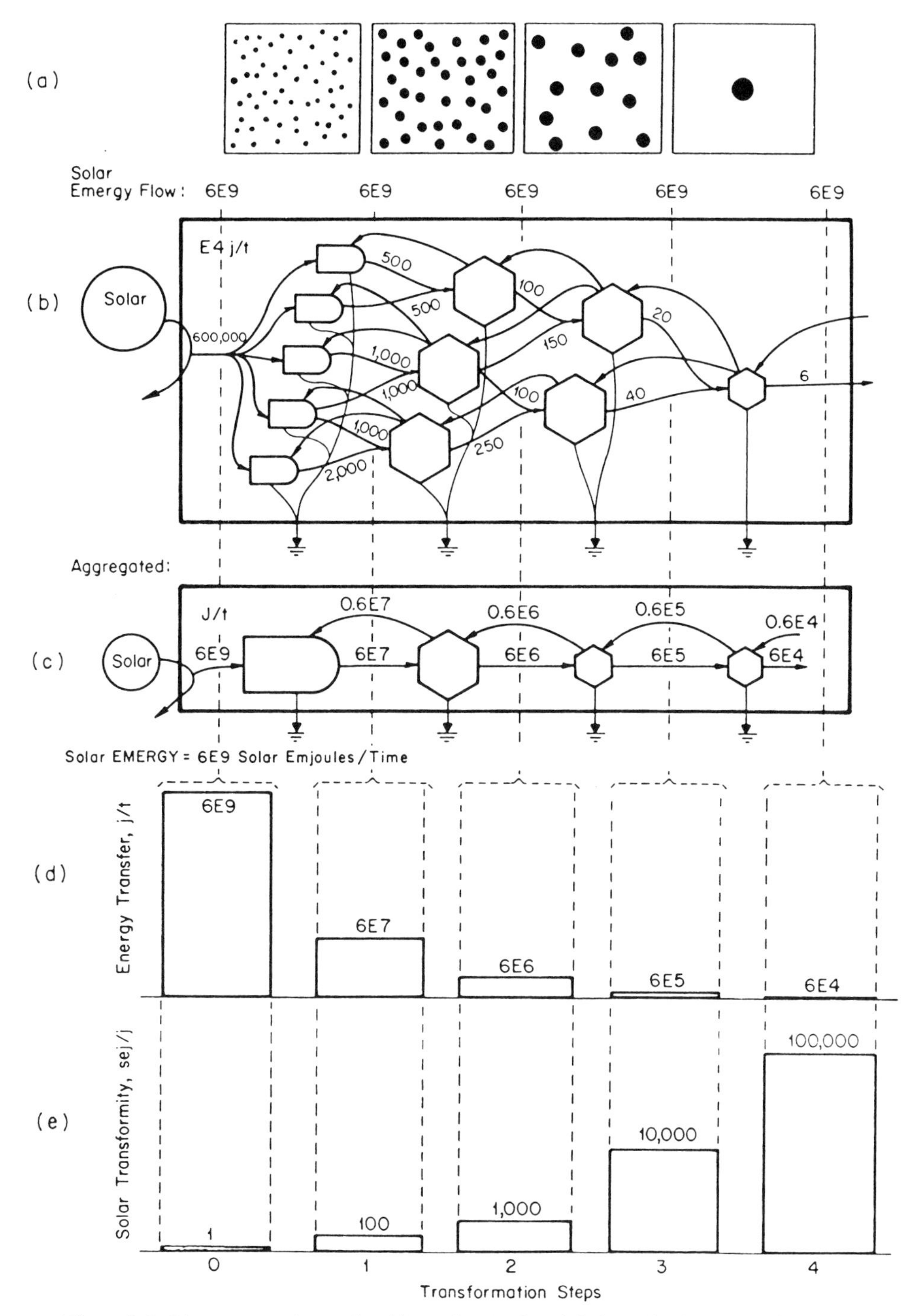

Figure 2.5. Energy transformation hierarchy. **(a)** Spatial view of units and their territories. **(b)** Energy network including transformation and feedbacks; hexagon-shaped symbol implied the contents given in Figure 2.3. **(c)** Aggregation of energy networks into an energy chain. **(d)** Bar graph of the energy flows for the levels in energy hierarchy. **(e)** Bar graph of solar transformities.

numbers on the pathways of Figure 2.5b decrease in energy flow through successive transformations within a one source network. If the components of similar transformity are aggregated, a web can be represented by a chain diagram (Figure 2.5c). If successive transformities are similar, the energy may decrease by a constant percentage with each stage in the chain. The successive decrease of energy, as it is divided by the transformation process, is illustrated by the graph of decreasing energy in Figure 2.5d. If the single source is solar insolation, the solar transformities increase by a constant percentage, as graphed in Figure 2.5e. From left to right in the chain in Figure 2.5c, the solar EMERGY is constant.

In this chain of transformations, energy decreased through successive transformations, but the source EMERGY is transmitted and the transformity increases (Figure 2.5e). Thus the transformity represents the position of flows and storages in the universal energy hierarchy. Transformations are accompanied by energy dispersal, which measures the work done in generating a smaller flow of higher transformity energy.

Because many joules of energy of one kind converge in any energy transformation, producing fewer joules of the next energy type, all the processes of the universe are apparently part of an energy hierarchy (Figure 2.2). Because energy transformation networks are apparently universal in all systems, hierarchical energy structure is also universal. Is this a fifth energy law?

SPATIAL DIMENSIONS OF ENERGY HIERARCHY

Another property of energy self-organization is the spatial clustering of structure and processes that include storage and energy transformation into a single unit. Examples are storms, organisms, stars, and industries. As suggested by Figure 2.5a, units at higher transformity levels have larger centers and operate over larger territories.

In order to explain the energy hierarchy concepts simply, Figure 2.5 was drawn with only one source, whereas most real situations have more than one. The fishery food chain in a Florida estuary is partly evaluated in Figure 2.6.

In the hierarchy of the universe, from the very small to the very large, there are storages at each level. Small storages affect small areas and turn over rapidly, whereas larger storages affect larger areas and have longer replacement times, lower depreciation rates, slower turnover times, larger sizes, and larger territories of support and influence that go with higher positions in hierarchies. On the energy systems diagrams in this text, items with small territory, fast turnover time, and low transformity are placed on the left. On the right, items have slow turnover time, large territory, and high transformity. A famous film, *The Powers of Ten,* shows the way each level is nested in the next (Morrison and Morrison, 1983).

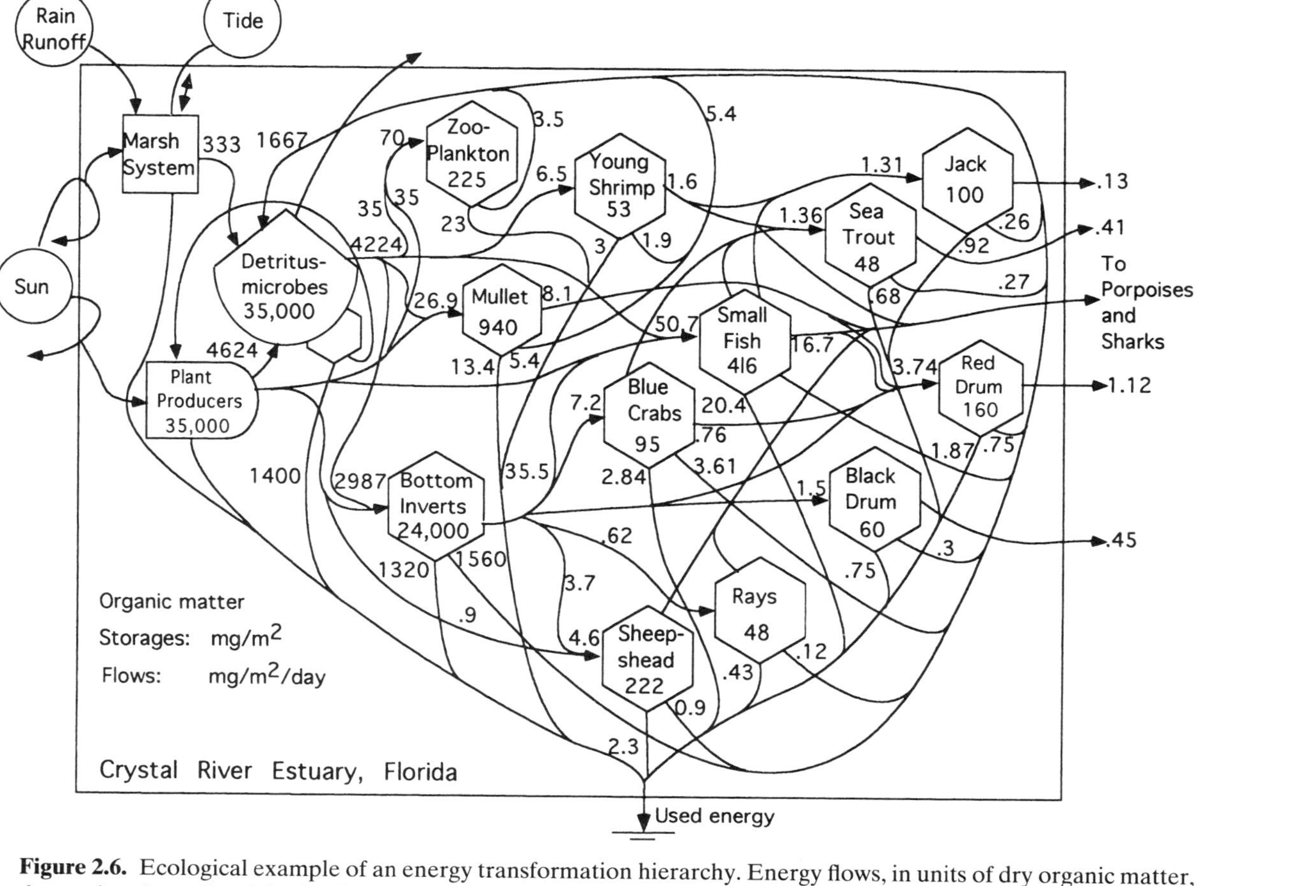

Figure 2.6. Ecological example of an energy transformation hierarchy. Energy flows, in units of dry organic matter, decreasing through a fisheries food chain in the estuary at Crystal River, Fla. Calculations by Kemp et al. (1975) based on weight data from Homer (1977). Here 1 mg of dry organic matter contains 4 to 4.5 small-calories of energy.

MAXIMIZING EMPOWER WITH REINFORCING FEEDBACK

Since self-designing processes are generally applicable to networks on any scale, they may be operating on all scales at the same time. The "maximizing power" concept of Lotka (1922a,b) is ambiguous because it might imply that transformations on a smaller scale with more energy but lower transformity prevail over those on the larger scale with higher transformity. Hence the following statement, which places energy of each level on a common basis using the concept "empower" (defined in Table 1.1, synonymous with EMERGY flow), is better:

> **Prevailing systems are those whose designs maximize empower by reinforcing resource intake at the optimum efficiency.**

This statement includes the maximizing of the resource intake *and* the operation at the optimum efficiency for maximum power. In network terms, the best use is to feed back as a multiplier, as shown in Figure 2.3. In other words, both *intake* and its best *use* are maximized.

Since any energy transformation has less energy in the output, to prevail, the lesser energy produced and stored must be able to feed back and reinforce the input production, which is only possible if it multiplies or otherwise amplifies that process. Thus, energy transformations that do not develop reinforcing design will not be reinforced and not long continued. After a symposium attended by 75 former students and associates, studies and theories on energy systems, EMERGY, and maximum power were assembled in a Festschrift volume dedicated to this author (Hall, 1995).

ENERGY MATCHING ACCORDING TO TRANSFORMITY

By reasoning that surviving designs will transform energy only with the products having a commensurate feedback effect, we can predict that the output of energy transformations that are continuing must be of a higher quality in the sense of doing more with less. However, that high quality is only useful if it interacts with and amplifies a lower-quality, higher-quantity energy source. Differences in energy quality were described by Tribus and McIrvine (1971), but without a way to measure them. The position in the energy-quality scale is measured by transformities that indicate what at one level is required for the next. The various energy flows in each unit of a network are all expressed on a common basis of what is required in EMERGY units.

The relationships between many small individuals and a few larger units are two-way—products going one way, with controls feeding back in the opposite way (Figures 2.3–2.5). The many flowers contributing to fewer bees are reinforced by the bees spreading out to pollinate many flowers. Well-adapted carnivores may be expected to provide such large-scale services as spatial organization and genetic control to the populations on which they depend. In aggregated diagramming, when one symbol (such as a "consumer" symbol) is used to represent many individuals (Figure 2.5b), more small units are represented by the symbols to the left than by those to the right.

GRAPH OF ENERGY QUANTITY AS A FUNCTION OF SOLAR TRANSFORMITY

The two variables, energy and solar transformity, graphed in Figure 2.5d,e, are the coordinates of the graph in Figure 2.7a. The energy-versus-transformity graph (Figure 2.7) is a general way of representing an energy hierarchy. As the solar transformity increases with successive transformations, the quantity of energy flow decreases. This plot represents power spectra in a generalized way, using data for any field of science or society. Sometimes this graph is called an *energy quantity-quality diagram,* since solar transformity is a measure of energy quality. Figure 2.7b shows the same graph on double logarithmic coordinates. Also see Figure 2.4.

Doherty (1990) fitted the data from four countries on national uses of different kinds of energy to a straight line with these coordinates. The R^2 values of his statistical regressions were 0.84 for Papua New Guinea, 0.84 for Liberia, 0.92 for Sweden, and 0.89 for the United States.

Energy Transformation Chain of a Large River

An example of the energy hierarchy is the network of small streams converging to form a river. The potential energy of water elevated by rain falling on highlands is used up in transformation steps, as small streams generate larger ones. The convergence of the small streams is an energy transformation chain. The flow of geopotential energy of water elevated above sea level is converted into kinetic energy, friction, and work on geological and biological systems. In Figure 2.8, available energy of elevated water in the Mississippi River (Diamond, 1984) is given as a function of increasing transformity downstream. The decline of geopotential energy at each stream junction is shown as streams converge downstream. Note the sketch of stream orders 1–6 in the center of the graph (Figure 2.8).

When rains fall on elevated lands, the geopotential energy of the elevated water carries the EMERGY of atmosphere, ocean, and land (evaluated in Chapter 4). Rains also bring to land the chemical potential energy of freshwater (distilled from the salty sea). Our theory is that self-organizing watersheds use the geopotential of their upper zones to develop stronger, higher transformity streams that can disperse the waters in floodplains and lower deltas, thus maximizing the biological productivity and other contributions there. Romitelli (1995) evaluated these characteristics and

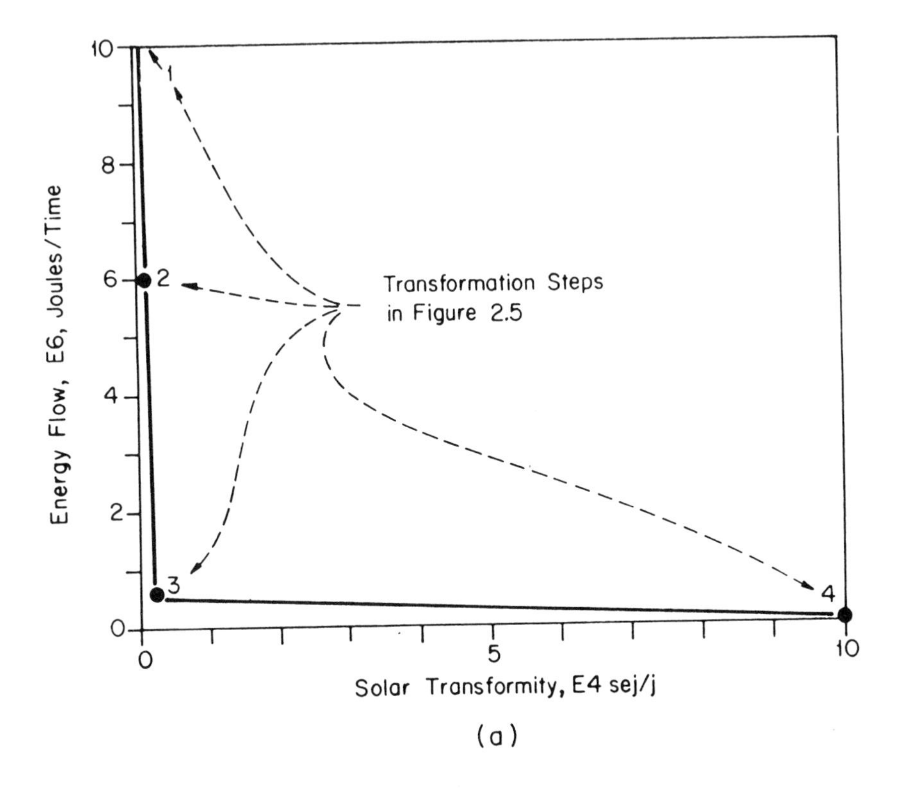

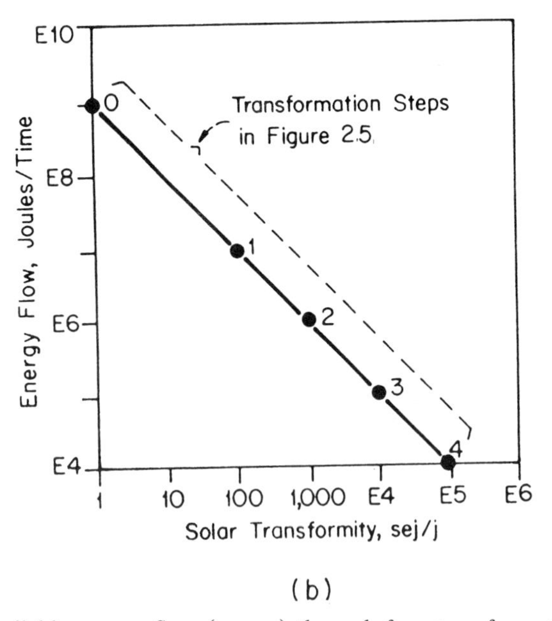

Figure 2.7. Available energy flow (power) through four transformation steps of the energy hierarchy in Figure 2.5 as a function of solar transformity. **(a)** Numerical coordinates. **(b)** Both coordinates logarithmic.

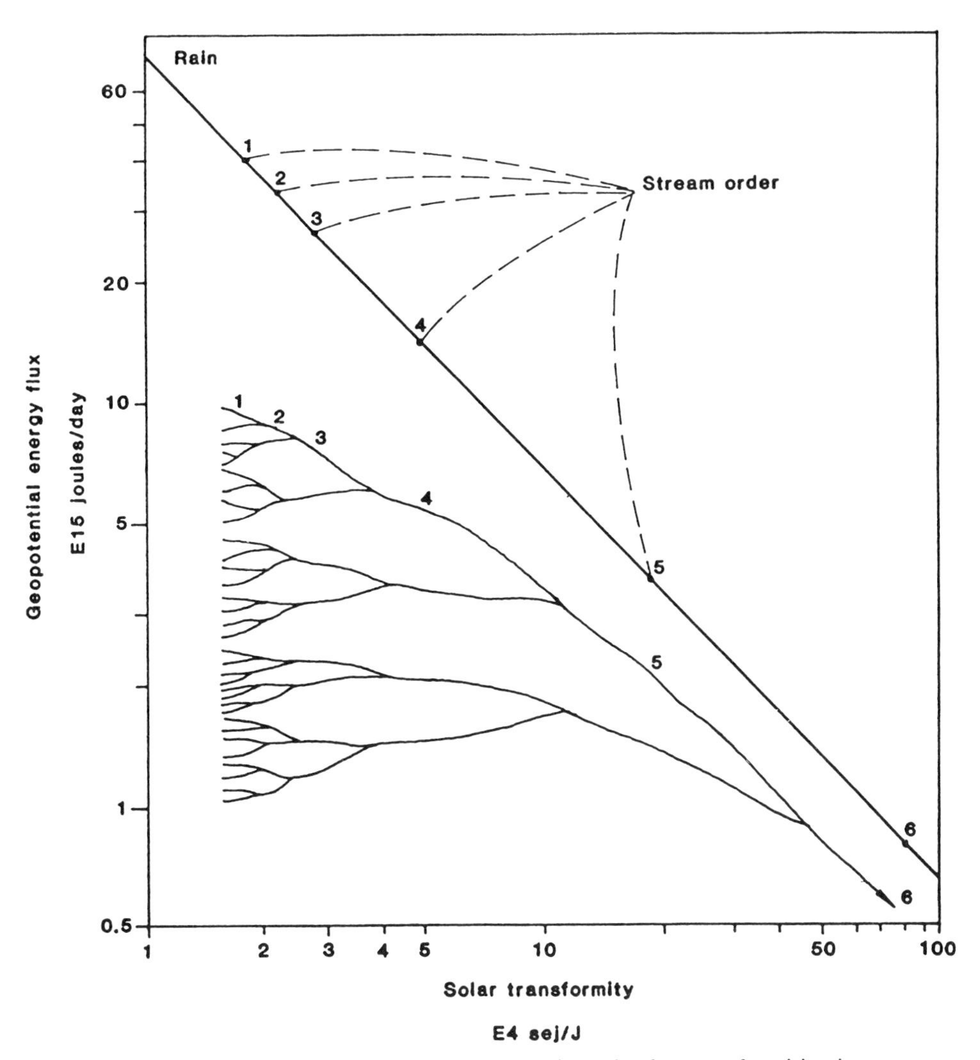

Figure 2.8. Geopotential energy flux as a function of solar transformities in stream hierarchies of the Mississippi River basin (Diamond, 1984). Transformation steps 0–6 are stream orders 0–6. Transformities were calculated by assigning 7.62×10^4 sej/g solar EMERGY to 2.57×10^{18} g rainfall of the Mississippi basin and dividing by joules of stream energy. Solar EMERGY per gram was obtained by dividing world solar EMERGY (9.44×10^{24} sej/yr) by world rain on land (1.05×10^{20} g/yr).

considered the loss of productivity when hydroelectric power developments divert the geopotential.

TRANSFORMITIES IN FLOWS OF SIMILAR OR DIFFERENT TYPE

The series of transformities increasing with successive converging of water in a river (Figure 2.8) is an example of the range of transformities to be found in flows of one type, in this case waters of streams. Sometimes we use a mean value from a reference table of transformities for a category such as stream water. Using averages is a rough procedure that can be improved by selecting an appropriate transformity from a graph of available energy as a function of transformity like that in Figure 2.8.

Other energy transformation series are heterogeneous, where each level in the hierarchy is occupied by a different kind of unit with recognizably different processes. In an ecological food chain there is an energy hierarchy transforming energy through plants, then small animals, then larger consumers, and so on. These levels in the hierarchy are called trophic levels in ecology, with each level occupied by different species (Figure 2.6). The transformities for ecosystem networks each refer to a different species. Part of the utility of the transformity concept is its ability to compare energy hierarchical positions of entirely different kinds of things. Contrast this heterogeneous energy chain with the homogeneous one in the Mississippi River system (Figure 2.8).

EQUATIONS DESCRIBING ENERGY TRANSFORMATION HIERARCHY

By making the simplifying assumption that the percent use of energy at each step in the energy chain is constant, a set of equations may be written for the graphs of energy and transformity, where energy distributions are a series of identifiable units. Here one step is an order of magnitude (power of 10) change of scale. Figure 2.9 gives these equations and the graphs they represent (H. T. Odum, 1984a). The slope of the natural logarithmic equation, k, is the efficiency of transformation per step. (K is the slope if logarithms to the base 10 are used.)

In these expressions E is the number of energy transformation steps (the number of units in a transformation series). Given the observed energy transformed in series, the number of steps is calculated from the transformity of the series. For example, Ricklefs (1979) used this procedure to calculate the steps in ecological food chains from the observed energy flows.

There are many published graphs of energy flows for series of transformations, where energy quantity is plotted as a function of stage of development. Examples are age graphs of trees or fishes, energy plots of storms of different size, plots of energy of molecules of different velocity, and so forth. It is possible to get transformities from these distributions if the baseline of EMERGY input to the series is known. From logarithmic graphs of energy quantity as a function of various measures of hierarchical position (energy intensity, age, size, territory, and so on), transformities R can be determined.

Transformity and Heat Engines

The amount of heat energy that can be converted into useful mechanical work depends on the temperature difference between hot and cold sources.

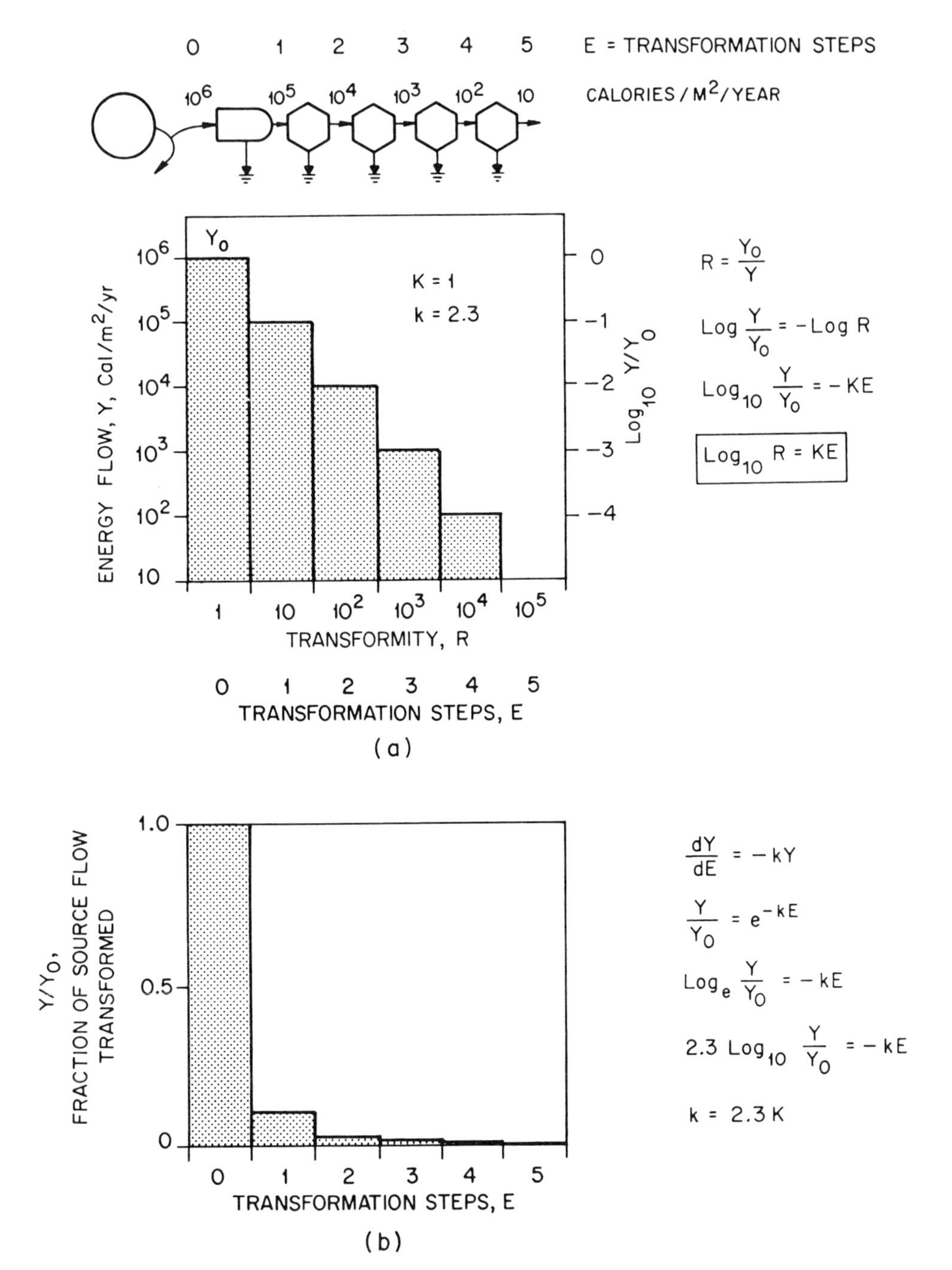

Figure 2.9. Mathematical equations relating energy flows, transformation steps, and transformities for a one-source energy hierarchy chain with constant transformity per transformation step (H. T. Odum, 1984a). **(a)** Logarithmic graph of energy flow as a function of transformation steps (E) with derived equations; a scale of equivalent transformities (R) is also given. **(b)** Linear plot of fraction of initial energy flow as a function of transformation steps.

The higher the temperature gradient between hot and cold (source and sink), the more work that can be transformed per unit of heat flowing. In other words, only the heat involved in the difference in temperature is available for work.

The theoretical maximum efficiency for converting heat energy to mechanical energy is given by the Carnot ratio, which is the ratio of the Kelvin temperature difference between hot and cold divided by the hot temperature. The Carnot ratio has been used as a measure of the "quality" of heat energy availability.

As the backforce loading of a device arranged to receive the work of the heat engine is increased, the conversion process approaches a stalled condition. A stalled condition is "reversible" (that is, thermodynamically at equilibrium, with no free energy availability and nothing happening). Thus, almost no power is being converted when operation is most efficient.

Operating with less load allows the work to go faster but at lower efficiency. More power is drawn from the heat. When operated at maximum power output (at steady state), the efficiency of energy conversion is about half of the Carnot reversible efficiency (Odum and Pinkerton, 1955; Tribus, 1961; Cruzon and Ahlborn, 1975; Fairen and Ross, 1981).

As the temperature differential between heat source and environmental sink increases (dT), the theoretical efficiency of conversion to mechanical work and the solar transformity both increase. In Figure 2.10, the temperature

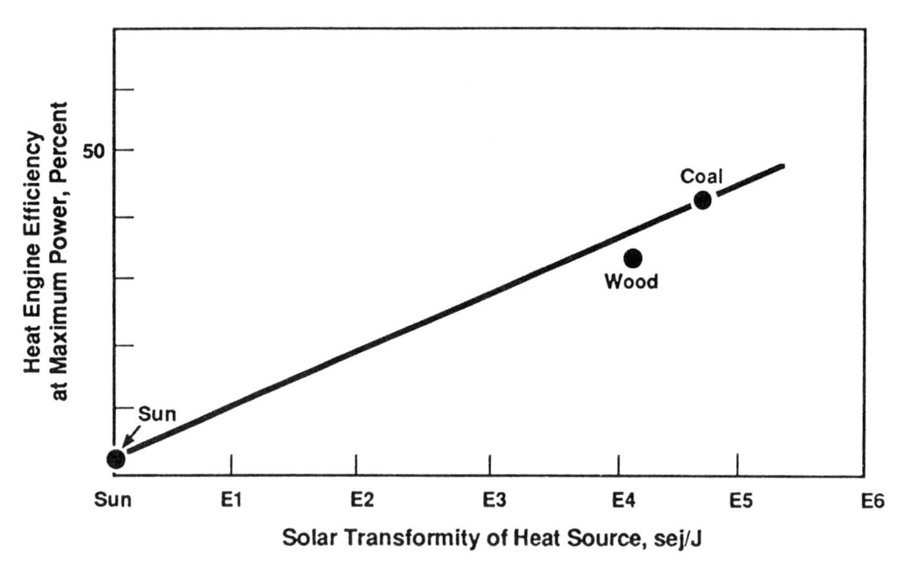

Figure 2.10. Heat-engine efficiency as a function of solar transformity of the source of heat. Efficiencies were calculated for observations on sustainable temperature gradient (ΔT) by calculating half the Carnot ratio (ΔT)/(hot T) in degrees Kelvin (°C + 273).

gradient (Carnot fraction) is graphed as a function of the solar transformity of the fuel used. Roberts (1975) provided a graph of efficiency of converting thermal energy difference into mechanical energy as a function of temperature for several kinds of conversion technologies.

Heat gradients in the biosphere due to differential heating by the sun are a few degrees, and are on the left side of the graph. Power plants with gradients of 800° k (wood power plants) to 5000° k (nuclear plants) occupy the right side of the diagram. Combustion of fuels in heat engines have a wide range of temperatures depending on the concentrations of fuels and oxygen (or other chemical reactants). Many industries are based on achieving high temperatures, and usually this means using fuels of higher transformity or further transforming these inputs in the process. The high temperatures deep in the earth, driving geologic convection, are also of relatively high quality.

An interpolation may be made to estimate solar transformities for heat sources by assigning 25,000 sej/J as the typical transformity for mechanical energy like that generated from typical power plants with 1000°C operating temperature. The Carnot ratio is about 700/1000 = 0.7, and the efficiency for maximum power is half of that: $(0.5 \times 0.7 \times 100) = 35\%$. Interpolating between a Carnot ratio of 0 and 0.7 for solar transformities of 0 to 25,000 sej/J can be done with the following expression:

$$\text{Solar transformity} = \frac{(\Delta T)\,(25{,}000\ \text{sej/J})}{(\text{hot } T)(0.7)}$$

DIFFICULTIES WITH PREVIOUS ENERGY VALUATIONS

Predating the scientific study of energy was the widely held belief that value results from work. Yet for a century of scientific development there was confusion on how the concept of prior work was to be measured quantitatively (see Table 14.1). Part of the confusion was caused because the original common-folk meaning of the word "energy"—as something required and used up during work—was cast aside by the physical sciences and engineering, which redefined the word as that which is conserved. The modern technical equivalent of the original folk concept is the second law, which states that "available" (potential) energy is used up during work as it is degraded. Aleksev (1986) reviews the roots of energy concepts in the history of philosophical thought.

As long as comparisons are made between energies of the same type, we may say that available energy measures the ability to do work, but as soon as we compare two different kinds of energy, the comparison becomes incorrect. Transformities must be used to relate the work of two energies of different types.

Previous efforts to evaluate environment and resources using energy are given in Chapter 14, where each method is compared with EMERGY accounting. Some of the controversy on energy-environmental accounting in recent history is included there. The failure to convert each kind of energy into each other kind before adding or comparing energy flows may be the main reason that energy theories of value have been initiated and dropped so many times. We can impove the previous attempts to use energy by multiplying energy-flow data by transformities, thus converting them into EMERGY flows.

SUMMARY

There is a universal energy hierarchy, and it provides transformities for quantitatively relating energy on one scale to that of another. Transformities can be used to put items at different scales on a common basis, which is the EMERGY of one kind required for a product or service of another kind.

CHAPTER 3

EARTH EMERGY

Since the real wealth of the environment comes from the work of the geobiosphere, EMERGY accounting starts with an evaluation of earth energy processes. Data on energy in sun, tide, and earth heat contributing to global processes are used to estimate the average EMERGY flows of the system of wind, water, and earth. This EMERGY flow is used as a baseline reference from which transformities of principal flows of the earth are calculated, thus making a start on quantifying main features of the energy hierarchy of the earth. Using these transformities, the EMERGY content in energy flows to local areas can be estimated (energy flows of sun, wind, rain, geologic cycles, and so forth). Where several environmental flows are by-products of the same source (the geobiosphere), double counting is avoided by diagramming the relationship of the local area to the larger earth.

A BASELINE EMERGY BUDGET FOR CALCULATING SOLAR TRANSFORMITIES OF THE EARTH

Whereas there appears to be a continuous series of energy transformations making up the energy network of the universe, represented schematically in Figure 2.2, a baseline from which to calculate transformities of the earth is a practical need for EMERGY accounting. For this purpose, we place the window

Readers not concerned with the reference baseline for calculating EMERGY flows of the earth may skip to the next chapter.

of evaluation around the geobiosphere and identify the main energy sources contributing over a long-run average (Figure 3.1). The sun, the tides, and the heat sources deep in the earth interact as a single coupled system with a network of processes that include the human economic systems and the production and maintenance of storages of globally shared information. In Figure 3.1 solar transformities increase from left to right along the series of successive energy transformations. Solar transformity measures position in the energy hierarchy and indicates the appropriate range of effective action (see Chapter 2).

Three main energy sources contribute to the global operation of the geobiosphere, each of a different energy form (sunlight insolation, tidal energy, and deep heat from inside the earth). An estimate of the solar transformity of each of these was obtained (Table 3.1) so that the total solar EMERGY inflow operating the earth could be represented in Figure 3.1. In the aggregate, the main renewable flows of the earth have an EMERGY flow of 9.44×10^{24} solar emjoules/year (sej/yr). On an EMERGY basis almost half is from the deep heat, half from the sunlight, and 15% from tidal energy. Explanation follows:

Crustal Heat

Part of the heat that operates the convection of the seafloor spreading cycle is from the independent heat sources deep in the earth (radioactive disintegrations and residual heat from earth formation when dispersed matter fell inward to the center of gravity, releasing heat). Another part of the heat comes down into the crust from energy transformations driven by the sun. The solar energy driving the ocean-winds-hydrologic cycle contributes to the terrestrial sedimentary cycle, which is coupled to the deep earth convection. Rivers move sediments from the mountains to the sea, and isostatic readjustment raises the mountains to replace the matter eroded. Heat energy is also added to the convection cycles from other surface processes that are solar driven; examples are the compression of sedimentary deposition under the river deltas, and the chemical potential energy deposited in the sediments that move downward in the continental and oceanic cycles (oxidized and reduced compounds deposited together), which are later released at higher temperatures and pressures.

Whereas the inflow of solar energy per area of earth is a low-transformity energy source (1 by definition), the high temperature concentrations deep in the crust are of higher quality. A solar transformity of 6055 sej/J was calculated (Table 3.1, note 2) as the quotient of the earth's solar energy absorption divided by the surface-originating part of the crustal heat estimated by difference.

Tidal Energy

Global tidal energy input was also estimated in Table 3.1. The tidal energy absorbed by the oceans was taken as the current energy. Then the solar transformity of the large sluggish river flow at New Orleans was used to estimate tidal EMERGY contributed.

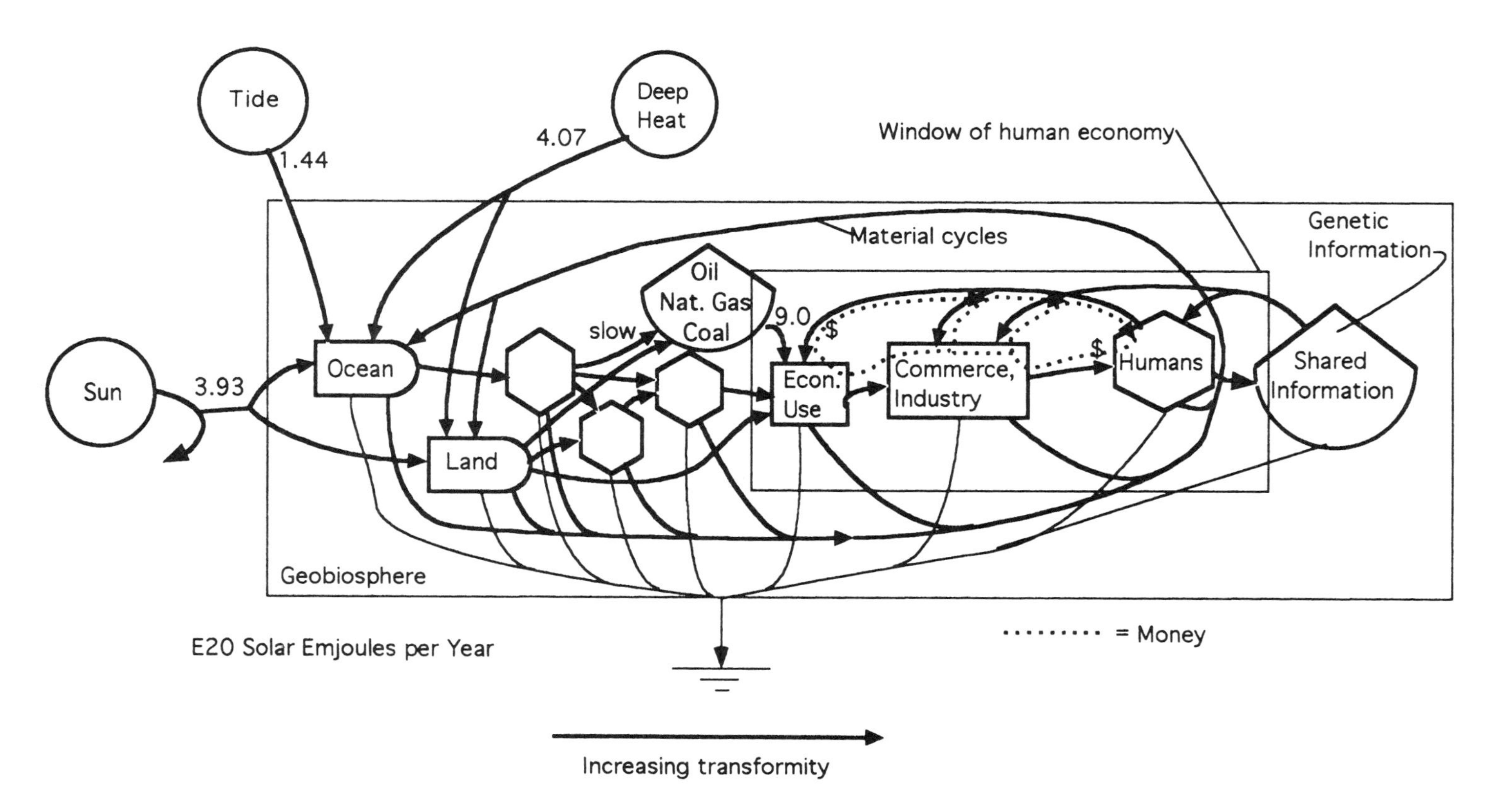

Figure 3.1. Overview of the global energy hierarchy and its baseline annual EMERGY budget.

TABLE 3.1. Annual Budget of Solar EMERGY for the Earth Geobiosphere*

Note	Source	Energy Flux (J/yr)	Solar Transformity (sej/J)	Solar EMERGY Flux (sej/yr)
1	Solar insolation	3.93×10^{24}	1	3.93×10^{24}
2	Deep earth heat	6.72×10^{20}	6,055	4.07×10^{24}
3	Tidal energy	8.515×10^{19}	16,842	1.44×10^{24}
4	Total			9.44×10^{24}

* See Figures 3.1 and 3.2.

[1] Sunlight: Solar constant, 2 cal/cm^2/min = 2 Langley per minute (Ly/min); 70% absorbed (Von der Haar and Suomi, 1969); earth cross section facing sun, 1.27×10^{14} m^2.

$$(2 \text{ Ly/min})(10 \text{ kcal/m}^2\text{/Ly})(1.278 \times 10^{14} \text{ m2})(5.256 \times 10^5 \text{ min/yr})(4186 \text{ J/kcal})(0.7) = 3.93 \times 10^{24} \text{ J/yr}$$

Solar transformity = 1 by definition.

[2] Heat release by crustal radioactivity (1.98×10^{20} J/yr) plus heat flowing up from mantle (4.74×10^{20} J/yr) is given by Sclater et al. (1980).

Solar transformity of earth heat driving deep geologic processes was evaluated from the deep heat contributed from the solar-driven sedimentary cycle passing energy downward as compression and chemical potentials (6.49×10^{20} J/yr) estimated by subtracting deep sources from total heat outflow (13.21×10^{20} J/yr). Solar EMERGY from surface is given in note 1.

$$\text{Solar transformity of deep heat} = \frac{3.93 \times 10^{24} \text{ sej/yr}}{6.49 \times 10^{20} \text{ J/yr}} = 6055 \text{ sej/J}$$

[3] Tidal energy received by the earth, 2.7×10^9 ergs/sec (Munk and Macdonald, 1960):

$$\frac{(2.7 \times 10^{19} \text{ ergs/sec})(3.153 \times 10^7 \text{ sec/yr})}{1 \times 10^7 \text{ ergs/J}} = 8.515 \times 10^{19} \text{ J/yr.}$$

Tidal energy transformed into ocean currents, 1.65×10^{19} ergs/sec (Miller 1966):

$$\frac{(1.65 \times 10^{19} \text{ ergs/sec})(3.153 \times 10^7 \text{ sec/yr})}{(1 \times 10^7 \text{ ergs/J})} = 5.2 \times 10^{19} \text{ J/yr}$$

Solar transformity of tidal currents assumed to equal that for stream currents in Table 3.2, item 5,27,764 sej/J.

Solar EMERGY of tidal currents:

$$(5.2 \times 10^{19} \text{ J/yr})(27{,}764 \text{ sej/J}) = 1.44 \times 10^{24} \text{ sej/yr}$$

Solar transformity of tidal energy received calculated from the energy received (8.515×10^{19} J/yr) and the solar EMERGY of the tidal currents:

$$\frac{(1.44 \times 10^{24} \text{ sej/yr})}{8.515 \times 10^{19} \text{ J/yr}} = 16842 \text{ sej/J}$$

[4] Summing is appropriate where sources are separate; see Figure 3.7.

The estimate of global annual EMERGY flow budget was used as a baseline reference standard in this book and in recent research evaluations. Since all the transformities are based on the same baseline, evaluations involving comparisons and differences will not be affected much by refinements in the baseline budget. The exact value of the reference does not have to be known. The situation is somewhat analogous to measurements of elevation made relative to sea level, where the absolute value of sea level height from the center of the earth is not important to evaluation of elevations from the baseline. Eventually geophysical specialists can determine a more accurate global EMERGY baseline.

Among the refinements possible and not yet adequately analyzed is the high-quality "solar wind" that brings high-energy radiation and particles to earth affecting the ionosphere, the distribution of earth charge, and by some theories, controls long-range aspects of weather convection. These inputs to the earth are related to sunspot cycles with peaks every 10 years or so. The insolation and the solar wind are subject to similar geometric divergence and dilution between the sun and earth. With large EMERGY and relatively small though intense energy flux, their transformity is large. On the longer time average, however, the solar wind is a by-product of the same processes in the sun that generate the regular solar insolation. To add more EMERGY for the solar wind would be double counting. See Chapter 6.

Earlier, different baselines were used. In order that older studies can be updated to the current baseline easily, Appendix E supplies the dates and parameters for corrections. The more recent changes and those to be expected when the baseline is refined do not affect EMERGY accounting very much.

SHARED EMERGY OF MUTUALLY NECESSARY SECTORS

Since the main features of the geobiosphere, the atmosphere, the ocean, and the earth cycles are each organized with their energies necessarily coupled with the others, an appropriate energy systems diagram shows closed loops where each interacts with the others (Figure 3.2). Each flow in this aggregated system is a by-product of the others (coproducts). All of the EMERGY flowing into this network is required for all of the flows. The total EMERGY is that of sunlight insolation, tidal energy absorbed, and heat entering the geobiosphere from sources deeper in the earth.

The different ways of aggregating a system for analysis and the effect on EMERGY accounting is the subject of Chapter 6. For the gross features of the geobiosphere where every sector is necessarily coupled to every other sector, all pathways have the same EMERGY requirement (9.44×10^{24} sej/yr in Figure 3.2). This estimate of EMERGY flow of the geobiosphere is used as the baseline for evaluating the transformities of the main-sector flows of wind, water, and earth.

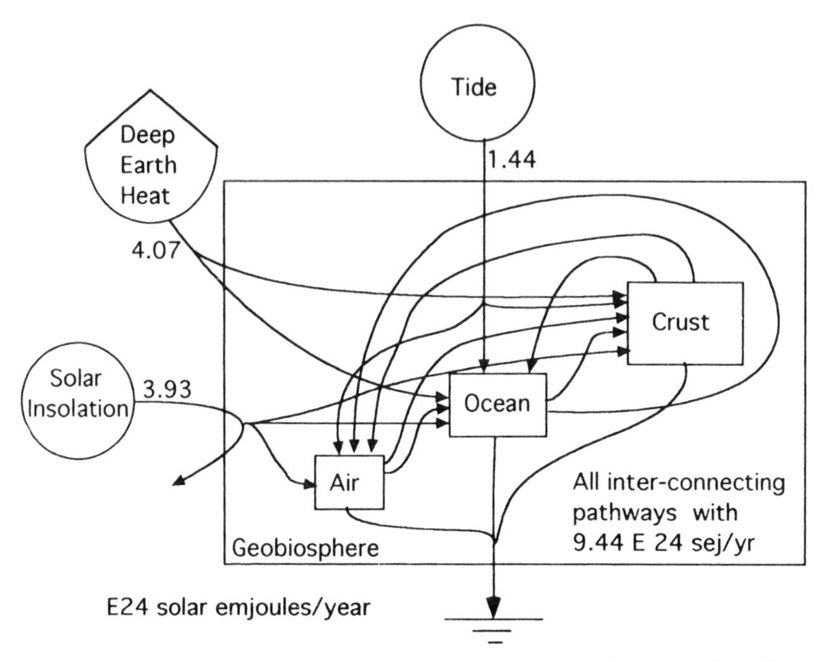

Figure 3.2. EMERGY basis for coupled atmosphere, ocean, and earth cycles, where all the pathways share the same baseline empower value of 9.44×10^{24} sej/yr.

EVALUATING EARTH PROCESSES

Calculating EMERGY and transformities for energy flows of the geobiosphere depends on correct representation of a system's hierarchy, where solar transformities increase from left to right. Independent evidence about hierarchical position comes from graphs of territorial size and turnover time now available in reference books of meterology, oceanography, and earth science. A sector or process at one level has a smaller territory and faster replacement time than a unit at the next higher level. Steele (1985) summarized data on hierarchical time-space position of items in the biosphere. Veizer (1988a,b) and Veizer et al. (1989) presented graphs for plate movements and other processes of geology.

Figure 3.3 is an overview of the hierarchy of global earth processes, enlarging somewhat the view in Figure 3.1. Because of their energy coupling, the main sectors of the earth network shown—including atmospheric circulation, oceanic circulation sedimentary cycle, and several processes of mountain main-

---

Figure 3.3. Geobiospheric hierarchy. **(a)** Correlation of turnover time and territory. **(b)** Landscape view of sea, land, and mountain-building centers. **(c)** Aggregated energy systems diagram. **(d)** Solar EMERGY per gram of recycling matter.

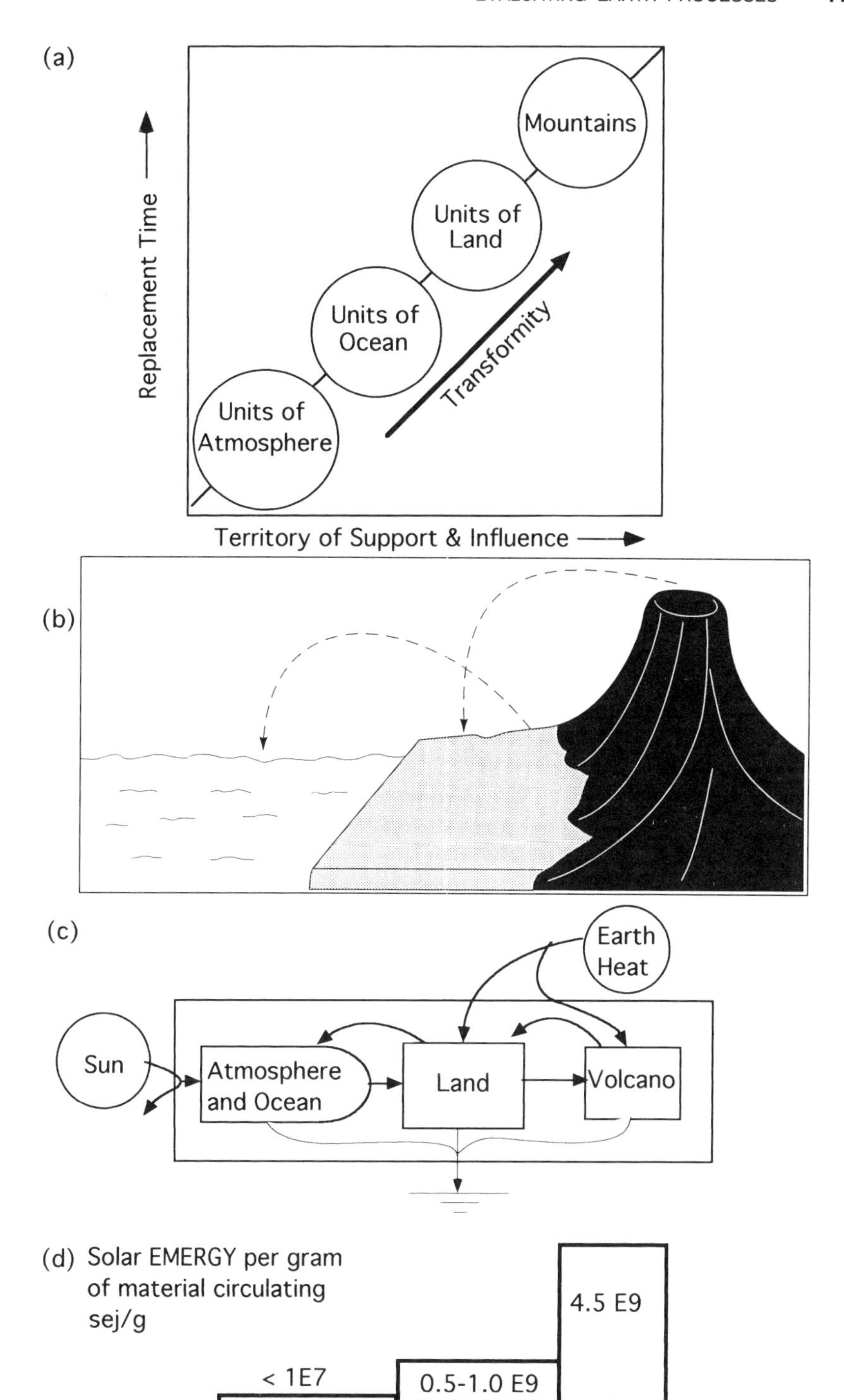
(a)
Replacement Time
Mountains
Units of Land
Units of Ocean
Units of Atmosphere
Transformity
Territory of Support & Influence
(b)
(c)
Earth Heat
Sun
Atmosphere and Ocean
Land
Volcano
(d) Solar EMERGY per gram of material circulating sej/g
4.5 E9
< 1E7
0.5-1.0 E9

TABLE 3.2. Solar Transformities of Global Energy Flows*

Note	Item	Solar EMERGY* (sej/yr)	Energy Flux (J/yr)	Solar Transformity (sej/J)
1	*Global Flows* (Figure 3.2)	9.44×10^{24}		
2	Surface wind		6.31×10^{21}	1,496
3	Physical energy, rain on land		9.0×10^{20}	10,488
4	Chemical energy, rain on land		5.187×10^{20}	18,199
5	Physical stream energy		3.395×10^{20}	27,764
6	Waves absorbed on shores		3.09×10^{20}	30,550
7	Earth sedimentary cycle		2.746×10^{20}	34,377
8	Chemical stream energy		1.948×10^{20}	48,459

* Based on global solar EMERGY flux in Figure 3.2.

[1] Sum of solar, geologic, and tidal EMERGY (see Figure 3.2).

[2] Wind used at surface of the earth—estimated as 10% of total flux of wind energy, or 2×10^{12} kW (Monin, 1972):

$$(2 \times 10^{12}\ \text{kW})(1\ \text{J/sec/W})(1 \times 10^{3}\ \text{W/kW})(3.154 \times 10^{7}\ \text{sec/yr})(0.10) = 6.31 \times 10^{21}\ \text{J/yr}$$

Solar transformity: ratio of total biosphere input, (9.44×10^{24} sej/yr) to surface wind energy:

$$\frac{9.44 \times 10^{24}\ \text{sej/yr}}{6.31 \times 10^{21}\ \text{J/yr}} = 1496\ \text{sej/J surface wind}$$

[3] Physical energy in rain on elevated land—World's rain on land, 105,000 km^3/yr; average elevation of land, 875 m (Ryabchikov, 1975):

$$(1.05 \times 10^{5}\ \text{km}^3)(1 \times 10^{12}\ \text{kg/km}^3)(9.8\ \text{m/sec}^2)(8.75 \times 10^{2}\ \text{m}) = 9 \times 10^{20}\ \text{J/yr}$$

$$\text{Solar transformity} = \frac{9.44 \times 10^{24}\ \text{sej/yr}}{9.0 \times 10^{20}\ \text{J/yr}} = 10{,}488\ \text{sej/J}$$

[4] Chemical potential energy in rain—continental rain 105,000 km^3/yr, 10 ppm rain compared with 35,000 ppm.

$$\text{Gibbs free energy/g} = \frac{(8.33\ \text{J/mole/deg})(300\ \text{K})}{18\ \text{g/mole}} \log e \left(\frac{999{,}990}{965{,}000}\right) = 4.94\ \text{J/g}$$

$$(1.05 \times 10^{5}\ \text{km}^3/\text{yr})(1 \times 10^{15}\ \text{g/km}^2)(4.94\ \text{J/g Gibbs free energy}) = 5.187 \times 10^{20}\ \text{J/yr}$$

$$\text{Solar transformity} = \frac{9.44 \times 10^{24}\ \text{sej/yr}}{5.187 \times 10^{20}\ \text{J/yr}} = 1.8199 \times 10^{4}\ \text{sej/J rain}$$

tenance—are necessary to each other. Figure 3.3a shows the way the earth hierarchy operates, with small, fast atmospheric cycles on the left driving somewhat larger and slower oceanic processes, which in turn drive the much larger slower earth cycles that converge on mountain-building zones. Figure 3.3b is drawn with elements increasing in solar transformity from left to right to reflect the way our diagrams are ordered.

To evaluate the transformities and EMERGY flows of large-scale earth processes, data on flows of energy or matter were obtained (Tables 3.1–3.6). Some flows were estimated by multiplying the stored quantities by turnover times. Then the solar transformities of these earth processes were obtained by dividing the annual global EMERGY flow (baseline budget) by the energy flow. For some calculations the solar EMERGY per unit mass flow was evaluated. Finally, the EMERGY flows supporting main processes of the earth were calculated by multiplying the flows of energy by solar transformities or by multiplying the flows of matter by EMERGY per unit mass. Results of these evaluations are given in Table 3.2.

---

[5] Physical energy in stream flow—Global runoff = 39.6×10^3 km^3/yr (Todd, 1970); average elevation = 875 m:

$$(39.6 \times 10^3 \text{ km}^3/\text{yr})(1 \times 10^{12} \text{ kg/km}^3)(9.8 \text{ m/sec}^2)(875 \text{ m}) = 3.395 \times 10^{20} \text{ J/yr}$$

$$\text{Solar transformity} = \frac{9.44 \times 10^{24} \text{ sej/yr}}{3.4 \times 10^{20} \text{ J/yr}} = 27{,}764 \text{ sej/J}$$

[6] Wave energy absorbed at shore—Estimated as the energy of average wave coming ashore (Kinsman, 1965) multiplied by facing shorelines:

$$(1.68 \times 10^8 \text{ kcal/m/yr})(4.39 \times 10^8 \text{ m})(4186 \text{ J/kcal}) = 3.09 \times 10^{20} \text{ J/yr}$$

$$\text{Solar transformity} = \frac{9.44 \times 10^{24} \text{ sej/yr}}{3.09 \times 10^{20} \text{ J/yr}} = 30{,}550 \text{ sej/J}$$

[7] Earth sedimentary cycle—Work of earth uplift replacing erosion without net change in elevation indicated by heat flow. From Sclater et al. (1980); continental heat flow is 2.746×10^{20} J/yr; solar from Table 3.1.:

$$\text{Solar transformity} = \frac{9.44 \times 10^{24} \text{ sej/yr}}{2.746 \times 10^{20} \text{ S/yr}} = 34{,}377 \text{ sej/J continent heat flow}$$

[8] Chemical potential energy in rivers—Rivers represent concentration over water dispersed as rain. A solar transformity for world average river is given: global runoff = 39.6×10^3 km^3/yr; typical dissolved solids = 150 ppm:

$$\text{Gibbs free energy per gram water: } \frac{(8.33 \text{ J/mole/deg})(300 \text{ K})}{18 \text{ g/mole}} \ln\left(\frac{999{,}850}{965{,}000}\right) = 4.92 \text{ J/g}$$

$$(3.96 \times 10^{19} \text{ cm}^3/\text{yr})(0.99985 \text{ g/cm}^3)(4.92 \text{ J/g}) = 1.948 \times 10^{20} \text{ J water/yr}$$

$$\text{Solar transformity} = \frac{9.44 \times 10^{24} \text{ sej/yr}}{1.948 \times 10^{20} \text{ J water/yr}} = 4.85 \times 10^4 \text{ sej/J river water}$$

To calculate a solar transformity of an energy flow within the web of the geobiosphere, divide the total solar EMERGY supporting the web (Figure 3.2) by the energy flow (Table 3.2). As further discussed in Chapter 6, long-operating self-organizing systems have their sectors coupled and mutualistic so that the EMERGY required for each necessary sector is the same, namely, the total EMERGY supporting the system.

Table 3.2 includes the cycle of earth materials. Continents are continuously eroded by winds, waters, and other climatic processes. As land is removed by erosion, the reduced weight allows land to rise up from below by isostatic adjustment so that the land is cycled. Garrels et al. (1975) suggested 2.4 cm/1000 yr as the average erosion-replacement rate for rock with an average density of 2.6 g/cm^3. The resulting solar EMERGY per gram of land cycle is:

$$\frac{9.44 \times 10^{24} \text{ sej/yr}}{(2.4 \text{ cm}/1000 \text{ yr})(2.6 \text{ g/cm}^3)(1.5 \times 10^{18} \text{ cm}^2 \text{ land area})} = 1.00 \times 10^9 \text{ sej/g}$$

The outward flow of earth heat energy processed through this system is known from extensive geologic data on thermal gradients. Using a summariz-

TABLE 3.3. Solar EMERGY per Unit Mass in Earth Processes, Based on Global EMERGY 9.44 × 10^{24} sej/yr

Item	Products of Earth Process (Figure 3.3)	Storage* ($\times 10^{24}$ g)	Material flux† ($\times 10^{16}$ g/yr)	Solar EMERGY/g‡ ($\times 10^9$ sej/g)
		Ocean Floor		
1	Oceanic basalt	5.4	6.34	0.15
2	Pelagic and abyssal sediments	0.63	0.97	0.97
		Continents		
3	Granitic rocks	9.8	1.89	0.50
4	Mountains on land	0.94	0.84	1.12
5	Metamorphic rocks	7.8	0.65	1.45
6	Continental sediment	2.2	0.50	1.88
7	Volcanic extrusion at surface (25.3% of land mountains)	0.24	0.21	4.5

* Storages (1–7) were obtained from Ronov and Yaroshevsky (1969) and Ronov (1982).
† Annual flux of material (1–7) was obtained by dividing storages by the turnover times. Turnover times were the half-lives from Veizer (1988b) divided by 0.693 with results as follows: pelagic and abyssal sediments, 64.9 × 10^6 yr, oceanic basalt, 85.1 × 10^6 yr; continental sediment, 432 × 10^6 yr; mountains and volcanoes at land surface, 112 × 10^6 yr; granitic rock, 519 × 10^6 yr; metamorphics, 1197 × 10^6 yr.
‡ Values for solar EMERGY/gram (1–7) were obtained by dividing 9.44 × 10^{24} sej/yr by the material flux.

ing estimate of the heat flow (Wylie, 1971), the solar transformity (solar EMERGY/heat energy flux) of the earth cycle is:

$$\frac{9.44 \times 10^{24} \text{ sej/yr}}{2.74 \times 10^{20} \text{ land heat J/yr}} = 3.4 \times 10^{4} \text{ sej/J}$$

Some of the main units of the earth's tectonic processes are listed in Table 3.3 with estimates of mass and annual flux, the latter calculated from turnover times given by Veizer (1988a,b) and Veizer et al. (1989). Since these major components of the earth system are mutually coupled and necessary, global annual solar EMERGY was divided by the estimates of material flow to obtain solar EMERGY per gram.

As described in Chapter 2, the energy hierarchy is quantitatively represented by territory, turnover time, and transformity. Tables 3.4 and 3.5 use turnover times to estimate orders of magnitude of transformities and EMERGY storage of main earth components. Figure 3.4 graphs solar transformity and turnover time showing the scales of energy hierarchy of the earth including information explained in Chapter 12. Figure 3.5 is a graph of size and turnover time (half-life) from Veizer (1988a). Table 3.4 shows that the larger the scale of hierarchy, the larger is the storage, which is consistent with a small energy flow but a slow turnover time.

TABLE 3.4. EMERGY of Some Global Storages (Natural Capital)*: Possible Orders of Magnitude (Odum, 1994)

Item	Replacement time (yr)	Stored EMERGY* (sej)	Em$ Value† (1992 $)
World infrastructure	100	9.44×10^{26}	\$ 6.3×10^{14}
Fresh waters, glaciers on land	200	1.89×10^{27}	\$ 1.26×10^{15}
Soil, terrestrial ecosystems	500	4.7×10^{27}	\$ 3.1×10^{15}
Cultural and technological information‡	1×10^{4}	9.44×10^{28}	\$ 6.3×10^{16}
Atmosphere	1×10^{6}	9.44×10^{30}	\$ 6.3×10^{18}
Ocean	2×10^{7}	1.89×10^{32}	\$ 1.25×10^{20}
Continents	1×10^{9}	9.44×10^{33}	\$ 6.3×10^{21}
Genetic information	3×10^{9}	2.8×10^{34}	\$ 1.86×10^{22}

* Product of annual solar EMERGY flux, 9.44×10^{24} sej/yr and order-of-magnitude replacement times in column 2. This calculation is appropriate where the category of storage is a mainline sector requiring and contributing to the the whole earth energy systems network.

† Solar EMERGY flow in column 3 divided by 1.5 sej/1992 \$. EMERGY per unit of money was obtained by extrapolating the EMERGY per year used by the United States in 1983 (Odum et al., 1987a) to 1992 by substituting estimated fuel use and gross economic product (see Appendix D).

‡ Replacement time for the shared human learned information taken as 10,000 yr—but even the order of magnitude is unknown.

TABLE 3.5. Solar Transformities of Sedimentary Cycle Processes in Parallel (Figure 3.6)

Note	Item	Material flux* (g/yr)	Solar EMERGY per Mass (sej/g)	Solar Transformity (sej/J†)
1	Global sedimentary cycle	9.36×10^{15}	1.0×10^{9}	—
	Proportions of Continental Area			
2	Shale, 42%	3.93×10^{15}	1.0×10^{9}	1.0×10^{7}
3	Limestone, 18%	1.68×10^{15}	1.0×10^{9}	1.62×10^{6}
4	Sandstone, 20%	1.87×10^{15}	1.0×10^{9}	2.0×10^{7}
5	Evaporites, 1%	0.94×10^{15}	1.0×10^{9}	3.3×10^{6}
6	Soil clay from shale	4.68×10^{15}	2.0×10^{9}	—
7	Coal	1.84×10^{11}	1.0×10^{9}	3.4×10^{4}
8	Sedimentary iron ore	4.85×10^{10}	1.0×10^{9}	6.2×10^{7}
9	Bauxite (aluminum ore)	8.83×10^{10}	1.0×10^{9}	1.5×10^{7}

* The global flux of the ore as part of the sedimentary cycle was calculated as follows: The known reserves of an ore were divided by the mass of the upper kilometer of land to obtain the fraction of the sedimentary cycle that this ore represents. Then this fraction was multiplied by the flux of the sedimentary cycle. (tn is the abbreviation for metric ton, also called the Tonne.) The mass of the upper kilometer of rock is:

$$(1\ \text{km})(1.5 \times 10^{14}\ \text{m}^2\ \text{land area})(1 \times 10^{3}\ \text{m/km})(2.62\ \text{tn/m}^3\ \text{density}) = 3.93 \times 10^{17}\ \text{tn}$$

The flux of the sedimentary cycle is:

$$(2.4\ \text{cm/1000 yr})(1.5 \times 10^{14}\ \text{m}^2\ \text{land area})(1 \times 10^{4}\ \text{cm}^2/\text{m}^2)(2.6\ \text{g/cm}^3) = 9.36 \times 10^{15}\ \text{g/yr}$$

† Solar transformity was calculated by dividing the fraction of global solar EMERGY assigned to this ore by the Gibbs free energy in the flux of that ore:

$$\text{solar transformity (in sej/J)} = \frac{(\text{fraction})(9.44 \times 10^{24}\ \text{sej/yr})}{(\text{Gibbs energy/g})(\text{flux in g/yr})}$$

Gibbs free energy for rock is the energy difference between rock and its weathered state after reacting with rainwater (Gilliland et al., 1978).

[1] Global EMERGY is from item 4, Table 3.1, 9.44×10^{24} sej/yr. Sedimentary cycle is from Garrels et al. (1975), 2.4 cm/1000 yr for rock of density 2.6; continental area, 1.5×10^{14} m^2:

$$\text{EMERGY/mass of global sediment: } \frac{9.44 \times 10^{24}\ \text{sej/yr}}{9.36 \times 10^{15}\ \text{g/yr material flux}} = 1.01 \times 10^{9}\ \text{sej/g}$$

[2] $(9.36 \times 10^{15}\ \text{g/yr})(0.42) = 3.93 \times 10^{15}$ g/yr; Gibbs free energy = − 100 J/g. Thus the solar transformity of shale is:

$$(1.0 \times 10^{9}\ \text{sej/g})/(100\ \text{J/g}) = 1.0 \times 10^{7}\ \text{sej/J}$$

[3] $(9.36 \times 10^{15}$ g/yr$)(0.18) = 1.68 \times 10^{15}$ g/yr; Gibbs free energy = – 611 J/g. Thus the solar transformity of limestone is:

$$(1.0 \times 10^9 \text{ sej/g})/(611 \text{ J/g}) = 1.63 \times 10^6 \text{ sej/J}$$

[4] $(9.36 \times 10^{15}$ g/yr$)(0.20) = 1.87 \times 10^{15}$ g/yr; Gibbs free energy = – 50 J/g. Thus the solar transformity of sandstone is:

$$(1.0 \times 10^9 \text{ sej/g})/(50 \text{ J/g}) = 2.0 \times 10^7 \text{ sej/J}$$

[5] $(9.36 \times 10^{15}$ g/yr$)(0.01) = 0.94 \times 10^{15}$ g/yr; Gibbs free energy relative to seawater = –300 J/g. Thus the solar transformity of evaporite is:

$$(1.0 \times 10^9 \text{ sej/g})/(300 \text{ J/g}) = 3.33 \times 10^6 \text{ sej/J}$$

[6] Soil from sedimentary cycle is in steady state; 42% is from shale; half of rock mass was lost in soil formation (Siegel, (1974); after Krauskopf, 1967 and Goldick, 1938. The solar EMERGY of weathered rock is:

$$\text{solar EMERGY per gram of clay} = \frac{(1.0 \times 10^9 \text{ sej/g})(1 \text{ g shale})}{0.5 \text{ g soil}} = 2.0 \times 10^9 \text{ sej/g}$$

[7] Coal reserve = 7.64×10^{12} tn; Gibbs free energy = –29,302 J/g; the fraction = $(7.64 \times 10^{12}$ tn$)/(3.93 \times 10^{17}$ tn$) = 1.96 \times 10^{-5}$. Then the material flux of coal is:

$$(1.96 \times 10^{-5})(9.36 \times 10^{15} \text{ g/yr}) = 1.84 \times 10^{11} \text{ g/yr}$$

The energy flux as coal is:

$$(29{,}302 \text{ J/g})(1.84 \times 10^{11} \text{ g/yr}) = 5.39 \times 10^{15} \text{ J/yr}$$

The solar transformity is:

$$(1.0 \times 10^9 \text{ sej/g})/(29{,}302 \text{ J/g}) = 3.4 \times 10^4 \text{ sej/J}$$

[8] Iron-ore reserve = 2.035×10^{12} tn; Gibbs free energy = –16.2 J/g; the fraction = $(2.035 \times 10^{12}$ tn$)/(3.93 \times 10^{17}$ tn $= 5.18 \times 10^{-6}$. Then the material flux of iron is:

$$(5.18 \times 10^{-6})(9.36 \times 10^{15} \text{ g/yr}) = 4.85 \times 10^{10} \text{ g/yr}$$

The energy flux as iron is:

$$(16.2 \text{ J/g})(4.85 \times 10^{10} \text{ g/yr}) = 6.89 \times 10^{11} \text{ J/yr}$$

The solar transformity is:

$$(1.0 \times 10^9 \text{ sej/g})/(16.2 \text{ J/g}) = 6.2 \times 10^7 \text{ sej/J}$$

[9] Bauxite reserve = 3.519×10^{12} tn; Gibbs free energy = –65.3 J/g; the fraction = $(3.519 \times 10^{12}$ tn$)/(3.93 \times 10^{17}$ tn$) = 8.95 \times 10^{-6}$. Then the material flux of bauxite is:

$$(8.95 \times 10^{-6})(9.36 \times 10^{15} \text{ g/yr}) = 8.34 \times 10^{10} \text{ g/yr}$$

The energy flux as bauxite is:

$$(65.3 \text{ J/g})(8.83 \times 10^{10} \text{ g/yr}) = 5.76 \times 10^{12} \text{ J/yr}$$

The solar transformity is:

$$(1.0 \times 10^9 \text{ sej/g})/(65.3 \text{ J/g}) = 1.5 \times 10^7 \text{ sej/J}$$

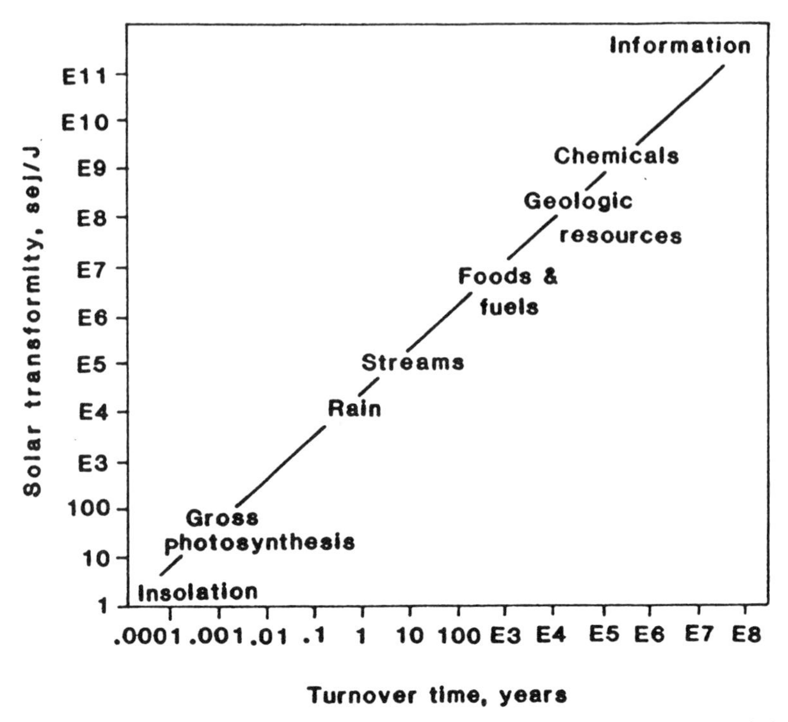

Figure 3.4. Solar transformity as a function of turnover time. The long-term information is genetic.

EARTH PRODUCTS GENERATED BY PARALLEL PROCESSES

Whereas the energy diagram of intercoupled earth sectors in Figure 3.2 was appropriate because the sectors were mutually necessary, for processes that are in parallel, a different diagram structure is required. The evaluation of the main sectors of the geobiospheric system required a diagram that includes only major, necessary, mutualistic divisions of the earth energy web. In Figure 3.6 a sector divides (splits) its flow into parallel flows, thus dividing the EMERGY among the several pathways.

For example, as part of the main sedimentary cycle of the earth, several kinds of sedimentary rock are formed and brought to the land surface in different places. In some places there are shales, limestones, sandstones, unconsolidated sediments, coal, iron ores, bauxite, and so forth. Many different processes are in parallel, as suggested in Figure 3.6. It may be appropriate to assign the total EMERGY flows of the sedimentary cycle to each of these parallel processes in proportion to the mass flows of each. Table 3.5 has some material flows and resulting transformities obtained by assigning EMERGY according to fraction of the total mass flow.

Some of the main units of the earth's tectonic processes are listed in Table 3.3 with estimates of mass and annual flux, the latter calculated from turnover

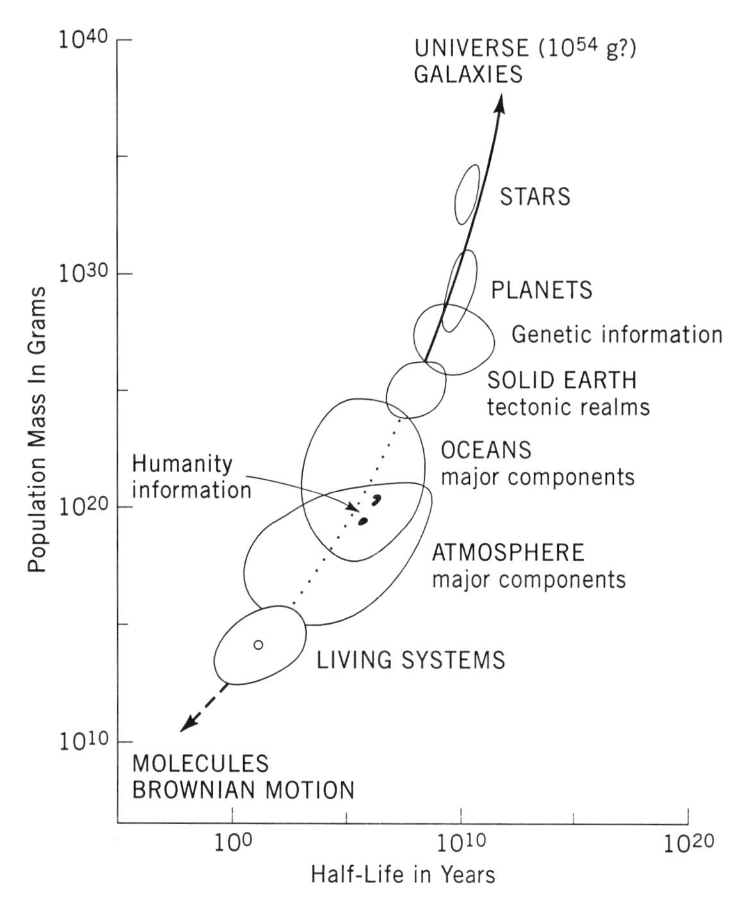

Figure 3.5. Generalized plot of mass and half-life from Veizer (1988a, reprinted by permission of Kluwer Academic Publishers). A circle for genetic information was added; see Chapter 12.

times given by Veizer (1988a,b) and Veizer et al. (1989). Since these major components of the earth system are mutually coupled and necessary, global annual solar EMERGY was divided by the estimates of material flux to obtain the solar EMERGY per gram.

HIGH MOUNTAIN CYCLES

As Figure 3.3 suggests, rapidly rising and eroding mountains such as those found in New Zealand, Taiwan, and Japan are at a higher hierarchical level than the general sedimentary cycle. They represent the convergence of energies of several kinds accumulating from larger plate areas. They are the hierarchical centers of larger sections of the land. Some are volcanic, while others are folded sedimentary rock and metamorphic. As the focus of high-energy

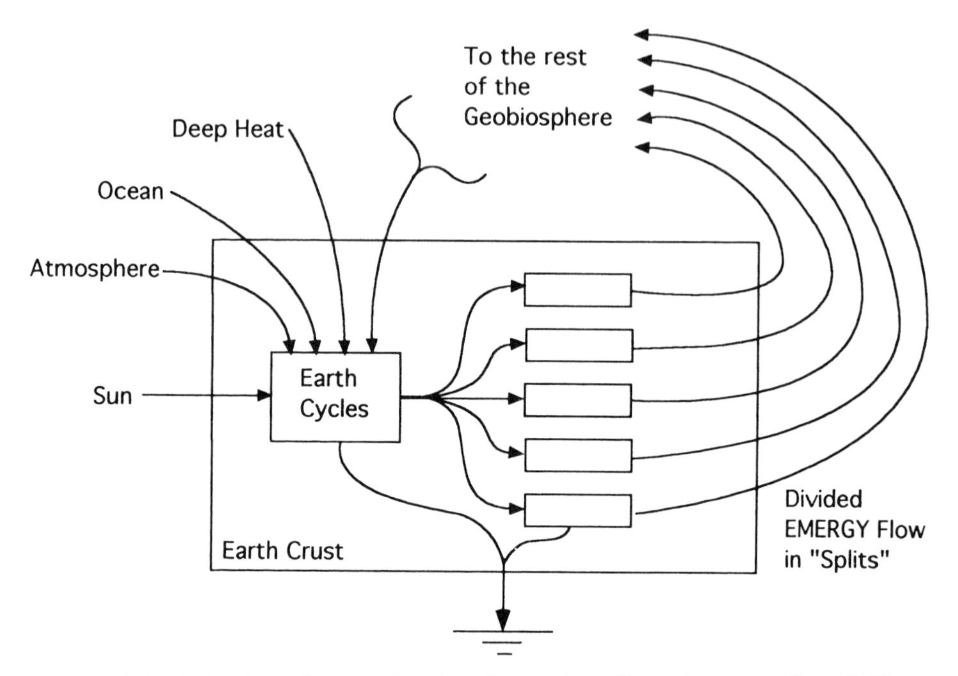

Figure 3.6. Evaluating solar transformity where a base flow of EMERGY flow divides (splits) into several parallel production processes.

TABLE 3.6. Solar EMERGY per Gram of Rock

Process	Turnover Time (×10⁶ yr*)	Solar EMERGY per Unit Mass† 10⁹ (sej/g†)	Products Contained
Ultramafic	2453	2.7‡	Nickel, chromium
Metamorphic	1948	1.45?	Manganese, zinc
Volcanic-sedimentary	1154	4.5	Zinc, copper, lead
Chemical-sedimentary	597	1.88	Uranium
Hydrothermal	124	2 ?	Tin, gold
Sedimentary placer	72	1+	Zirconium
Weathering residual	45	1.09	Iron, aluminum

* Half-lives from Veizer (1988a, b) and Veizer et al. (1989) divided by 0.693 to obtain steady-state turnover times.

† An EMERGY per unit mass of a related process was assumed to be appropriate where it was a shared part of a group of processes in parallel at the same position in the earth hierarchy. See related values in Table 3.3.

‡ Mantle contribution to continental cycle, 16% (Veizer, 1988a); solar EMERGY contribution 4.07×10^{24} sej/yr (Table 3.1); continental cycle, 9.36×10^{15} g/yr (Table 3.5). EMERGY per unit mass:

$$(0.16)(9.36 \times 10^{15} \text{ g/yr})(4.07\ 10^{24} \text{ sej/yr}) = 2.7 \text{ E9 sej/g}$$

phenomena, they have sharply pulsing transformations but over long periods of time. Mountains, volcanoes, and earthquakes exert autocatalytic controls from hierarchical centers. Global solar EMERGY is assigned to the material flows in Table 3.3.

Just as we divided the solar EMERGY of the sedimentary cycle among its parallel sectors, it may be appropriate to apportion the total solar EMERGY of the mountain centers among their processes in parallel. Some of the mineralogical products of earth processes may be evaluated with Table 3.6. Processes are arranged in order of their turnover times, which correlate with their solar EMERGY per unit mass, since longer times involve more environmental work.

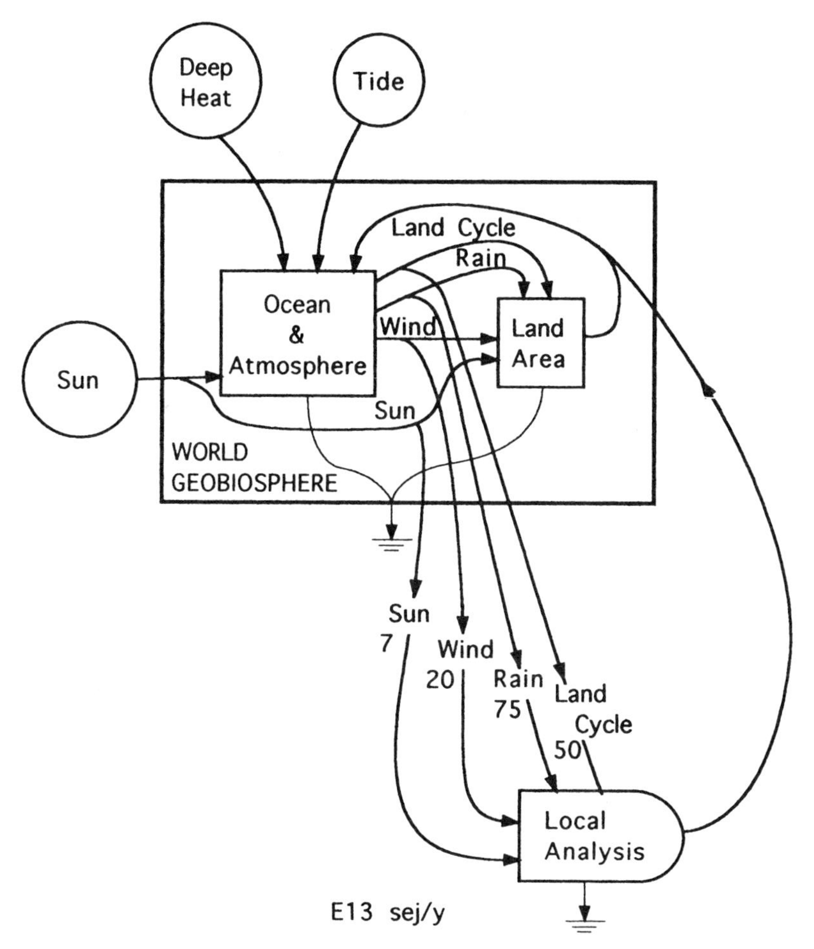

Figure 3.7. Evaluating the EMERGY flows from the geobiosphere to a local area. Since the flows to the local area are coproducts from the same source, their EMERGY contents are not independent. To avoid double counting, utilize only the largest one (in this example, rain).

If a rock or mineral can be identified as a product of one of these processes with a similar turnover time, a similar value of solar EMERGY per gram may be inferred.

EVALUATING SOLAR EMERGY FROM THE EARTH SYSTEM SUPPORTING A LOCAL AREA

The main geobiospheric process of the earth contributes several inputs of sun, wind, rain, waves, land cycle, etc., to each area of the earth (Figure 3.7). Globally, the EMERGY required for each is the same since they are coproducts of each other. To add the EMERGY flows calculated separately would double count the original sources' contribution to an area.

The direct sun on every area contributes back to the whole world system in which it is a part. The wind, rain, and land cycle are all by-products of the world process, but different areas receive different percentages of earth's EMERGY in various inputs. Deserts may have more EMERGY in the land cycle, whereas rainforests may receive more EMERGY in rain.

To determine the total geobiospheric input, we could calculate and subtract out the double counting from each input so they could be added. A simpler way to determine how much solar EMERGY the earth system has contributed is to use the largest of the geobiospheric inputs and ignore the rest (as double counting).

The example in Figure 3.7 has the solar EMERGY inputs of each pathway written on the pathway. Rain is the largest (75×10^{13} sej/yr), and this is the correct one to use for the earth contribution of solar EMERGY flow to that local analysis.

SUMMARY

A baseline annual budget for global solar EMERGY was defined from which solar transformities of some main flows of air, water, and the earth cycle of the geobiosphere were calculated. These tables of transformities can be used for EMERGY evaluations of various smaller areas and subsystems of the planet.

CHAPTER 4

ENVIRONMENTAL PRODUCTION AND ECONOMIC USE

In this chapter both environmental production and its economic use are included within the window of systems attention. Calculating EMERGY flows into mining, agriculture, forestry, and fisheries evaluates the real wealth contributed by local environmental products. Calculating EMERGY of the purchased fuels, materials, goods, and services evaluates the resources brought in with economic investment and use. Indices relate EMERGY flow and the circulation of money, contrasting EMERGY value and market value. Two numerical examples, silk manufacture and shrimp aquaculture, illustrate EMERGY and dollar flows at the economic use interface.

As with all systems, we approach the environmental-economic interface by examining how the subject fits into the larger scale (the top-down approach). Figure 4.1 is the world energy system aggregated into three sectors; an environmental sector (natural areas), an environmental-economic production sector, and the main economy (including commerce, industry, urban consumers, and so forth). Dilute materials released from products as waste are shown recycling back into environmental production. Money (dashed line) is circulating in closed loops, somewhat conserved in the short term. What money can buy depends on the real-wealth production of the environmental processes on the left (note the two producer symbols) deriving their EMERGY from the baseline of earth inputs (see Chapter 3). By increasing the real wealth in the economy, the buying power of the circulating currency is increased.

CIRCULATION OF MONEY

Money circulates among humans and their economic activities. Money is a type of information that flows as a countercurrent in exchange for real goods

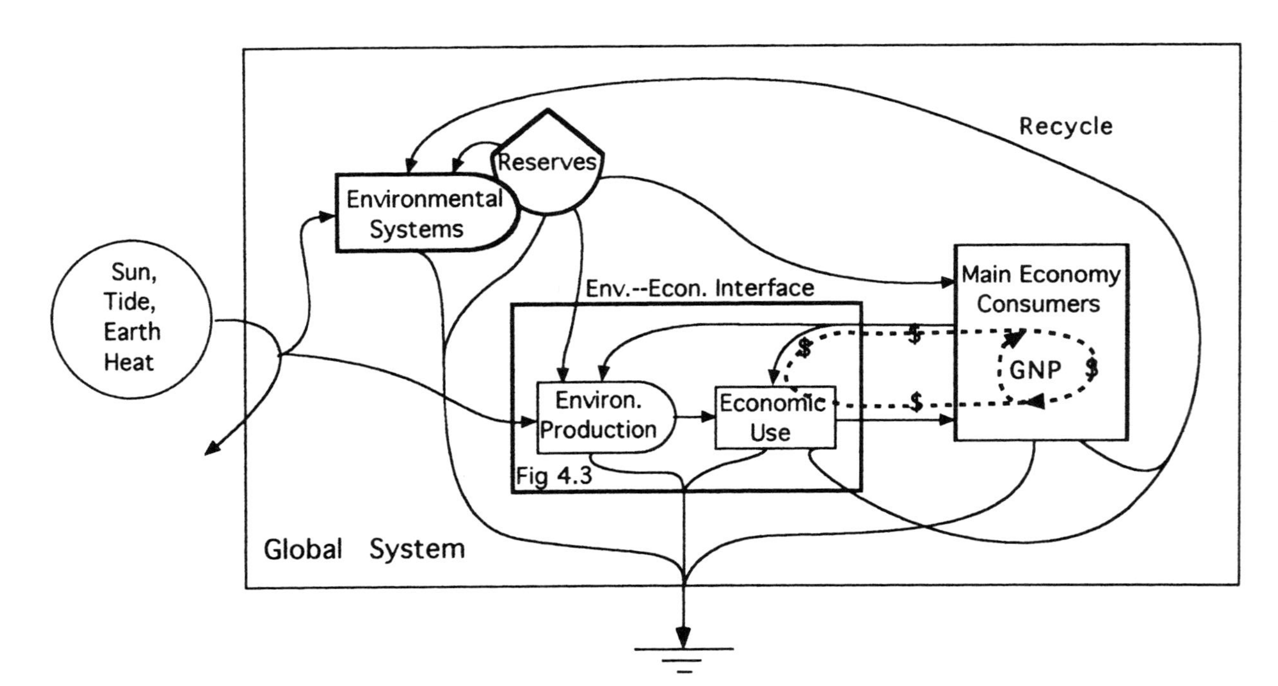

Figure 4.1. Systems overview of the earth showing its environmental production and the human economic system.

and services. Understanding the economy requires that both money circulation and the pathways of real wealth be represented together but separately. (Money is shown as dashed lines in systems diagrams such as (Figure 4.1.) Money is only paid to people and never to the environment for its work.

In a systems web (Figure 4.1), energy flows are large on the left, decreasing with successive transformation to the right. Environmental sectors on the left may have no dollar flow if humans are not involved. A few dollars are paid to people in environmental use sectors in the middle (i.e., forestry and fisheries). More money is circulated in sectors to the right (those concerned with paid human work).

The ratio of energy flow to money flow is an index with large values on the left and small values on the right. See for example the tables for the regional system of Gotland island (Sweden) by A. M. Jansson and J. Zucchetto (1978a,b). The ratio is an indication of the relative role of human service to environmental service within that sector. The distribution shows why energy has been important in environmental study and money in human thinking. However, neither of these properly measure work until their EMERGY contents are determined.

SOLAR EMERGY PER UNIT MONEY

Wealth directly and indirectly comes from environmental resources measured by EMERGY (not energy). Food, clothing, housing, information, health services, and other real wealth are measured by EMERGY. What the money circulating in the economy buys depends on the solar EMERGY production and the money circulation. Although the price of any particular product or service depends on its cost, local scarcity, and the willingness of people to pay, the buying power of money on the average depends on how much real wealth there is to buy. Therefore, the buying power of money within an economy may be calculated by dividing EMERGY use by the money circulation to obtain the EMERGY/money ratio. For 67 counties of Florida, Regan (1977, 1978) found a high correlation between measures of economic product and an estimate of EMERGY use in those counties, including rain, tides, and uses of soil and fuel.

If the total solar EMERGY used in a year by a state or nation is divided by the gross economic product expressed in local money units, an EMERGY-money index results (solar EMERGY/unit money with units in solar emjoules per dollar). Calculations for the United States are shown in Figure 4.2. The EMERGY/money quotient is a measure of buying power of the money. If more money is circulated for the same EMERGY flow, or less EMERGY is produced for the same money, there is inflation, defined as a decrease in the EMERGY/money ratio.

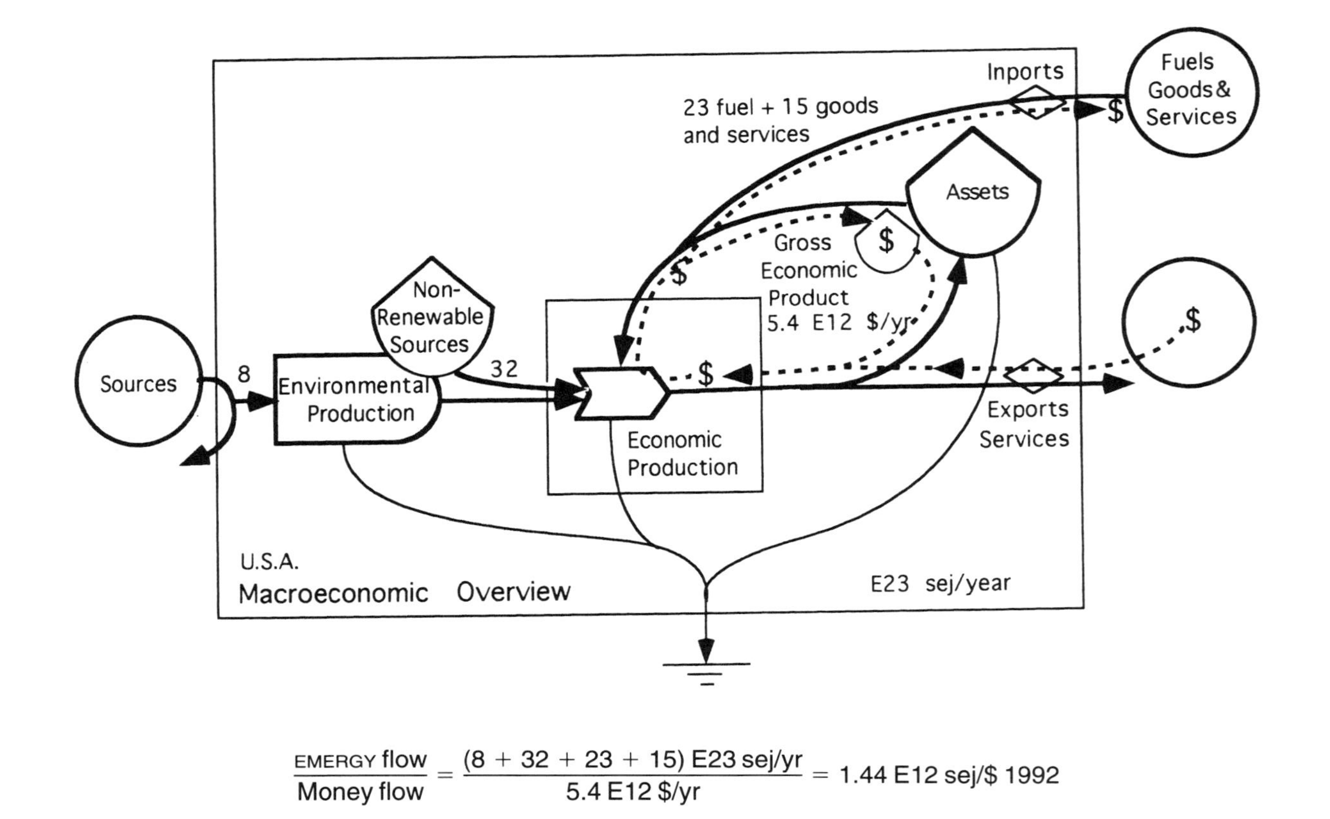

$$\frac{\text{EMERGY flow}}{\text{Money flow}} = \frac{(8 + 32 + 23 + 15)\ \text{E23 sej/yr}}{5.4\ \text{E12 \$/yr}} = 1.44\ \text{E12 sej/\$ 1992}$$

Figure 4.2. Energy systems overview of the United States for 1992 showing main EMERGY contributions, the gross economic product measuring currency circulating, and the EMERGY/currency ratio.

World Ratio of EMERGY per Money

For the world system (Figure 10.6), we can divide the world EMERGY flow by the world gross economic product that it supports, to obtain the solar EMERGY per unit U.S.$. For 1980 the EMERGY/$ ratio was 2.0×10^{12} sej/$:

$$\frac{9.44 \times 10^{24} \text{ sej/yr renewable} + 10.8 \times 10^{24} \text{ sej/yr nonrenewable}}{10.1 \times 10^{12} \text{ \$/yr world gross economic product}} = 2.0 \times 10^{12} \text{ sej/\$}$$

Since EMERGY flows can be estimated for environmental conditions for centuries in the past, it is possible to relate currencies of the past to present-day money values. A dissertation in accounting by Boyles (1975) explored EMERGY measures for giving long-range stability to accounting (under the name *embodied energy*).

EMDOLLAR (EM$)

If a flow of EMERGY is responsible for a portion of the real wealth of an economic system, we can infer that this proportion of the systems buying power is due to this EMERGY flow. For example, if agriculture is contributing 10% of the total annual EMERGY budget, then we may infer that 10% of the buying power of the gross economic product is due to agriculture.

The proportion of gross economic product due to an EMERGY flow is the Emdollar, abbreviated Em$:

> **The *emdollar* (Em$) value of a flow or storage is defined as its EMERGY value divided by the EMERGY/money ratio for an economy for that year.**

In this book, solar EMERGY is used, and emdollars are obtained by dividing by the solar EMERGY/money ratio. For example, an EMERGY flow of 1×10^{12} sej/yr divided by the EMERGY/money quotient for the U.S. economy in 1993 (1.4×10^{12} sej/$) = 0.71 1993 Em$/yr:

$$\frac{1 \times 10^{12} \text{ sej/yr}}{1.4 \times 10^{12} \text{ sej/1993 \$}} = 0.71 \text{ 1993 Em\$/yr}$$

The dollars of an economy identified with EMERGY of a resource contribution have been called *macroeconomic value,* referring to real wealth measured in emdollars. However, in this book the phrase is not used because it can be confused with economic indices.

Emdollars are the appropriate measure for discussing large-scale considerations of an economy, including environment and information, as well as human goods and services. Making decisions that will maximize emdollars is the same as maximizing EMERGY production and use, making an economy sustainable and competitive.

Evaluating Emdollars of Storages

Flows and storages may be expressed in solar EMERGY and emdollars. Flows are in solar emjoules per unit time and emdollars unit time. The EMERGY of a storage is the EMERGY flow used in developing the storage. An overview of the production of stored EMERGY value is shown in Figure 1.4.

For example, a rough estimate of the EMERGY in solar emjoules stored in a forest hectare that required 200 years to develop is the product of the annual EMERGY use (5×10^{14} sej/ha/yr) times the time of development:

$$\text{Stored EMERGY:} \quad (5 \times 10^{14}\ \text{sej/ha/yr})(200\ \text{yr}) = 1 \times 10^{17}\ \text{sej/ha}$$

When the EMERGY storage is divided by the 1993 EMERGY/money ratio, the quotient is the emdollar value in 1993 emdollars:

$$\text{Em\$ value:} \frac{1 \times 10^{17}\ \text{sej/ha}}{1.4 \times 10^{12}\ \text{sej/1993 \$}} = \$71{,}428\ \text{1993 Em\$/ha}$$

For another example, see the evaluation of peat storage in a swamp in Table 5.4.

Evaluating Average Services

Since money is paid to people for their services, the money can be given an average solar EMERGY value by multiplying their income by the average EMERGY/money ratio of the economy from which the services were purchased. This procedure gives an average EMERGY/money to the purchased services. Thus, using an average sej/$ ratio for evaluating services is most appropriate for overall aggregate calculations. For a particular service, it may be more accurate to multiply the metabolic energy of the human service by the solar transformity of that service. See Chapter 12.

ECONOMIC USE INTERFACE

Within Figure 4.1 is the box "Env.-Econ. interface," which contains the environmental production–economic use subsystem. This box is enlarged as Figure 4.3. Notice that the EMERGY entering the frame is partly from the renewable and nonrenewable earth sources on the left and also that from purchased, high-transformity fuels, electricity, capital goods, labor, and services from the right. Thus, EMERGY value is added from free and purchased inflows. Money

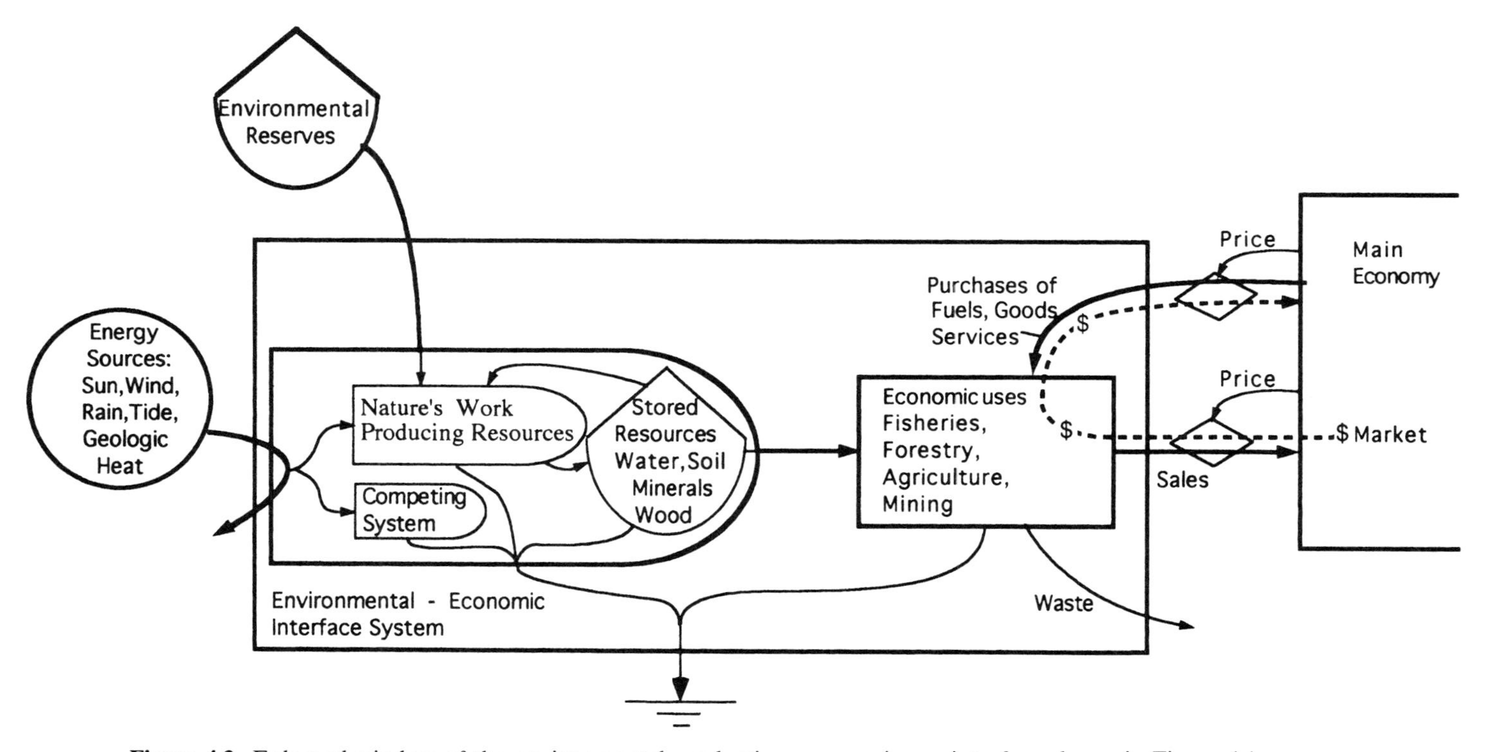

Figure 4.3. Enlarged window of the environmental production–economic use interface shown in Figure 4.1.

is only paid for the purchased inputs and only to people who are in the sectors to the right.

Environmental-economic interfaces like this one connect environmental products with human use. As shown in Figure 4.3, environmental products interact with items bought from the economy. The environmental product becomes an economic product that is sold to pay for the goods and services necessary for the processing. The money paid for the product goes back to the right, to people. Money is not paid to environmental systems on the left that did the production work.

As shown in Figure 4.3, two EMERGY sources contribute to the interface process: part from the work of nature on the left and part bought from the economy to the right. If the main economy on the right has a separate source, the two EMERGY flows may be added to obtain the EMERGY of the output of the product sent out for sale. The total EMERGY flow is a measure of that system's output. The perceptive reader comparing the smaller window (Figure 4.3) with the global picture (Figure 4.1) will see that the purchased inputs are indirectly supplied from environmental resources elsewhere in the system. If questions about double counting of the same environmental sources arise, turn to Chapter 6, which deals with details of assigning EMERGY to pathways. Much of the monied economy runs on past fuel-energy reserves, an independent source by virtue of its different time of formation from contemporary environmental inputs. Where small sectors are evaluated, double counting involved in adding renewable environmental EMERGY and purchased EMERGY is negligible.

MARKET VALUES NOT FOR ENVIRONMENTAL EVALUATION

Since money is paid only to people for their contribution, and never to the environment for its contribution, money and market values cannot be used to evaluate the real wealth from the environment. When the resources from the environment are abundant, little work is required from the economy, costs are small, and prices low. However, this is when the net contribution of real wealth to the economy is greatest: this is when everyone has abundant resources and a high standard of living.

When the resources are scarce, obtaining costs are higher, supply-and-demand principles cause higher prices, and the market puts a high value on the product. However, this is also when there is little net contribution of the resource to the economy, real wealth is scarce, and standards of living are low. Market prices are not proportional to the contribution that resources make to the economy; prices are low when EMERGY contributions are greatest:

> **Market values are inverse to real-wealth contributions from the environment and cannot be used to evaluate environmental contributions or environmental impact.**

With EMERGY evaluation all pathways are measured on the same basis, namely the solar EMERGY of the available energies previously used up. Contributions from nature and those by humans are evaluated in common units (EMERGY or emdollar units).

Since money can only evaluate the contributions of people, using the money paid to evaluate an environmental resource greatly underestimates its contribution. The EMERGY of environmental resources contributes more real wealth than is paid for.

NET EMERGY BENEFIT IN PURCHASES OF ENVIRONMENTAL PRODUCTS

When a commodity is purchased, money is paid for the goods, as diagrammed in Figure 4.4a. The EMERGY and emdollar values may be calculated for both, as has been done for the oil purchase in Figure 4.4b. The EMERGY benefit to the purchaser can be inferred by calculating the ratio of the EMERGY in the product to that in the buying power of the money paid.

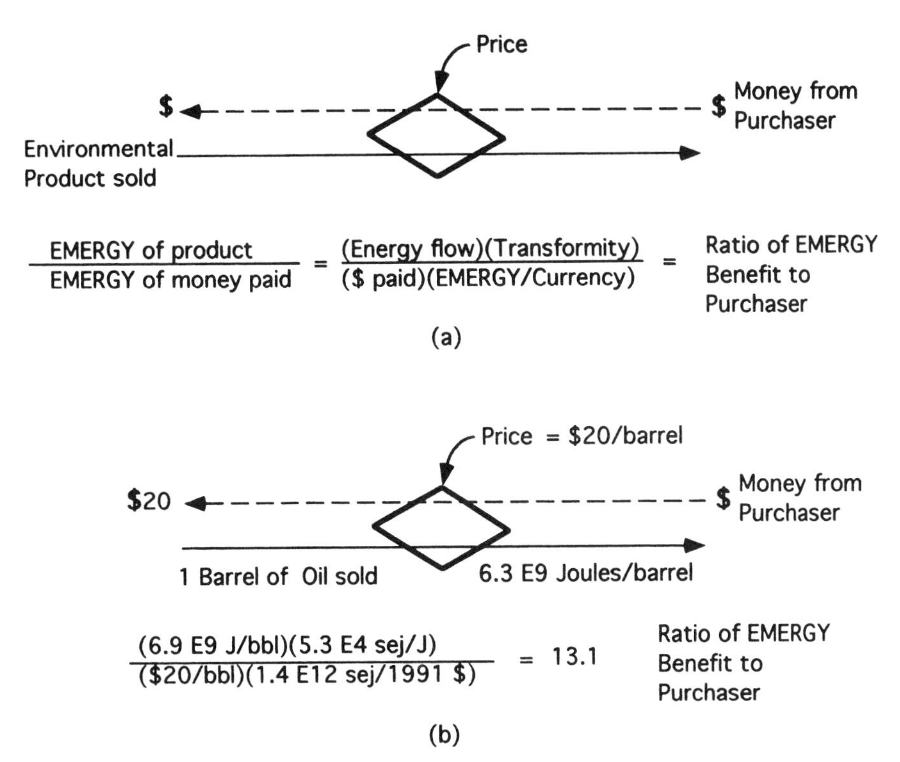

Figure 4.4. EMERGY flows in a purchase process. **(a)** Diagram defining an exchange ratio. **(b)** An example, the purchase of oil (bbl = barrels).

In the oil example (Figure 4.4b), the solar EMERGY of the oil was obtained by multiplying the energy in the purchased oil by its solar transformity. The solar EMERGY in the money's buying power was obtained by multiplying the money paid by the solar EMERGY/money ratio. Thirteen times more EMERGY was delivered to the purchaser than was available in the buying power of the payment.

Table 4.1 shows prices and EMERGY contributions of many environmental products. Almost always the EMERGY value of an environmental product is much larger than that in the payment.

VALUE ADDED AND INCREASED TRANSFORMITY

As illustrated by Figure 4.5, the output from the environmental interface may be processed through successive stages (left to right). For example, wood goes to cutter, transporter, wholesaler, retailer, and so forth. At each stage additional inputs of EMERGY from fuels, electricity, goods and services, and so on, are shown connecting from above, each accompanied by a counterflow of money in payment. Each stage has "value added." The value added by human services is often expressed by the dollar flow paid for the added services. However, since there are other EMERGY inflows besides services added, EMERGY evaluation is required to be complete. For example, the EMERGY in furniture is that of the wood from environmental work, EMERGY of human services added, and the EMERGY of the electricity, equipment, and varnish used in manufacture.

As Figure 4.5 shows, the money paid for the first step of economic processes is much less than the total money circulation due to that product, which may be 5 or 10 times larger (see the last money loop on the right end of Figure 4.5). Economic studies are often made to determine the overall contribution of a resource input by tracking the item through the economy. An easier way to estimate the overall contribution is to evaluate the EMERGY and divide by the EMERGY/money ratio to obtain the emdollar (EM$) value.

The chain of economic process increases the transformity of the products moving to the right to be processed further by the economy. Transformity increases in ecological and economic energy transformation chains.

REINFORCEMENT OF ENVIRONMENTAL PRODUCTION

In the environmental-economic interface of Figure 4.3, products are sold, and if the prices are good, money accumulates and buys more inputs to harvest more of the environmental product (wood, fish, crops, and so forth). The economic process is mathematically autocatalytic and tends to accelerate and grow. As the environmental product gets scarce, prices rise, which encourages those using the products to go after more. By pulling down the environmental

TABLE 4.1. EMERGY Advantages of Buyers of Agricultural Products

Note	Item	Unit	1983 Price ($)	EMERGY Ratio*
1	Water	Acre-foot	50	1.9
2	Potatoes	100 lb	8.50	2.0
3	Fuel (1984)	Gallon	1.00	3.3
4	Wheat	Bushel (bu)	3.55	3.5
5	Cotton	Pound	0.59	3.9
6	Beef	100 lb	55	6.5
7	Fertilizer	Ton	164	11.8
8	Wool	Pound	0.83	16.7

* Ratio of EMERGY in commodity to EMERGY of money paid. Prices from Texas Livestock, Dairy and Poultry Statistics for 1983, Texas Dept. of Agriculture (Odum et al., 1987a).

[1] 1 acre-foot of water:

$$(1 \text{ ac-ft})(4.05 \times 10^3 \text{ m}^2/\text{acre})(0.3 \text{ m/ft})(1 \times 10^6 \text{ g/m}^3)(5 \text{ J/g})(4.1 \times 10^4 \text{ sej/J}) = 2.49 \times 10^{14} \text{ sej}$$

$$\text{EMERGY ratio: } (2.49 \times 10^{14} \text{ sej})/[(\$50)(2.6 \times 10^{12} \text{ sej/\$})] = 1.9$$

[2] 100 lb of potatoes:

$$(100 \text{ lb})(454 \text{ g/lb})(0.22 \text{ dry})(4 \text{ kcal/g})(4186 \text{ J/kcal})(2.6 \times 10^5 \text{ sej/J})/[(\$8.50)(2.6 \times 10^{12} \text{ sej/\$})] = 2.0$$

[3] 1 gallon of fuel:

$$\text{EMERGY ratio: } (1.3 \times 10^8 \text{ J/gal})(6.6 \times 10^4 \text{ sej/J})/[(\$1)(2.6 \times 10^{12} \text{ sej/\$})] = 3.3$$

[4] 1 bu of wheat:

$$(1 \text{ bu})(27.2 \times 10^3 \text{ g/bu})(3.3 \text{ kcal/g})(4186 \text{ J/kcal})(8.6 \times 10^4 \text{ sej/J})/[(\$3.55)(2.6 \times 10^{12} \text{ sej/\$})] = 3.5$$

[5] 1 lb of cotton:

$$(1 \text{ lb})(454 \text{ g/lb})(3.7 \text{ kcal/g})(4186 \text{ J/kcal})(8.6 \times 10^5 \text{ sej/J})/[(\$.59)(2.6 \times 10^{12} \text{ sej/\$})] = 3.9$$

[6] 100 lb of beef:

$$(100 \text{ lb})(454 \text{ g/lb})(2.82 \text{ kcal/g})(4186 \text{ J/kcal})(1.73 \times 10^6 \text{ sej/J})/[(\$55)(2.6 \times 10^{12} \text{ sej/\$})] = 6.5$$

[7] 1 ton of fertilizer:

$$(1 \text{ ton})(9.07 \times 10^5 \text{ g/ton})(5.55 \times 10^9 \text{ sej/g})/[(\$164/\text{ton})(2.6 \times 10^{12} \text{ sej/\$})] = 11.8$$

[8] 1 lb of wool:

$$(1 \text{ lb})(454 \text{ g/lb})(5 \text{ kcal/g})(4186 \text{ J/kcal})(3.8 \times 10^6 \text{ sej/J})/[(\$.83)(2.6 \times 10^{12} \text{ sej/\$})] = 16.7$$

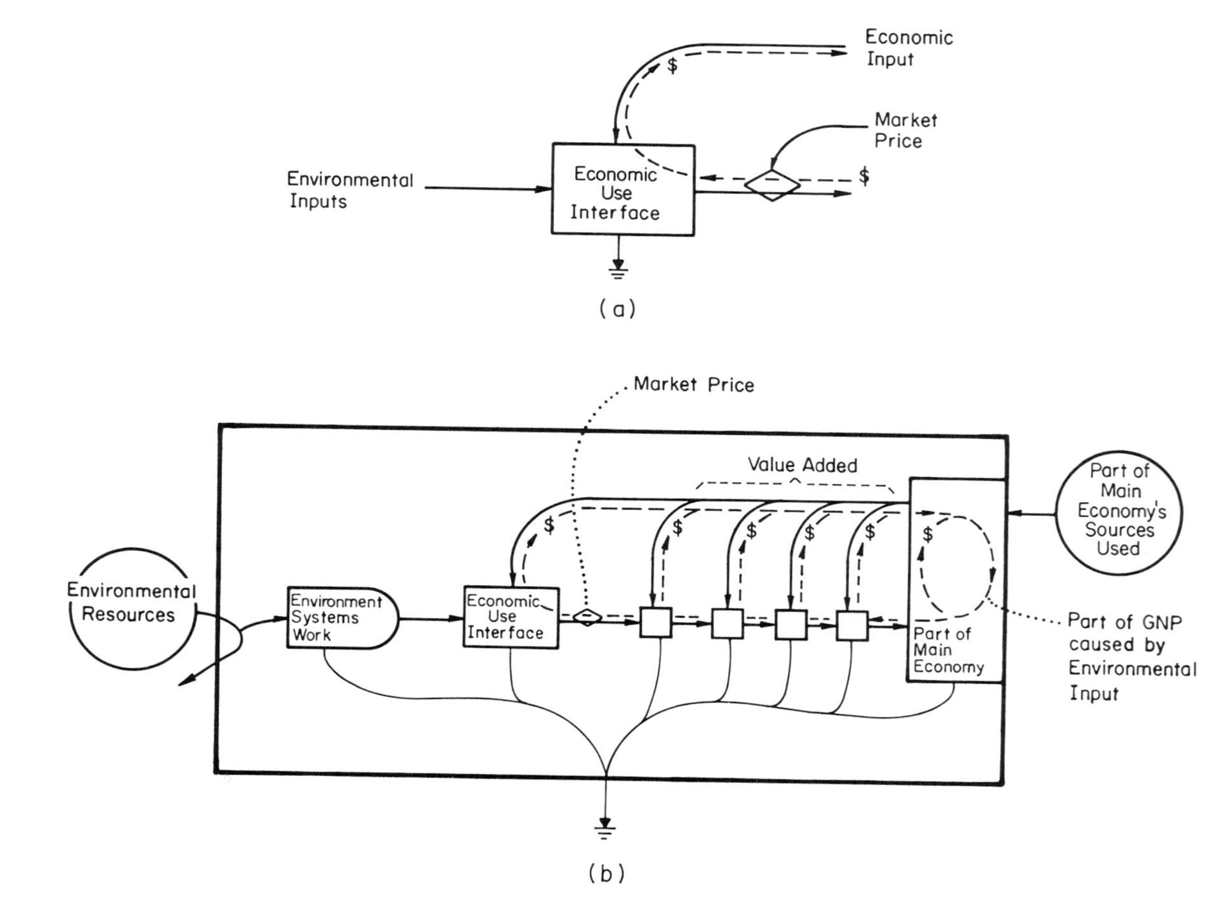

Figure 4.5. EMERGY flow with successive economic-energy transformations that increase transformity of products in which EMERGY and dollar values are added.

stocks that are part of the production process, the environmental producers are diminished and tend to be replaced by their competitors. Cutting the desirable trees causes weed trees to replace them. Harvesting the desirable fish causes their inedible competitors to replace them. In other words, the interface design shown in Figure 4.3 is not sustainable.

Figure 4.6 is better because it has a reinforcement back from the economic system. Agriculture tends to be sustainable, because the economy and the farmers feed back goods, services, fertilizers, and seeds to reinforce and encourage the environmental system that is in economic use. Notice in Figure 4.6 the special flow of money to pay for the feedbacks to reinforce the environmental production process.

However, when economic competition is severe, the reinforcement efforts may be omitted, causing eventual collapse of the environmental basis, and thus of the economic production as well. Many of the major fisheries of the world have now collapsed in this way.

Some authors use the monetary cost of reinforcing nature as a measure of its value. This is incorrect, underestimating the wealth required for replacement, because the major free environmental contributions are not included. Notice in Figure 4.6 the way the EMERGY for the economic feedback to help restore natural capital is only part of the EMERGY required. The rest is derived from the energy sources on the left.

EXAMPLES OF PRODUCTION BY ENVIRONMENTAL ECONOMIC INTERFACE SYSTEMS

Results of evaluating the EMERGY flows of two environmental production systems are given as examples next. Each is based on an EMERGY evaluation table prepared according to the procedures described in Chapter 5.

Indian Silk Manufacture

An environmental-economic EMERGY analysis of Indian silk manufacture is shown in Figure 4.7 and Table 4.2. This system has considerable labor within the system but not many purchased goods and services from the main economy. In Figure 4.7a, inputs contributing solar EMERGY are written on the pathways. These are summarized in a three-arm diagram (Figure 4.7b).

Shrimp Aquaculture in Ecuador

In Figure 4.8 is an overview of shirmp aquaculture in Ecuador (Odum and Arding, 1991). For purposes of this chapter the pathways have been much aggregated; the original diagram had 14 input pathways, but with the same totals for inputs and outputs.

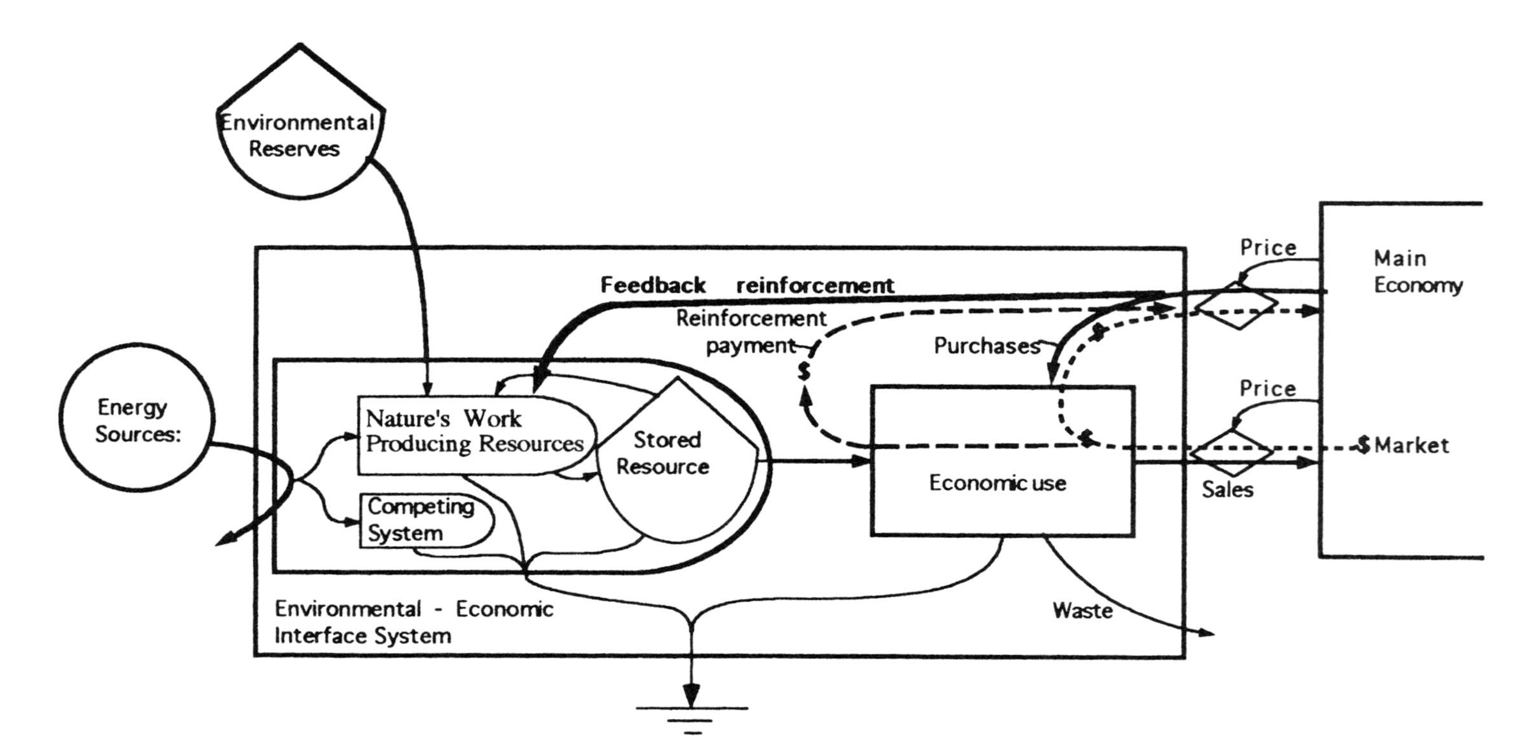

Figure 4.6. Environmental-economic interface showing the competitors that displace desired production when feedback reinforcement is missing.

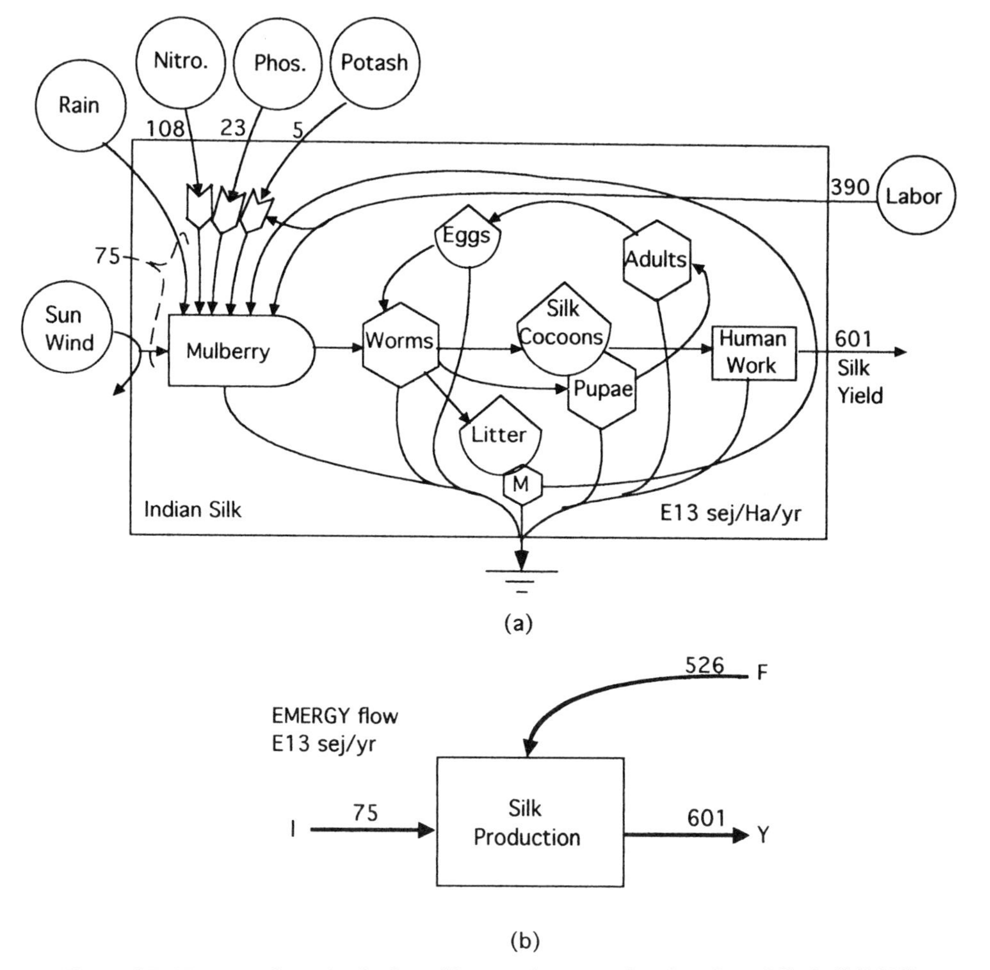

Figure 4.7. EMERGY flows for Indian silk manufacture using data from Mitchell (1979). **(a)** Systems diagram with contributing EMERGY flows. **(b)** Aggregated summary of EMERGY flows. See Table 4.2.

PRODUCTION INDICES AND THE THREE-ARM DIAGRAM

As shown in the two previous examples, EMERGY flows of environmental-economic systems can be summarized with a three-arm diagram (Figures 4.7b and 4.8b). Free environmental inputs are combined in the pathway from the left (environmental input I). Inputs from the economy are combined in the pathway from the top right (feedback F). The yield of product outflow is on the lower right (Y). Several indices calculated from this diagram give perspective on the type and efficiency of the environmental use.

TABLE 4.2. EMERGY Evaluation of Indian Silk Production, 1 ha (See Figure 15.2)

Note	Item, Unit	Data (units/yr)	Unit Solar EMERGY (sej/unit)	Solar EMERGY ($\times 10^{13}$ sej/yr)	Em$ value* ($/yr)
1	Direct sun, J	7.32×10^{13}	1.0	7.3	—
2	Rain, J	5.0×10^{10}	1.5×10^{4}	75	375
3	Nitrogen (N), g	2.57×10^{5}	4.19×10^{9}	108	540
4	Phosphorus (P_2O_5), g	5.1×10^{4}	4.6×10^{8}	23	115
5	Potassium (K_2O), g	5.1×10^{4}	9.5×10^{8}	5	25
6	Labor, J	4.81×10^{10}	8.1×10^{4}	390	1950
7	Sum of inputs omitting sun to avoid double counting			601	3005

Transformities of products derived from sum of EMERGY input (601×10^{13} sej/yr) divided by energy flows:

Note	Item	Energy Flow (J/yr)	Transformity (sej/yr)
8	Mulberry leaves	2.5×10^{11}	2.4×10^{4}†
9	Pupae formed	3.0×10^{9}	2.0×10^{6}
10	Litter	2.22×10^{11}	2.7×10^{4}
11	Cocoon silk	1.75×10^{9}	3.4×10^{6}

* Solar EMERGY/yr divided by 2×10^{12} sej/U.S. $ for 1990.

[1] From Mitchell (1979): 1.75×10^{10} kcal/yr

$$(1.75 \times 10^{10}\,\text{kcal/yr})(4186\,\text{J/kcal}) = 7.32 \times 10^{13}\,\text{J/yr}$$

[2] Assumed:

$$(1.0\,\text{m}^3/\text{m}^2/\text{yr})(1 \times 10^6\,\text{g/m}^3)(5\,\text{J/g})(1 \times 10^4\,\text{m}^2/\text{ha}) = 5 \times 10^{10}\,\text{J rain/yr}$$

[3–5] Fertilizer used per hectare. EMERGY per gram from Odum and Odum (1983). EMERGY per gram for P_2O_5 was 0.23 of that for phosphorus. EMERGY values used here are different from embodied energy given by Mitchell based only on embodied fuel.

[6] Energy per person was supplied by Mitchell (1979) and multiplied by 4186 kcal J. Solar transformity for primitive (uneducated) labor is 8.1×10^4 from Odum and Odum (1983).

[7] Main EMERGY inputs are the rain, fertilizer, and labor:

$$75 + 108 + 23 + 5 + 390 = 601 \times 10^{13}\,\text{sej/yr}$$

[8–11] Solar transformities are ratios of input solar EMERGY (601×10^{13} sej/yr) divided by energy flow of each product from Mitchell (1979). Price of silk used to calculate EMERGY exchange ratio: $12,835/t 1983 U.S.$. For transformities see p. 71.

$$\text{Solar EMERGY per gram: } \frac{601 \times 10^{13}\,\text{sej}}{8.36 \times 10^4\,\text{g}} = 7.19 \times 10^{10}\,\text{sej/g}$$

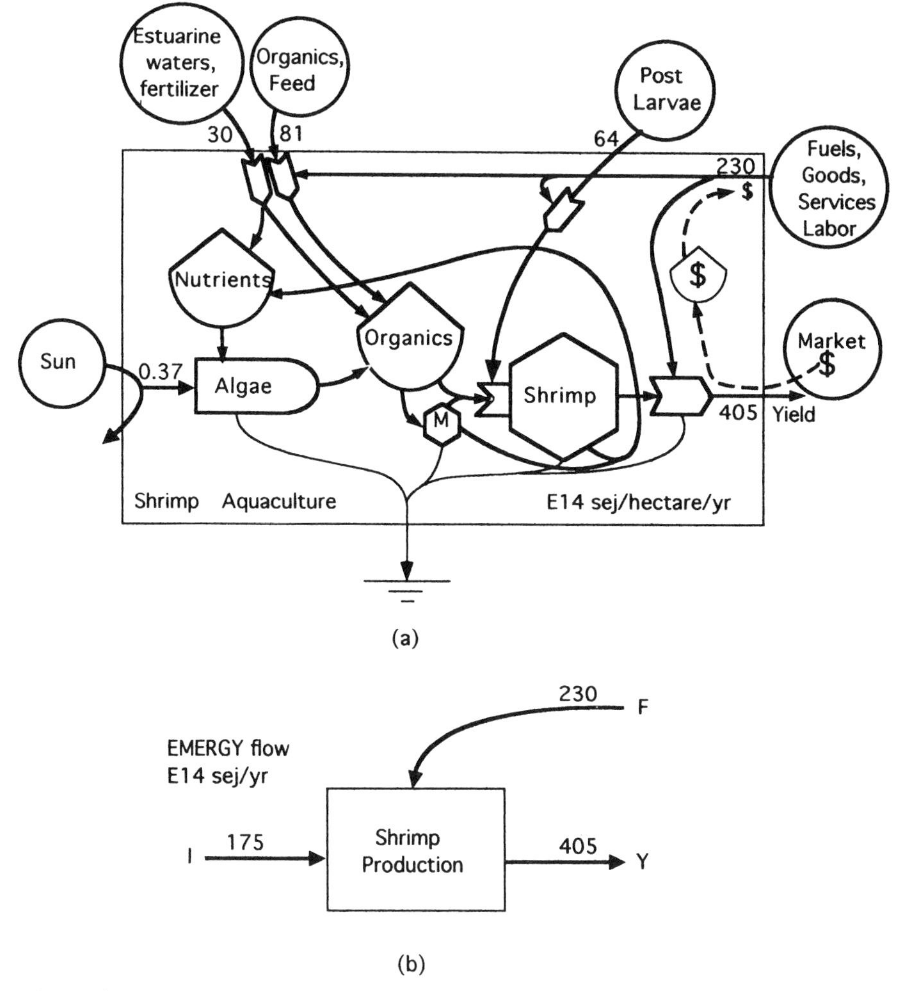

Figure 4.8. EMERGY flows for shrimp aquaculture in Ecuador (Odum and Arding, 1991). **(a)** Systems diagram with contributing EMERGY flows. **(b)** Aggregated summary of EMERGY flows.

EMERGY Investment Ratio

EMERGY investment ratio is the purchased EMERGY (F) feedback from the economy (services and other resources), divided by the free EMERGY inflow from the environment (I). This index measures the intensity of the economic development and the loading of the environment. Examples are given in Table 4.3. In highly developed countries like the United States, the investment ratio tends to be 7 or higher (see Chapter 10). Depending on how some of the pathways are assigned to the three-arm diagram, there are several variations (see Chapter 5). For the silk example (Figure 4.7b) the ratios are as follows:

TABLE 4.3. EMERGY Investment Ratios of Useful Environmental Products

Note	Item	Economic/ Environment Ratio*
1	Harvested mature tropical forest trees, Jari, Brazil	0.14
2	Tropical plantation wood, Jari, Brazil	0.85
3	Temperate plantation wood, New Zealand	0.97
4	Forage, Italy	1.32
5	Rice, Italy	2.7
6	Wheat, Italy	3.2
7	Shrimp from aquaculture, Ecuador	3.4
8	Mutton, new Zealand	3.7
9	Olives, Italy	4.1
10	Silk, India	6.9
11	Sugarcane alcohol, Bahia, Brazil	7.0
12	Cotton, Texas	9.6
13	Oranges and lemons, Italy	11.2
14	Corn, Illinois	12.5
15	Palm oil, Bahia, Brazil	17.0
16	Cacao dried seed for export, Brazil	17.0
17	Sunflowers, Italy	26.3

* Defined as purchased empower divided by free empower (Table 5.3).
Source: Items 1, 2, 11, 15, 16, Odum et al. (1986); 12, H. T. Odum et al. (1987b); 7, Odum and Arding (1991); 4–6, 9, 13, 17, Ulgiati et al. (1993); 3, Table 5.2, 5.3 10, Table 4.2; 8, Odum and Odum (1983); 14 Odum (1984a).

$$\text{Investment ratio of silk:} \quad \frac{(390 + 108 + 23 + 5) \times 10^{13}\ \text{sej/yr}}{75 \times 10^{13}\ \text{sej/yr}} = 7.0$$

$$\text{Investment ratio of shrimp:} \quad \frac{230 \times 10^{14}\ \text{sej/ha/yr}}{(30 + 81 + 64) \times 10^{14}\ \text{sej/ha/yr}} = 1.3$$

The EMERGY investment ratio for silk is much higher than the average ratio within the Indian economy. The silk is comparable in development intensity with intensive agriculture in developed countries. Intensive Illinois corn (Odum, 1984d) has an investment ratio of 12.5. For the shrimp aquaculture in Ecuador the components of environment drawn into the production are large compared to the purchased investment.

EMERGY Yield Ratio

The EMERGY exchange ratio (Figure 4.4) is the ratio of EMERGY of yield to the EMERGY of the money received *only.* In contrast, the *EMERGY yield ratio* is the EMERGY of yield divided by the EMERGY of *all* the feedbacks from the economy including fuels, fertilizers, and services.

The EMERGY yield ratio of each system output is a measure of its net contribution to the economy beyond its own operation. This index is important for evaluating fuels (Chapter 8), since they must support more than their own system. A good EMERGY yield ratio for domestic fuels in 1991 was 6. This net yield was less than if purchased from the Near East (Figure 4.4) with an EMERGY yield ratio of 13.1. No wonder that the percentage of foreign oil brought into the United States increased at that time.

For the silk example in Figure 4.7:

$$\text{Net EMERGY yield ratio of silk: } \frac{601 \times 10^{13} \text{ sej/yr}}{(390 + 108 + 23 + 5) \times 10^{13} \text{ sej/yr}} = 1.14$$

For the shrimp example in Figure 4.8:

$$\text{Net EMERGY yield ratio of shrimp: } \frac{405 \times 10^{14} \text{ sej/ha/yr}}{230 \times 10^{14} \text{sej/ha/yr}} = 1.76$$

For silk, the ratio is barely larger than 1, which means that the silk uses almost as much resource of the economy as it contributes. It is not a primary energy source that supports other activities.

For shrimp aquaculture in Ecuador, the net contribution to the economy is higher, a stimulus to the economy that is able to purchase it.

In these examples the output EMERGY was calculated by summing EMERGY inputs to the same system. No net EMERGY yield ratio of less than 1 can be obtained using this method. Using inputs for outputs assumes *a priori* that there are not unnecessary wastes and losses. If the system is being examined for efficiency and practicality, evaluating output with an independently determined transformity may be desirable for comparison.

Determining Transformity of Products

An environmental-economic system may be used to determine solar transformities of the products. The total EMERGY input flow (solar emjoules per time) is divided by the output energy flow (joules per time). For example, from Figure 4.7 a total solar EMERGY flow of 601×10^{13} sej/ha/yr into silk production generated 2.5×10^{11} J/ha/yr of collected mulberry leaves and 1.75×10^{9} J/ha/yr of silk. These are co-products of the same input EMERGY:

$$\text{Solar transformity of leaves: } \frac{601 \times 10^{13} \text{ sej/ha/yr}}{2.5 \times 10^{11} \text{ J/ha/yr}} = 2.40 \times 10^{4} \text{ sej/J}$$

$$\text{Solar transformity of silk: } \frac{601 \times 10^{13} \text{ sej/ha/yr}}{1.75 \times 10^{9} \text{ J/ha/yr}} = 3.43 \times 10^{6} \text{ sej/J}$$

For the shrimp aquaculture example (Figure 4.8), where energy of shrimp yield was 3.16×10^9 J/ha/yr:

$$\text{Solar transformity, harvested shrimp:} \quad \frac{4.05 \times 10^{16}\ \text{sej/yr}}{3.16 \times 10^{9}\ \text{J/yr}} = 1.28 \times 10^{7}\ \text{sej/J}$$

SUMMARY

By considering the circulation of money within the global system and through the environmental-economic interface, money was related to energy and EMERGY. Properties and indices of EMERGY production and use were studied with two numerical examples. Emdollars were assigned to products according to their proportion of the economy's EMERGY budget. Much more EMERGY is contained in environmental products than in the paid services used to process the products. Using market values (money paid to people) to evaluate wealth of environmental resources was found to be many times too small.

CHAPTER 5

EMERGY EVALUATION PROCEDURE

After the definitions, concepts, and examples relating energy, EMERGY, and money in previous chapters, we next set out the procedure for environmental accounting with EMERGY. Evaluation starts with energy systems diagramming to obtain an overview of the system, its parts and processes, the problems, the contributing factors, and alternatives for management. The pathways of the overview diagram determine the line items in an EMERGY evaluation table. Solar EMERGY and emdollars are calculated for each inflow, product, or item of special interest. On all scales involved, choices are made on the basis of maximum EMERGY production and use. Indices from evaluation tables are used to make policy recommendations. Evaluation procedures are illustrated for annual EMERGY production and use of a New Zealand pine plantation and for EMERGY storage in a Florida swamp.

ENERGY SYSTEMS DIAGRAMMING

To understand a problem, we need to understand both the mechanisms and the way the problem is controlled by the larger surrounding system. For understanding, evaluating, and simulating, procedures start with diagramming the system of concern. This initial diagramming may be done in detail with everything mentioned put on paper, even if thought to be minor. The first complex diagram is like an inventory (examples in Figures 2.6 and 5.1a). After a complex initial diagram is drawn, it may be simplified by aggregation, without eliminating any of the EMERGY flows. George Knox (1986) has many examples of diagramming estuarine systems.

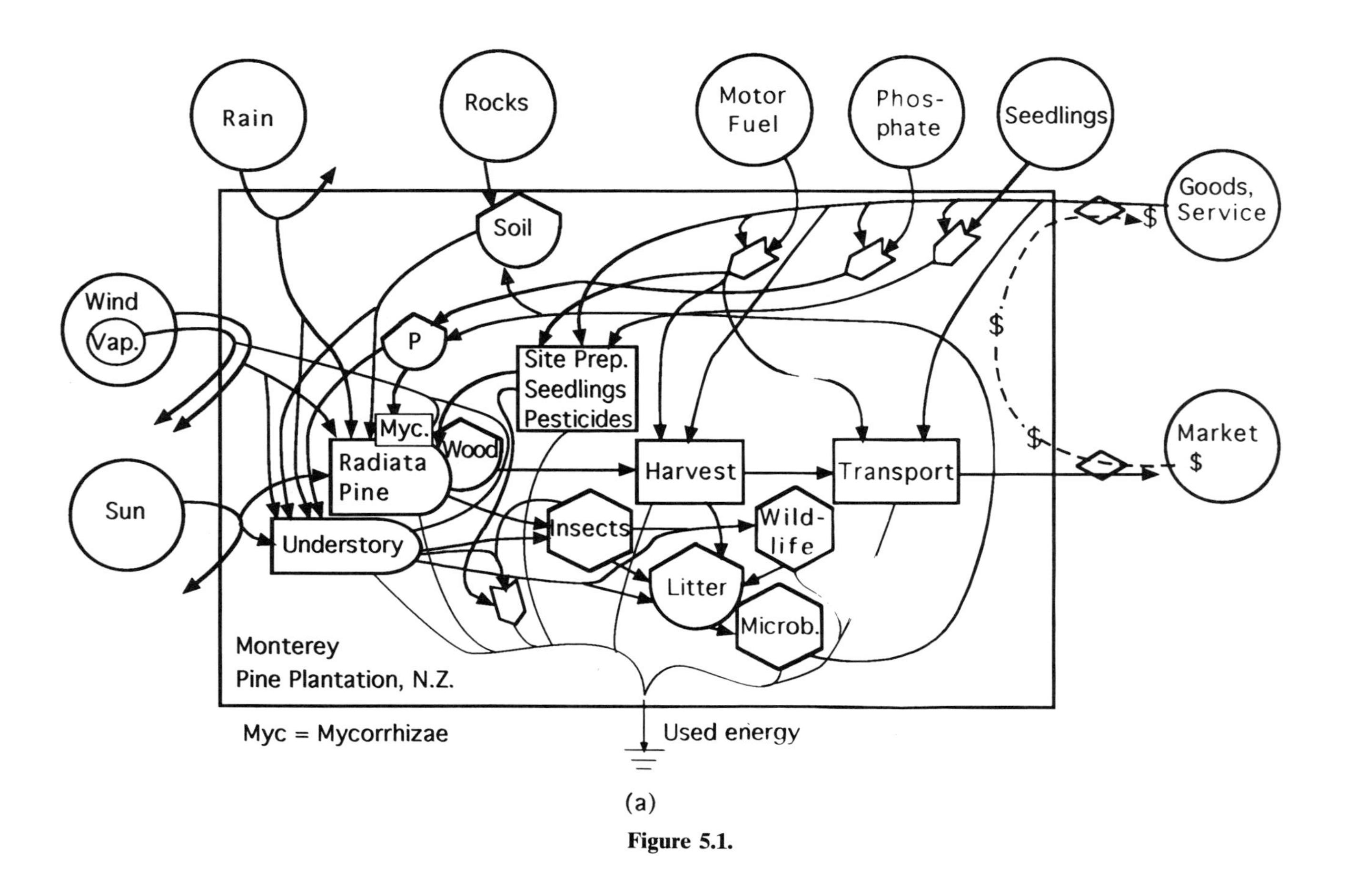
Rain
Rocks
Motor Fuel
Phos-phate
Seedlings
Goods, Service
Market $
Wind
Vap.
Sun
Soil
P
Site Prep.
Seedlings
Pesticides
Myc.
Wood
Radiata Pine
Understory
Harvest
Transport
Insects
Wild-life
Litter
Microb.
$
Monterey
Pine Plantation, N.Z.
Myc = Mycorrhizae
Used energy

(a)

Figure 5.1.

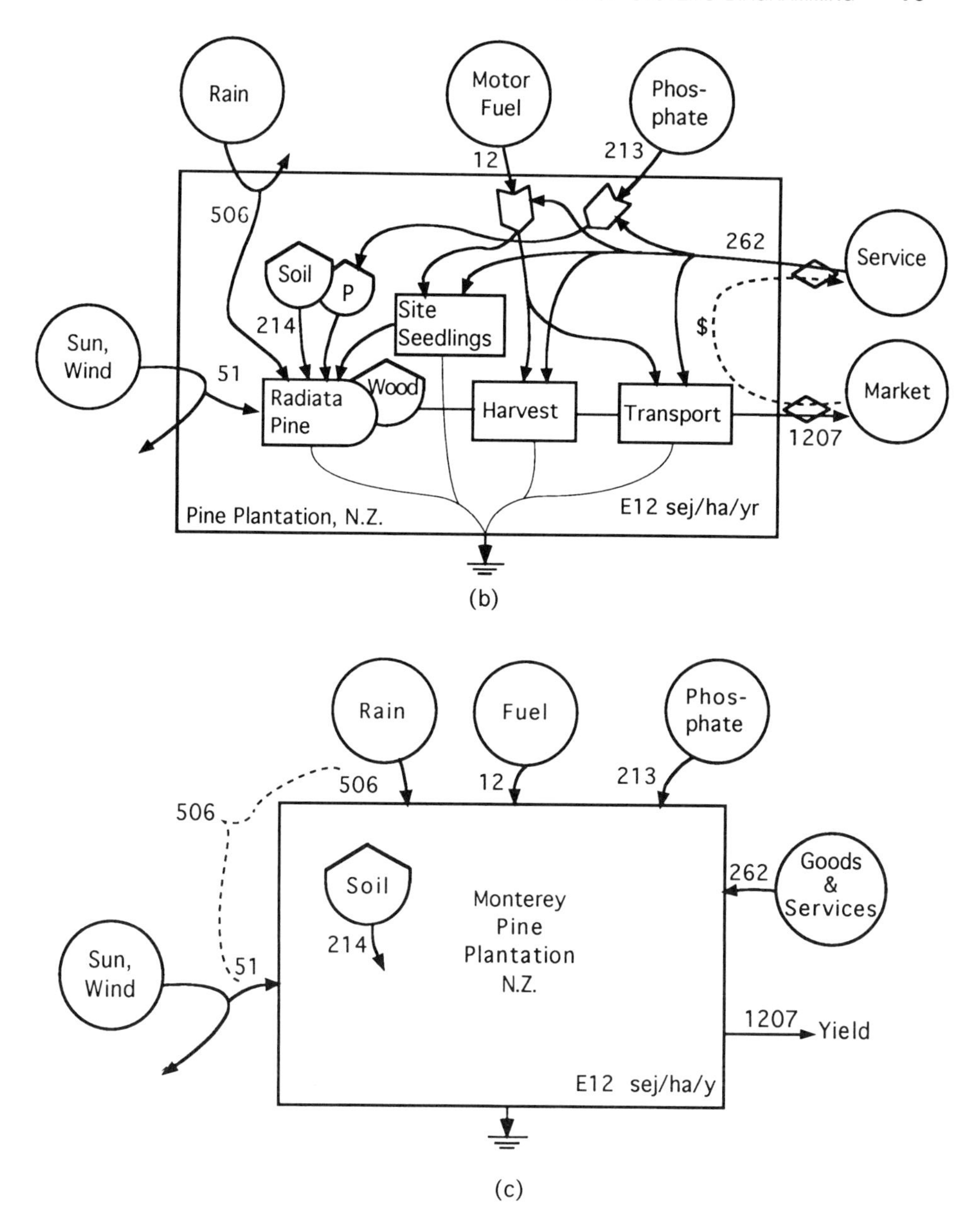

Figure 5.1. Energy systems diagram of Monterey pine plantations in New Zealand evaluated in Table 5.2. **(a)** Complex drawing, **(b)** Diagram with some details, **(c)** Aggregated summary.

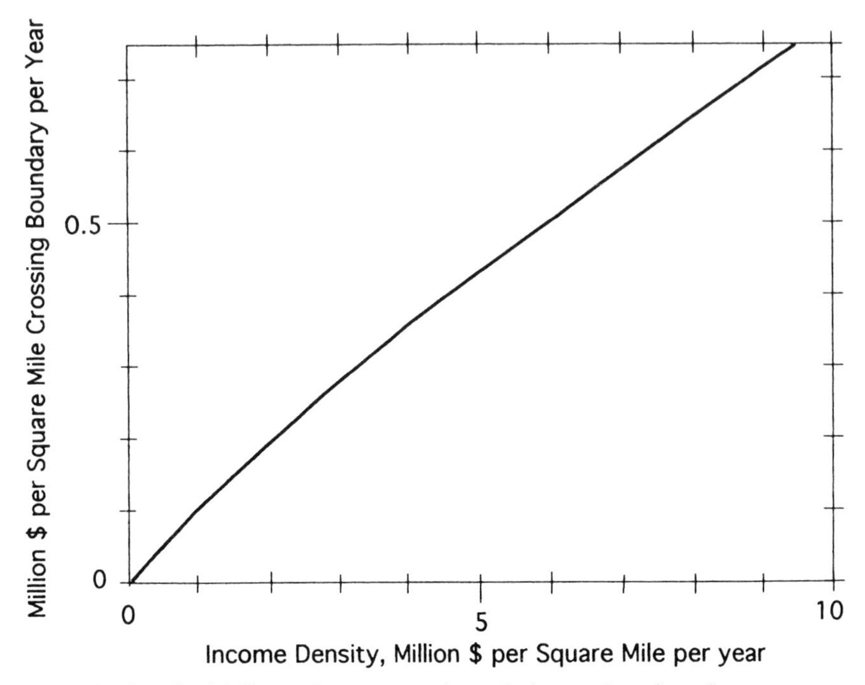

Figure 5.2. Graph of dollar exchange across boundaries as a function of area concentration of gross economic product (M. T. Brown, 1980).

Often this overview diagramming is done by a group of people around a table, sharing knowledge and making the diagram together. Sometimes the leader draws on a blackboard while another person makes a permanent version on poster-sized paper. Pencil is used since it is necessary to erase often, relocating symbols so as to fit everything in that is mentioned. The diagram on paper is two-dimensional but has to represent a system that has many dimensions (many variables). When the first draft of a detailed diagram is completed, the group is usually satisfied that all their ideas and inputs are incorporated in the result. Since the diagram usually includes the environment and the economy, it is an organized impact statement.

Using the Monterey pine plantations (Figure 5.1) as an example, these are the usual steps in diagramming:

(1) Define the boundary of the window of the systems overview, thus separating the internal components and processes from the influences from outside that boundary.

(2) List the important sources (external causes, external factors, and forcing functions). Importance means than an effect is suspected to be 5% or more of the total system function.

(3) List the principal components within the boundary and units believed to be important considering the scale of the system defined.
(4) List the processes (flows, relationships, interactions, production and consumption processes, and so on). Include flows and transactions of money believed to be important.
(5) Draw the systems diagram, starting with the external sources arranged around the rectangular frame marking the boundary. Draw in the symbols for components. Arrange sources and components in order of transformity from left to right. Then connect pathways between the symbols.

Symbols

The symbols (Figure 1.2) each have energetic and mathematical meanings that are given in great detail elsewhere (H. T. Odum, 1983b, 1994a). Appendix A has a brief account that should facilitate diagramming for purposes of overviewing and EMERGY analysis.

Color

It is often desirable later for someone to redraw the complex diagram neatly for wall display. If color is desirable, the following color scheme is suggested:

Sunlight and heat dispersal—yellow
Producers—green
Water, nutrients, and other material resources—blue
Consumers, cities, high transformity units—red
Money—purple

Reducing Complexity

The first diagram often is more complex (having more parts and pathways) than the examples given in this chapter, especially when the contributing people are knowledgeable about the science of the smaller-scale parts of the system. A poster may be required to show the 20–50 parts. Where the window of interest is on a larger scale of time and space, the smaller, faster components such as chemical reactions and microbes may need to be aggregated into group categories. For example, in a particular case if it is established that time scale for the matter of interest is years, then items with a turnover time shorter than 1 yr are aggregated. Conversely, if the window of time is a small one, for phenomena lasting hours or minutes, then those items of large scale such as geologic inputs need to be aggregated. The first diagram may be kept as an inventory of what was considered.

Storages

Whether to include a storage for something depends on its turnover time. If the concern covers a scale of years, then storage tanks are not included for items with faster turnover time than 1 yr.

Diagram Aggregation

The initial detailed diagram for a system can be simplified by aggregation. This kind of simplification retains all resources and components, but combines them to make fewer pathways and symbols. An example of aggregation without omitting sources is Figure 5.1b, derived from Figure 5.1a.

For an EMERGY analysis, aggregation is made so that the flows remaining are those to be evaluated in an EMERGY analysis table. All outside sources and large inside storages that serve as nonrenewable sources must be included. The pathways to be evaluated are usually the following:

(1) Pathways contributing resources from outside crossing into the system. These include environmental inputs, fuels, minerals, money, goods, and services.
(2) Storages within the system that are large enough to act as nonrenewable sources for sustained short-time use. These are the storages with longer time constants than the time of study. For state and national studies, storages with longer turnover times than 1 yr are included.
(3) Pathways subject to change in alternatives under consideration.
(4) Pathways of special interest because of the problems being considered. Pathways that are evaluated may include main flows and smaller component flows if they are of interest.

At first, little concern need be given to problems of double counting. Evaluations and comparisons are useful whether one includes another item or not. Including an item in the EMERGY analysis table does not mean that it is to be added into totals.

Figure 5.1a, representing the New Zealand pine plantation system (Monterey pine, *Pinus radiata*), is an example of energy systems diagramming in preparation for EMERGY analysis. The inputs from the environment and from the economy are arranged from left to right in order of solar transformity. Making the diagram identifies the pathways to be evaluated in the EMERGY table.

EMERGY EVALUATION TABLE

Next, set up the EMERGY analysis table with the six-column format of Table 5.1.

Column 1: Line Number. List the line items to be evaluated. For each line item evaluated, indicate the source of raw data and calculations with a

TABLE 5.1. Tabular Format for EMERGY Evaluation

Note*	Item	Data Units (J, g, or $)	Solar EMERGY/unit (sej/unit)	Solar EMERGY (sej/yr)	Em$ 1991$ ($/yr)†
	(One line here for each source, process, or storage of interest.)				

* Footnotes for each line of the table go here.
† Solar EMERGY in column 5 divided by 1.4×10^{12} sej/$ for 1991.

footnote with the same number as the line in the table. Table 5.2 is an example of energy flows for pine plantations in New Zealand.

Column 2: Item. Next, a name of the item is written. It may be necessary to give a longer description in the footnote if there is not enough space in the main table. It may be convenient to include at the end of this caption the units of the data of column 3, for example joules, grams, or dollars (J, g, or $).

Column 3: Data. For each line item, evaluate the flow in raw units: energy, grams, or dollars, and place the number in column 3. Appendix B has formulae for calculating energy flows. These are standard expressions from physics, chemistry, engineering, economics, and geology. Helpful computer programs and spreadsheets are available for expressing raw data in units of joules, grams, or dollars. For some materials where the potential energy values are not readily calculated, data may be expressed in grams in place of joules. Data on human services and labor may be in dollars paid or time spent.

Column 4: Transformity or Other Unit EMERGY. A transformity (or a value of solar EMERGY per unit weight or solar EMERGY per dollar) goes in this column. A suitable transformity may be found in Appendix C. If a transformity is not available, calculate one from subsystem data assembled for the purpose. Nine methods are listed in Table 14.3. Examples of calculating transformities from observed systems are included in Chapter 4 and Appendix C.

Column 5: EMERGY (Flow or Storage). This value is calculated in column 5 by multiplying the data in column 3 by the EMERGY per unit of data in column 4. Numbers in this column should be made comparable by expressing them with similar powers of 10. For example, in Table 5.2 column 5 has all its numbers times $\times 10^{12}$ (1,000,000,000,000, or 1 trillion). We usually use the computer notation where E12 = 10^{12}.

Column 6: Emvalue (Emdollars). Finally, divide the EMERGY in column 5 by the EMERGY/money ratio for a particular currency of a particular year to get the emdollar value for column 6. EMERGY/money ratios for

TABLE 5.2. EMERGY Evaluation of Pine Plantation in New Zealand. Annual Flows per Hectare (see Figure 5.1)

Note	Item	Data	EMERGY/unit (sej/unit)	Solar EMERGY ($\times 10^{12}$ sej)	Em$* (1984 U.S.$)
1	Sunlight	5.14×10^{13} J	1	51.4	23.
2	Rain transpired	3.16×10^{10} J	1.6×10^4/J	506	230.
3	Soil used	3.39×10^9 J	6.3×10^4/J	214	97.
4	Phosphate added	4.84×10^6 J	4.4×10^7/J	213	97.
5	Fuel used	1.79×10^8 J	6.6×10^4/J	12	5.
6	Services	57 1978$	4.6×10^{12}/$	262	119.
7	Annual Yield	1.507×10^{11} J	8×10^3/J	1207	548

Solar Transformities

$$\text{Wood standing in forest: } \frac{(506 + 213 + 214 + 199) \times 10^{12} \text{ sej/yr}}{1.507 \times 10^{11} \text{J}} = 7511 \text{ sej/J wood}$$

$$\text{Harvested wood: } \frac{(506 + 213 + 214 + 12 + 262) \times 10^{12} \text{ sej/yr}}{1.507 \times 10^{11} \text{J wood/yr}} = 8009 \text{ sej/J wood}$$

* EMERGY flow in column 3 divided by 2.2 sej/$ for U.S. in 1983; analysis from IIASA report (H. T. Odum and Odum, 1983). Harvest after 24 yr of growth. Costs for 1978 from D. J. Mead.

[1] mean insolation 333.65 Langleys (Ly) per day (Lisle, 1960); 10 kcal/m²/Ly; 365 days:

$$(5.1 \times 10^9 \text{ J/m}^2\text{/yr})(1 \times 10^4 \text{ m}^2\text{/ha}) = 5.14 \times 10^{13} \text{ J/ha/yr}$$

[2] Rain used per year: 2.06 m rain − 1.42 m runoff = 0.64 m (Toebes, 1972):

$$(0.64 \text{ m}^3\text{/m}^2)(1 \times 10^4 \text{ m}^2\text{/ha})(1 \times 10^6 \text{ g/m}^3)(4.9 \text{ J/g}) = 3.16 \times 10^{10} \text{ J/ha/yr}$$

[3] Soil eroded:

$$(3 \text{ tn/ha/yr})(1 \times 10^6 \text{ g/tn})(0.05 \text{ organic})(5.4 \text{ kcal/g})(4186 \text{ J/kcal}) = 3.39 \times 10^9 \text{ J/ha/yr}$$

[4] Phosphate used, 2 tn/ha/24 yr:

$$(0.083 \text{ tn/ha/yr})(1 \times 10^6 \text{ g/tn})(58.3 \text{ J/g}) = 4.84 \times 10^6 \text{ J/yr}$$

[5] Liquid fuels used per cubic meter of wood from N.Z. Ministry of Forestry for 1981: logging, 52×10^6 J/m³; transport 130×10^6 J/m³; loading, 33×10^6 J/m³; and total 215×10^6 J/m³.

$$(215 \times 10^6 \text{ J/m}^3)(20 \text{ m}^3\text{/ha/24 yr}) = 1.79 \times 10^8 \text{ J/ha/yr}$$

[6] personal communication from D. J. Mead, 1978, Univ. of Canterbury, N.Z.: Fertilizer, $205; roads, $73; land preparation, $60; planting, $60; stock, $33; restock, $20; first thinning, $60; second thinning, $45; administration, $480; cutting and roads, $334; total $1370/ha/24 yr; $57/ha/yr. Evaluated with EMERGY/money for New Zealand:

$$(\$57\text{/ha/yr})(4.6 \times 10^{12} \text{ sej/\$}) = 262 \times 10^{12} \text{ sej/yr}$$

For planting and fertilizing only:

$$(\$1036/24\ \text{yr})(4.6 \times 10^{12}\ \text{sej/\$}) = 199 \times 10^{12}\ \text{sej/yr}$$

[7] Yield is 10 tn dry/ha/yr (1×10^{7} g/ha/yr) when averaged over one 24-yr cutting cycle:

$$(10\ \text{tn/ha/yr})(1 \times 10^{6}\ \text{g/tn})(3.6\ \text{kcal/g})(4186\ \text{J/kcal}) = 1.507 \times 10^{11}\ \text{J/yr wood harvest}$$

Total EMERGY of yield is taken as the sum of the inputs omitting sunlight.

different years estimated for the United States are included in Appendix D.

Evaluating Solar EMERGY Coming from the Earth System

Some of the line items in the table may be for sources from the earth's system of earth, air, and water, which we evaluated as the geobiospheric baseline in Chapter 3. As illustrated there with Figure 3.7, the main geobiospheric processes of the earth contribute several inputs, such as sun, wind, rain, waves, and land cycle, to each area of the earth. Because the transformities of these coproduct processes were calculated from the baseline, to add the EMERGY flows calculated separately would double-count the original sources contribution to an area. A simple way to determine how much solar EMERGY the basic earth system has contributed is to use the largest of the geobiospheric inputs and ignore the rest (assuming they are already included in the largest one).

Purchased Resources

The inflows that are purchased outside and brought into an area have two components, each of which has to be evaluated. One is the EMERGY contained in the available energy that is brought in. The other is the EMERGY that supported the human services. For example, if fuels are brought into an area, the EMERGY of the area includes the EMERGY in the fuel itself (energy multiplied by transformity) plus that in the services in mining, processing, and delivering (dollar cost multiplied by the EMERGY/money ratio appropriate for that economy and time). These are entirely separate sources and must be evaluated separately. If, however, the transformity used to calculate the EMERGY of a product already includes all the services that went into its delivery, then it would be double counting to also evaluate the services separately.

Evaluating Services

Payments of money represent services and the dollar data go in column 3 as raw data. An appropriate EMERGY/money ratio is written in column 4. The EMERGY/money ratio should be for the year of the data. If the money is from a foreign source, the data may be in international U.S. dollars. Therefore, the EMERGY/money ratio for that country should be used if available, or one from a country with a similar degree of economic development. Chapters 12 and 14 (especially Table 14.2) contain more on evaluating human services.

When a commodity is evaluated using a transformity, the services involved in generating that commodity were possibly included in that transformity. In many evaluations, total dollar flow may be used to evaluate services, at the same time that some commodities are evaluated separately. In this situation there would be some double counting of services, which can be corrected by subtracting from the dollar evaluation those dollars paid for the commodity. In practice these corrections are usually small, 1% or less. When analyses are made of large sectors of an economy such as all of agriculture, the correction may reach 10%.

Evaluating EMERGY Inflows to an Area

Whereas there are often statistics available on imports to countries, data are not usually available for "imports" to smaller areas (such as states, regions, cities, or development project areas). Usually there are data on income and population in a study area. The denser the population and economic activity, the higher the percentage of necessary resources that have to be imported from outside of the area. In New York City practically everything has to be purchased from outside. M. T. Brown (1980) used a base-nonbase occupational analysis from economics to develop nomograms relating the imports to the spatial concentration of income. One of these graphs is Figure 5.2. Given the areal income density (dollars per time) on the x-axis, the money spent on imports can be read from the curve on the y-axis (dollars per time). To obtain the EMERGY of the imported goods and services, the dollar expenditure can be multiplied by the EMERGY/money ratio *of the area from which imports come.*

Print EMERGY for Emdollar Values on an Aggregated Diagram

After evaluations are made with the EMERGY table, it may give perspective to print these values on pathways and storages of an aggregated systems diagram. An EMERGY diagram results if EMERGY values are used; see example in Figure 5.1. An emvalue diagram results if emdollars are used.

Comparing Line Items in the EMERGY Analysis Table

When all the desired flows (or storages) of the EMERGY analysis table have been evaluated, comparisons of EMERGY or the equivalent emdollars show immediately which ones are most important and contribute most to the combined economy of nature and humanity. If a line item is part of another, comparison shows how much of the larger one is due to the smaller one. More complex comparisons involve summations and ratios.

EVALUATE INDICES

In order to calculate some indices useful for interpretation, we further aggregate the EMERGY flows into a three-arm diagram (environmental inputs, purchased feedbacks, and output products in Figures 4.7b, 4.8b). Figure 5.3 shows the local environmental inputs with two parts (Renewable and non-renewable sources) and the feedbacks also with two parts (purchased resources and human services). Calculate the following:

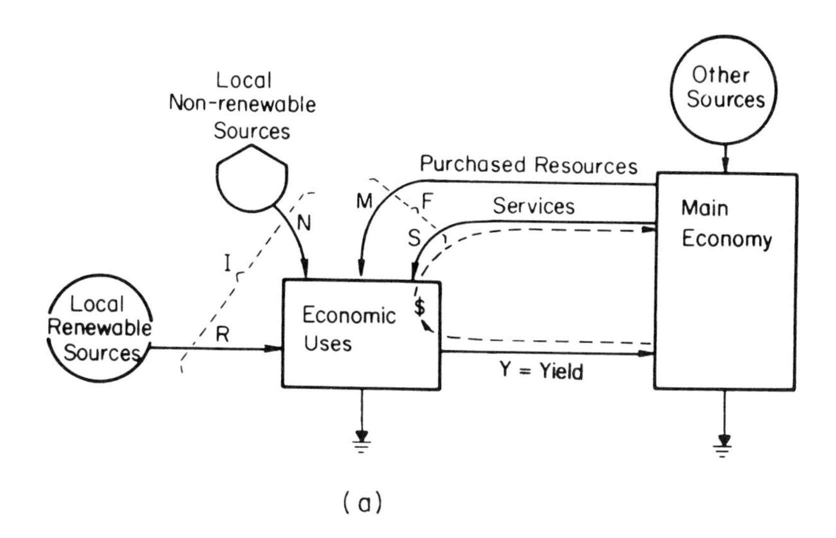

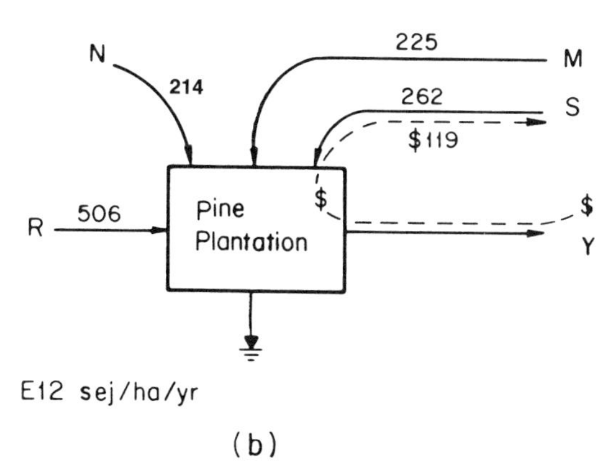

Figure 5.3. Diagrams illustrating EMERGY ratios in Table 5.3 for evaluating investment in a local resource. **(a)** Main pathways categories. **(b)** Example of data used for calculating investment ratios, Monterey pine plantations in New Zealand (See Table 5.2).

Solar Transformity of Products

Divide the total solar EMERGY inputs required by the energy of the yield.

EMERGY Yield Ratio of Products

Divide the yield output EMERGY flow by the sum of the feedback EMERGY from the economy.

Exchange Ratio

Divide the solar EMERGY flow of the yield product by the solar EMERGY of the money paid by the buyer (Figure 4.4).

Investment Ratios

Depending on how one includes nonrenewable inputs and services, there are several ratios of interest that relate the economic contributions to the environmental contributions. These are defined in Table 5.3 with examples from the pine plantation. As shown in Figure 5.3a, EMERGY contributions to an economic activity in a local area include the following (from left to right in the diagram):

R. Free *renewable* EMERGY of environmental inputs from such as sun, wind, rain

N. Free *nonrenewable* resource EMERGY from the local environment such as soil, forest wood, and minerals when used faster than produced

M. Purchased EMERGY of *minerals,* fuels, and raw materials brought to an area by the economic system

S. Purchased EMERGY in *services and labor,* the paid work of people

TABLE 5.3. Useful Ratios for Evaluating Economic Uses of Resources

Name of Index	Definition (Fig. 5.3a)	Pine Plantation (Fig. 5.3b)
Purchased/free	(M + S)/(R + N)	487/720 = 0.67
Nonrenewable/renewable	(N + M)/R	439/506 = 0.87
Service/free	S/(N + R)	262/720 = 0.36
Service/resource	S/(R + N + M)	262/945 = 0.28
Developed/environmental	(N + M + S)/R	701/506 = 1.39

For various comparisons, ratios may be made with different combinations of these four categories of inflow, each expressed in solar EMERGY units (sej/time). Indices combining all four kinds of inputs in Table 5.3 use data from the summary diagram of pine plantation EMERGY flow in Figure 5.1b. Other examples are given for Indian silk (Figure 4.7) and shrimp aquaculture (Figure 4.8).

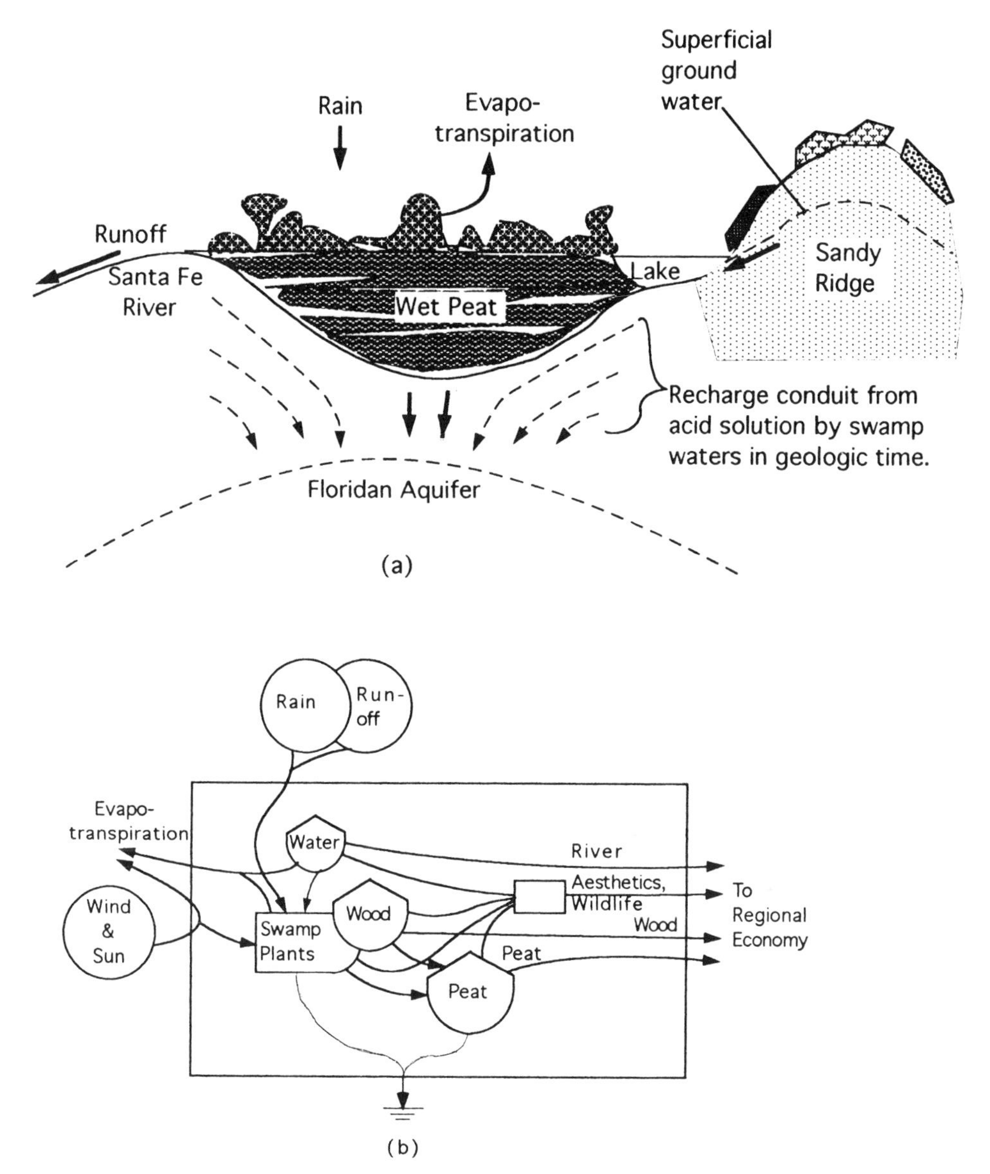

Figure 5.4. Santa Fe Swamp, Bradford County, Fla. **(a)** Lateral view. **(b)** Energy diagram of the system with three main storages.

EMERGY Yield per Money Spent

For some perspectives the EMERGY yield of a system to the main economy is related to the dollar costs. This index may be useful in deciding on alternatives where there is a question of where to spend the money. Whereas the EMERGY yield ratio gives the yield per unit of all feedbacks from the economy including

TABLE 5.4. EMERGY Evaluation of Storages in Sante Fe Swamp, Florida: 1958 ha (4896 acres) (see Figure 5.4)

Note	Item	Stored Energy ($\times 10^{14}$ J)	Solar Transformity ($\times 10^4$ sej/J)	EMERGY ($\times 10^{20}$ sej)	Em$* ($\times 10^9$ $)
1	Water storage	3.14	4.1	0.121	0.006
2	Wood storage	352.	3.2	10.6	0.4
3	Peat storage	1400.	1.9	26.6	1.2

* Solar EMERGY divided by 2.2×10^{12} sej/$ for 1984.

[1] Volume of water taken as volume of peat, 8.65×10^7 cu yd; 89.6% moisture; 5 J/g Gibbs free energy in soft water:

$$(8.65 \times 10^7 \text{ cu yd})(0.729 \text{ m}^3/\text{yd}^3)(1 \times 10^6 \text{ g/m}^3)(5 \text{ J/g}) = 3.15 \times 10^{14} \text{ J}$$

Solar transformity is that for stored water.

[2] Peat storage, 8.65×10^7 cu yd; dry wt, 10.4%; heat content, 9.2×10^3 btu/lb dry.

$$(9.2 \times 10^3 \text{ btu/lb dry})(0.254 \text{ kcal/btu})(4186 \text{ J/kcal})/(454 \text{ g/lb}) = 2.15 \times 10^4 \text{ J/g}$$
$$(6.56 \times 10^{12} \text{ g})(2.15 \times 10^4 \text{ J/g}) = 1.4 \times 10^{17} \text{ J}$$

Solar transformity of peat based on radio-carbon dating: 3 m deposition in 1500 yr (2 mm/yr). Annual EMERGY inflows in rain and net water inflow transpired:

$$(1.37 \text{ m}^3/\text{m}^2/\text{yr})(1 \times 10^6 \text{ g/m}^3)(5 \text{ J/g})(1.8 \times 10^4 \text{ sej/J}) = 1.23 \times 10^{11} \text{ sej/m2/yr}$$
$$(2.34 \text{ E7 m}^3/\text{yr})(1 \times 10^6 \text{ g/m}^3)(5 \text{ J/g})(4.8 \times 10^4 \text{ sej/J})/(1.98 \times 10^7 \text{ m2}) = 2.84 \times 10^{11} \text{ sej/m}^2/\text{yr}$$

Energy of peat deposited:

$$(2 \times 10^{-3} \text{ m})(1 \times 10^6 \text{ g/m}^3)(0.50 \text{ dry})(5.2 \text{ kcal/g dry})(4186 \text{ J/kcal}) = 2.18 \times 10^7 \text{ J/m}^2/\text{yr}$$

Solar transformity is:

$$((1.23 + 2.84) \times 10^{11} \text{ sej/m}^2/\text{yr})/(2.18 \times 10^7 \text{ J/m2/yr}) = 1.9 \times 10^4 \text{ sej/J}$$

[3] Wood storage based on timber survey in 1981; cord = (4 ft)(4 ft)(8 ft):

$$\frac{(2040 \text{ cords})(128 \text{ ft}^3/\text{cord})(0.027 \text{ m}^3/\text{ft}^3)(0.7 \times 10^6 \text{ g dry/m}^3)(3.5 \text{ kcal/g})}{(101 \text{ plots})(0.1 \text{ acre/plot})(4.05 \times 10^3 \text{ m}^2/\text{acre})} = 1.768 \times 10^9 \text{ J/m}^2$$
$$(4896 \text{ acres})(4.05 \times 10^3 \text{ m}^2/\text{acre})(1.78 \times 10^9 \text{ J/m}^2) = 3.52 \times 10^{16} \text{ J}$$

Solar transformity of cutover wood in the field: 3.2×10^4 sej/J.

Source: Modified from report to Georgia Pacific Corporation (Odum, 1984d).

human services, the EMERGY yield per money spent relates the yield to the paid human services *only*.

EMERGY OF STORAGES

To understand many systems and answer some policy questions, a separate table of EMERGY storages may be needed. Appendix B has formulae for calculating energy storages, similar to those for calculating energy flows in Appendix B. For the window of normal concern about resources, an EMERGY analysis table for storages might include those categories with a longer replacement time than 1 yr.

For example, the solar EMERGY and emdollar value of the storages of water, wood, and peat in the Santa Fe Swamp in northern Florida were important in making decisions about the use of that swamp. Figure 5.4 is a systems diagram of the Santa Fe Swamp. Table 5.4 contains the EMERGY evaluation of the storages in the Santa Fe swamp. Peat, water and wood were included because they have a turnover time longer than 1 yr.

Those storages that fill and discharge within the period of time of the analysis (usually 1 yr) are included in the flow table but not in the storage table. For example, in analyzing the pine plantation with annual data, most of the storages (sun, rain, nutrients, harvest) have a shorter replacement time than 1 yr.

The emdollar evaluation of the Santa Fe peat storage (Table 5.4) is 1.2 billion 1984 Em$, a much higher value to the public good than any market value that might use the peat for fuel. The peat deposit maintains and stabilizes water quantity and quality of the headwaters of the Santa Fe River and Lake Santa Fe.

SUMMARY

In this chapter instructions are given for making an EMERGY evaluation. Starting with verbal concepts in knowledgeable people, composite complex diagrams are made to aid systems overviews. Then simpler aggregate diagrams are drawn and used to define the line items in EMERGY evaluation tables. Then data are assembled and energy computations made using transformities to determine EMERGY of these items. Tables provide values for EMERGY flows (empower), stored EMERGY in reserves, and EMERGY indices. Emdollars (Em$) are used for comparisons with economic values.

CHAPTER 6

EMPOWER THROUGH NETWORKS: EMERGY ALGEBRA

This chapter contains the "algebra" of EMERGY accounting in networks.[1] Models may be aggregated differently, according to different concepts, and scales. The way networks are defined for different purposes affects EMERGY flow on pathways. Numerical data for a system may be given as numbers on network pathways, as tables and matrices, or as raw data for systems described only in words. Whatever the model, systems diagrams showing energy flow, EMERGY flow and transformity can be used to guide pathway accounting.

Different people aggregate networks emphasizing different properties. To use the quantitative work of others for EMERGY accounting may mean evaluating empower of networks aggregated according to someone else's concepts and systems view. The procedures in this chapter allow you to use energy systems language to translate someone else's data and models to EMERGY flows without necessarily endorsing their models or defending their numbers. Other approaches to energy and environmental accounting will be covered in Chapter 14.

TYPICAL EMERGY FLOW IN A SIMPLE SYSTEM

A typical basic configuration commonly observed in our window on systems usually contains at least one source, a producer, a consumer, a heat sink and the connecting pathways including a feedback reinforcement. Figure 6.1 shows the typical flows of energy, EMERGY, and transformity, with numbers placed

[1] Readers who are not concerned with evaluating flows on network energy diagrams may want to skip this chapter.

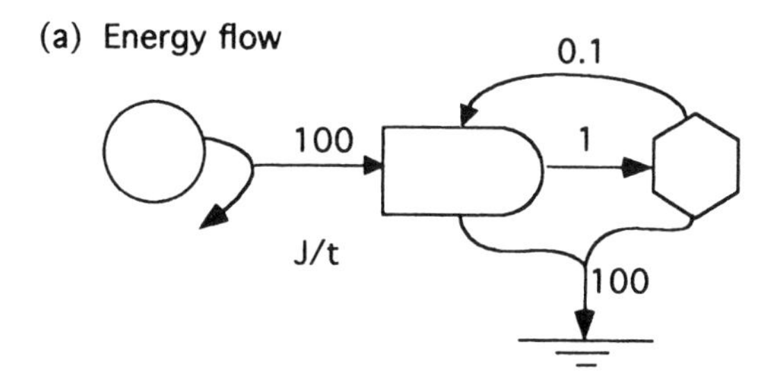

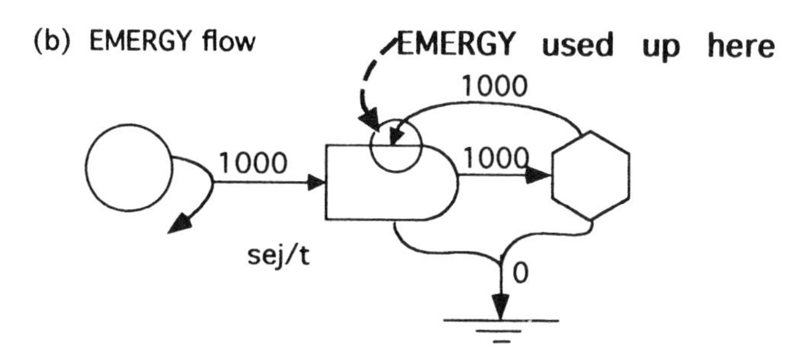

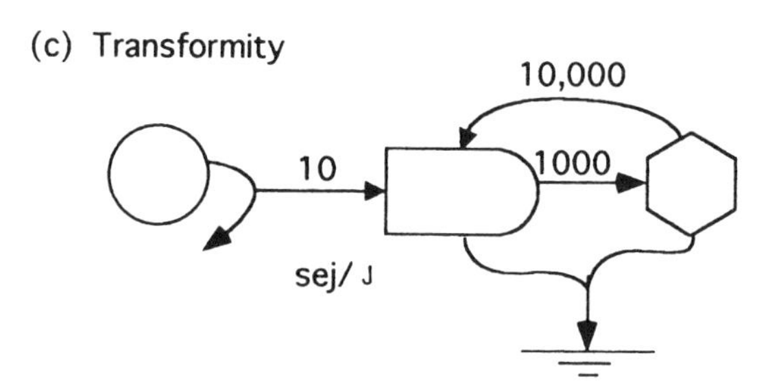

Figure 6.1. EMERGY flow through a typical network disappearing as the last available energy in its chain is used up in a feedback control action. **(a)** Energy flow. **(b)** EMERGY flow. **(c)** Transformity. In this and later figures t is an abbreviation for time.

on the pathways for the steady-state condition. (A system is in a steady state if the storages that may be in the system are constant because input flows equal output flows.) Whatever unit is on the left may be designated a primary producer ("producer" symbol) and those to the right designated consumers ("consumer" symbols). Or the general-purpose "box" symbol may be used, as in Chapter 2.

In the example (Figure 6.1b), EMERGY flowing in from the steady source on the left (1000 sej/t) continues through producer and consumer and finally is used up in the feedback control process as the last available energy carrying that EMERGY is transformed. In accounting jargon, the inflowing EMERGY ends when it *intersects its own tail* (dashed arrow in Figure 6.1b).

If the energy flows are known (Figure 6.1a), and the EMERGY inputs are known (Figure 6.1b), then the transformities are the quotients (EMERGY flow/energy flow) in Figure 6.1c. As always, each successive transformation increases the transformity in the energy hierarchy.

BRANCHES AND INTERSECTIONS

In systems networks there are several kinds of branches and intersections. Accounting of EMERGY flows depends on the type. Much of the confusion in network accounting comes when the kinds of branches and connections are not identified for the appropriate evaluation procedure. The real world has a mixture of types, but in the quest for simplicity, people may build their concepts around the principles that are appropriate for only one kind of systems aggregation. Another person who diagrams and thinks with a different type of branch intersections sees the same real system quite differently.

EMERGY Flow through Branches

In Figure 6.2, branches that split into two flows of the same type (Figure 6.2a) are distinguished from branching transformations that produce flows of two different types (coproducts in Figure 6.2c). In a *split branching,* a pathway divides into two branches of the same kind, each with the same transformity. For example, an island in a stream may separate the river into two branches; a steam supply in a pipe may be divided into two branches going to two different locations. The flow in the branches may be little changed from that in the undivided pathway. Notice the way splits are diagrammed as pathway branching without associated energy transformations.

Split-type branching may also take place through a storage (Figure 6.2b). The two outflows are drawing from the same storage and therefore are of the same kind and carry the transformity of the stored quantity. As explained in Chapter 1 (Figure 1.6), the transformity of outflows from a storage is higher than flows going in.

In a *coproduct branching* (Figure 6.2c), the flow in each branch is a different kind and has a different transformity. Paynter (1960), using "bond graph" network diagrams, showed the importance of distinguishing the two kinds of branching by designating coproduct branching with a "2" and split branching with a "1." Coproduct branching occurs with energy transformation. For examples, agricultural operations involving sheep produce wool and meat; harvest of a crop may yield primary product and by-product litter (silk and litter in Figure 4.7). With *co-product transformations* the branch flows are different in kind from the flow that entered the transformation. Each pathway has different transformities. Correct diagramming requires that co-product branching come from one of the symbols that can indicate an energy transformation (producer, consumer, interaction, or miscellaneous box).

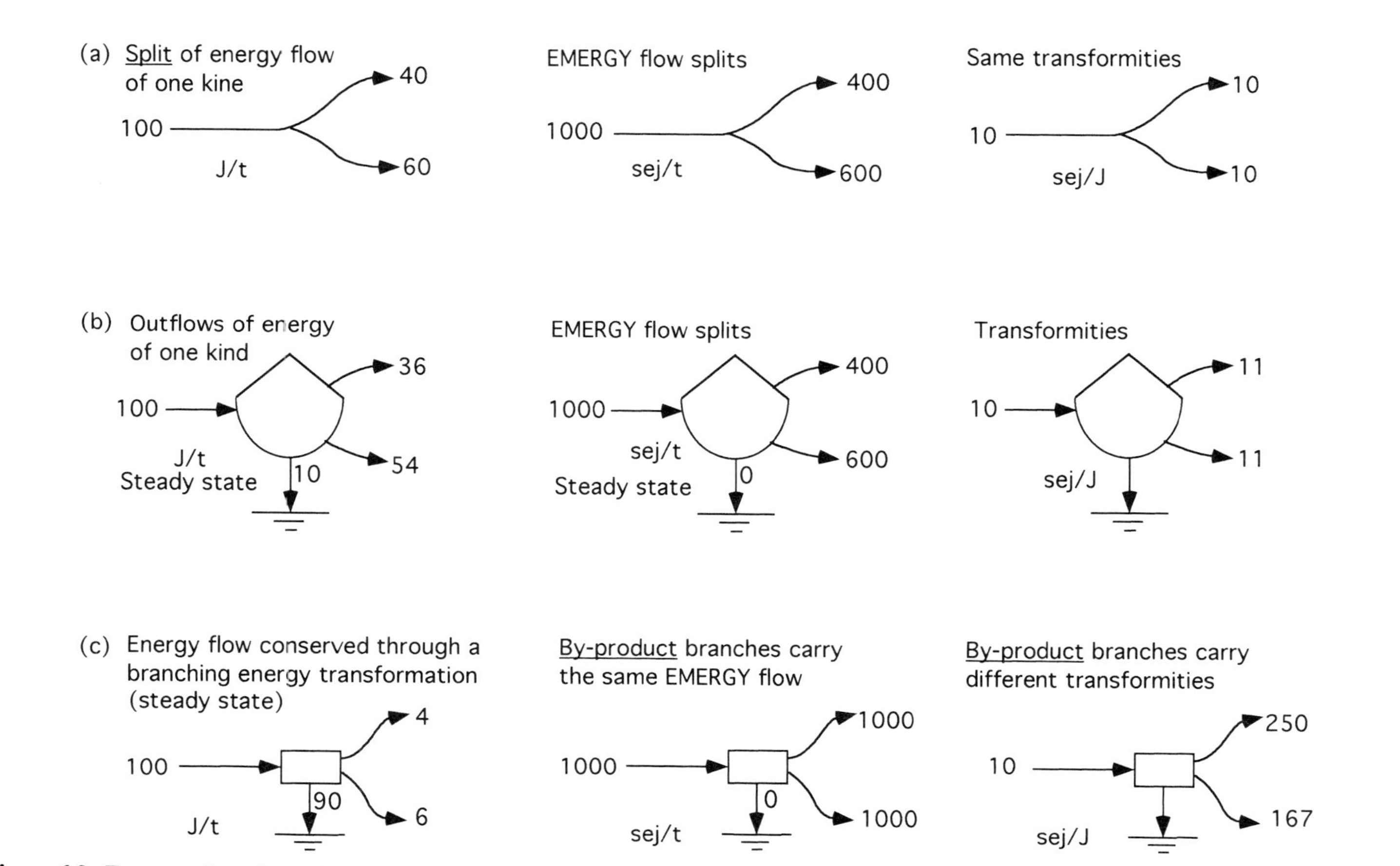

Figure 6.2. EMERGY flow through network branches. **(a)** Splitting of EMERGY into flows of the same kind. **(b)** Separation of EMERGY flows of the same kind outflowing from a storage. **(c)** Co-product energy flows of different type branching from energy transformations.

Since EMERGY is required for a product, all coproduct branches coming out of a transformation carry the same EMERGY (the full amount coming in). This concept is confusing to people at first. It is not that EMERGY was increased in the transformation. EMERGY on both pathways records the total input to the process. If the branches come together again, the two branches are not to be added because it is realized that the two sources are not independent. (See below.)

EMERGY Flows through Intersections

Figure 6.3 contrasts kinds of network intersections—those that join flows of the same kind, and those where flows of different kinds interact in energy

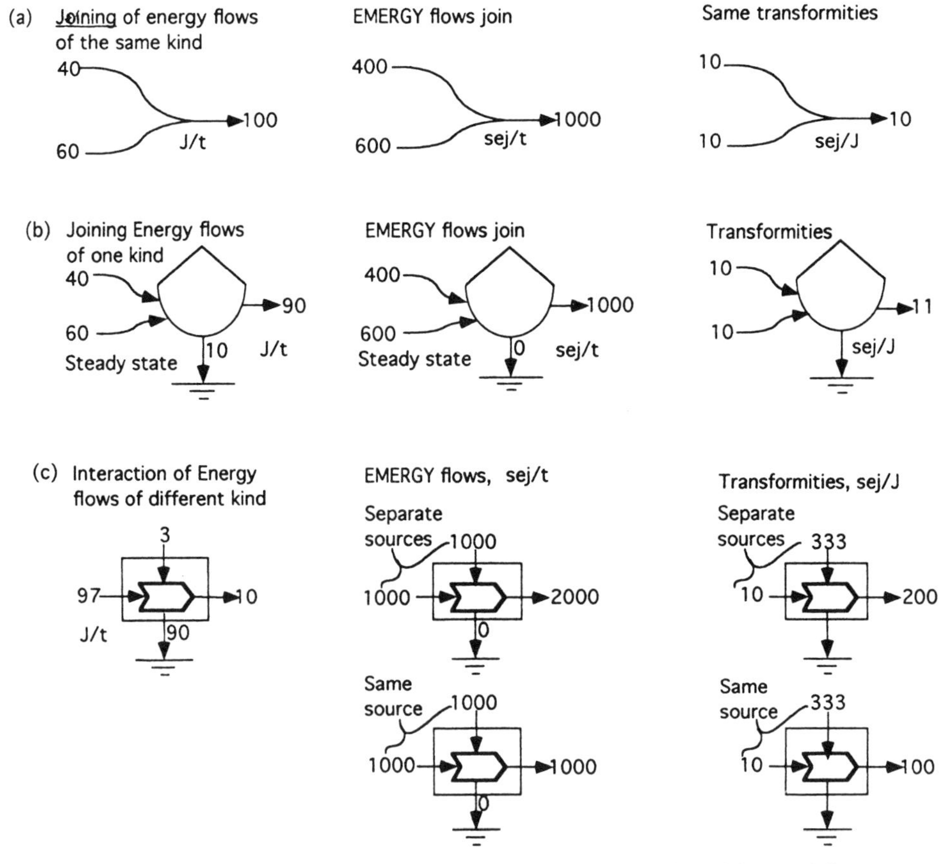

Figure 6.3. EMERGY flow with network intersections. **(a)** Joining of flows of the same kind. **(b)** Joining of flows of the same kind going into storage. **(c)** Interactions of flows of different kinds in productive energy transformations.

transformation. Where two flows of the same kind intersect, similar kinds of energy and transformity join (Figure 6.3a). Examples are the joining of two flows of stream water previously divided by an island and the rejoining of two flows of electricity of the same voltage. Correct diagramming shows a simple joining of the pathways without any energy transformations required.

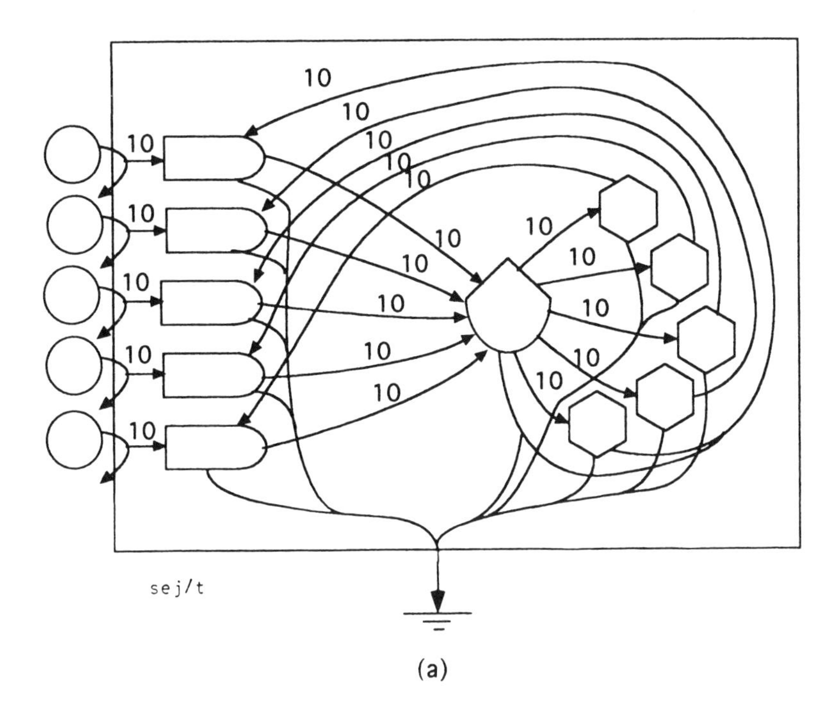

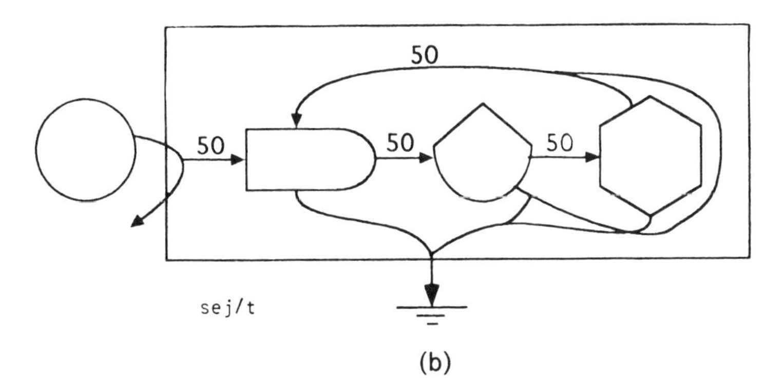

Figure 6.4. Environmental system with rural production, product storage, consumption, and feedbacks. **(a)** Energy flows disaggregated to show area sources and production contributing energies of the same kind in parallel with joins and splits. **(b)** Energy flows aggregated into mutually necessary units connected by co-product flows.

Two flows of the same kind may also join by flowing into the same storage (Figure 6.3b). For example, a silo may receive corn from many different farms; a water tank may receive water piped in from two lakes. The joining of the flows combines quantities of the same type and transformity. If inflows equal outflows, a steady state exists (Figure 6.2b), and the sum of the EMERGY inflows equals the EMERGY outflow.

Figure 6.3c shows intersections of flows of different kinds, with different transformities. In this type of intersection, interactions occur in which both inputs are required for energy transformations that make one or more output products. Diagrams use the "interaction" symbol, with the inputs to the symbol arranged from left to right in order of transformity (Figure 6.2c). Examples are chemical reactions and industries where two or more commodities form a factory product. We have already defined production as the output of an energy transformation (see Chapter 1). Most (if not all) energy transformations involve the interaction of two or more inputs of different transformity (see Figure 2.1).

TYPES OF NETWORK MODELS

Although the kinds of branches and intersections of energy flows shown in Figures 6.2 and 6.3 are found together in the real world, in systems windows sometimes we simplify our views of the environment and economy by using only one kind. Figure 6.4a compares a network of producers that "join" energies of a similar type from farm areas as they converge to a storage. Then branches of the same kind supply consumer households. From the central storage, the diagram shows energy flows dispersing as splits through units of consumption and then as feedbacks back to the rural area again (to the left). Much of Figure 6.4 consists of similar loops of production and consumption in parallel.

In contrast, Figure 6.4b shows the same network with energy flows aggregated into different kinds of blocks each necessary to the system function where energy branches are coproducts. For EMERGY accounting we often aggregate flows in diagrams this way, eliminating the flows that split and join. Figure 6.5 is an example of a system with almost all units necessary to all other units by their pathway connections of the interaction type. The only splits are from the initial energy inflow from source S.

SOURCES AND THEIR INDEPENDENCE

When we place our systems window on some part of our environmental-economic system, we arbitrarily define the parts and processes within the window as the system, and the inputs crossing into the systems window as *sources,* indicated with a circle. As diagrammed, Figure 6.4a has duplicate sources of the same kind, whereas all energy inflows of one kind are aggregated into a single source in Figures 6.1 and 6.4b.

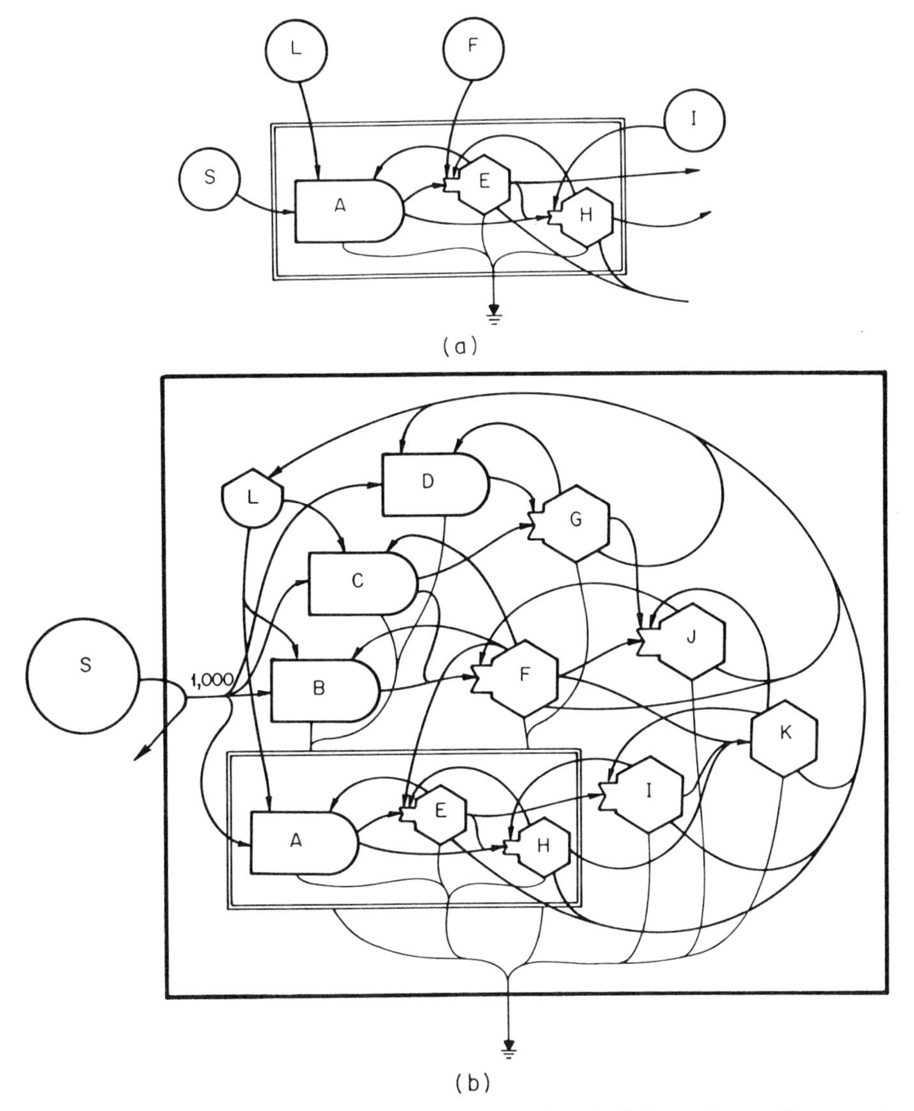

Figure 6.5. Inspection of the larger scale to determine the independence of a system's sources. **(a)** A systems window with four sources and three sectors. **(b)** A larger view of the system showing that the four sources are by-products of one source; their EMERGY inflows should not be added.

Looking Outside to Judge Source Independence

EMERGY accounting requires knowledge of the sources of inputs. Because EMERGY is a property of sources, it is a measure of the larger system, an outward-looking network property. Figure 6.5a is a typical system with several kinds of sources, energy transformation units, storage-containing symbols, and feedbacks. We might start to calculate the total EMERGY input in Figure 6.5a as the sum of the EMERGY inflows of the four sources (S, L, F, and I).

However, since EMERGY is a property of previous transformations, we cannot tell from the small systems window which sources are independently different and which are really by-products branching from one common source outside of the systems window. To determine whether the sources are independent requires a larger window. The larger view in Figure 6.5b shows that all four sources in Figure 6.5a are coproduct flows from the same source. For example, suppose an automatic chemical industry had various sectors of its plant all operating on the same source of fuel and that each sector was necessary to every other sector. To evaluate the EMERGY flow would require evaluating the fuel use only. The EMERGY for the small window is the same as that for the large window (1000 sej/t).

Parallel Sources of One Kind

In Figure 6.4a there are five different farming areas contributing EMERGY of a food product. Although the types of energy are the same, each involves a different EMERGY input. In other words, the sunlight, wind, and rains in one farm are a different set from those on another farm. It is appropriate to add these in the more aggregated overview (Figure 6.4b). (We could think of the separate flows of the same kind as coming from a split branching of the earth's total environmental energy flows.)

"Nonrenewable" Reserves and Contemporary Renewable Sources

Figure 6.6 has two sources that can be added as independent even though they came from the same source. One source is the contemporary daily inflows of environmental energies (such as sunlight or rain), which over long periods contribute to environmental storages. The other source is a previously stored resource in the time window of consideration that is being used faster than it is being stored. Since this use represents the EMERGY of an earlier time rather than the EMERGY coming in from the environmental sources now, we can add the storage contribution to the contribution from the external inflow. Examples are the using up of soil storages in some current agriculture and the pumping out of geothermal heat from hot springs areas faster than it is being supplied to the earth from below. In Figure 6.6b the total EMERGY of the product flow is the sum of 1000 (from the renewable source) and 750 (from the nonrenewable source), corrected by subtracting 20 for the part of the flow from storage that is not from the past storage.

Two External Sources

When there are two or more independent sources of different kinds, it may be expected that each source could make a maximum contribution

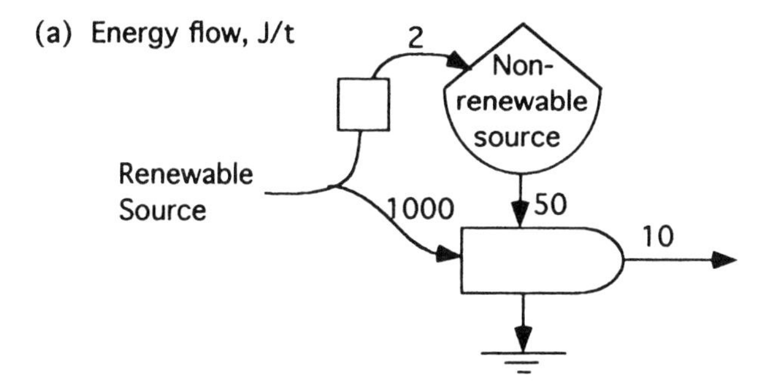

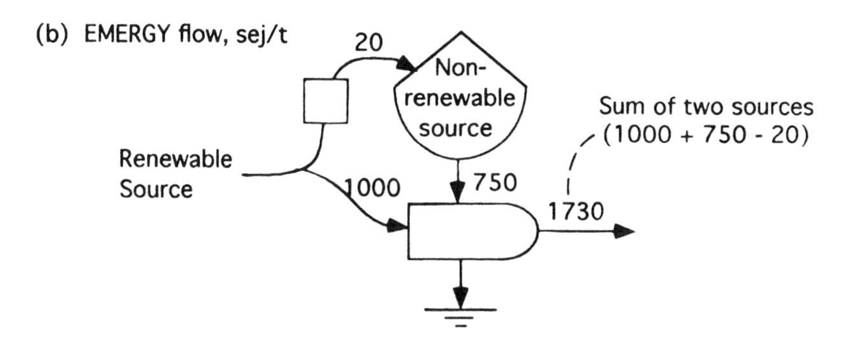

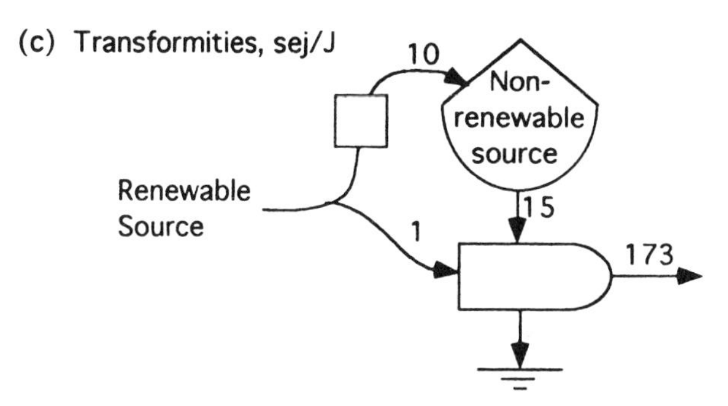

Figure 6.6. Diagram of a system with a storage reserve being used up faster than it is being renewed (in the time window of consideration). The "nonrenewable reserve" is an internal source that adds to the external source. In this and later figures t is an abbreviation for time.

if the network used each flow to amplify the other. Figure 6.7 shows a common pattern in which two sources contribute to a system of coupled producer and consumer sectors. Tracing each source to the point where it interacts with its own tail means that the closed loop carries both EMERGY flows.

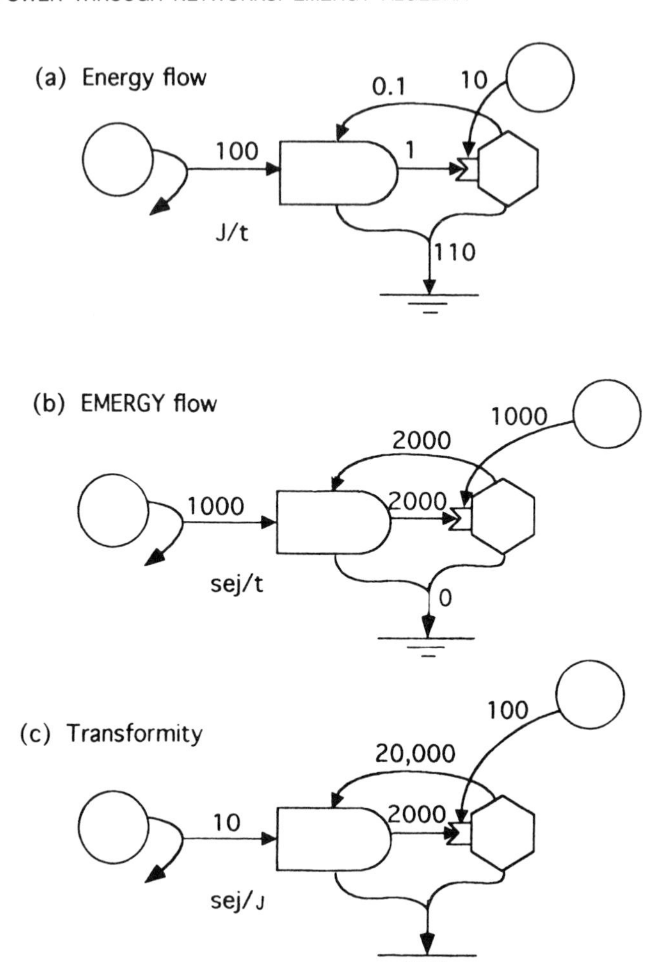

Figure 6.7. Energy and EMERGY flows in a production consumption network with two independent sources in steady state. **(a)** Energy flows. **(b)** Flows of solar EMERGY, summing on the loop pathways. **(c)** Transformities.

Interdependent Aggregation of Sources and Sectors

As expected from maximum empower concepts (Table 15.1), many if not most systems have developed networks that use sources of different kinds to reinforce and amplify each other. Therefore, we often aggregate components into mutually dependent sectors without omitting the available sources (two in Figure 6.7).

For example, the basic geobiospheric system evaluated in Chapter 3 has three main sources for the current window of time. In Figure 3.2 the system of Figure 3.1 was aggregated into three units (sectors), each necessary to the others. Pathways are co-products, each of which carries the same empower, the total of all three source inflows. Because of the way the system is aggregated, all of the EMERGY is required for all of these processes. All the internal loops are coproduct pathways with the same solar EMERGY flow (9.44×10^{24} sej/yr).

EVALUATING EMERGY FLOW BY TRACK SUMMING

The procedure used to evaluate the two-source networks shown in Figures 6.6 and 6.7 was to sum the result of tracking the EMERGY flow from each source separately. Tennenbaum (1988) developed a general procedure for track-summing each input source along every pathway until that pathway reaches the end of its contribution, where it connects with its input. Sources contribute along all the pathways of a network until they intersect with their own pathway, at which point the EMERGY disappears with the EMERGY use. The total EMERGY flow on any pathway is obtained by summing the separate EMERGY flows tracked through that segment. After total EMERGY flows are found for each pathway segment, the total EMERGY flow is divided by the energy flow to obtain the transformity.

Tennenbaum (1988) included a microcomputer program for the summing procedure, evaluating the EMERGY contributions to each sector of a network. The mathematics was written to sum all the separate pathways converging on a sector from the various inputs. A general statement for the summing was written in matrix-algebra form and adapted to a computer program in BASIC that calculates the transformities. The study included an application to a food-chain network for the Mississippi River estuary based on data of Bahr et al. (1982).

Figure 6.8 is an example of track summing of a complex network diagram containing energy-flow data with both split and coproduct branching. The figure legend details how each source and each pathway segment was evaluated. Starting with each source's input, EMERGY was assigned to all pathways to which it contributed until it "crossed its own tail" or left the system. When the diagram showed a coproduct branch, then the full value of EMERGY flow continued on each branch. When the diagram showed a split of energy of the same kind, then the EMERGY flow was split in the same ratio as the energy.

The summing and tracking method can be used objectively to designate EMERGY flows and transformities for any network that has been drawn and its energy flows evaluated even if data are incomplete, the diagrams are incomplete, and the diagram is a mix of splits and sector co-product flows. To make this evaluation is to show what that diagram represents, whether we think the diagram is representative of nature or not. All we have is our aggregate model perceptions of nature, and each person's experience and discipline may lead him or her to see it differently. However, given a network with data on sources and their transformities, we can use track summing automatically to write the EMERGY flow and transformity for each pathway.

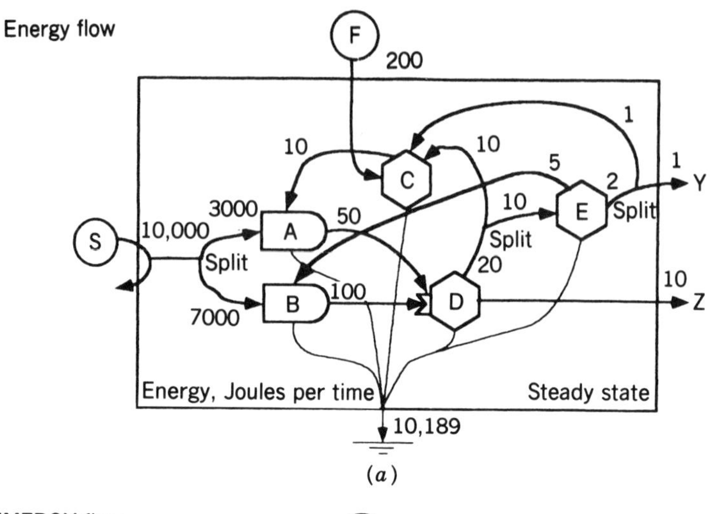

(a)

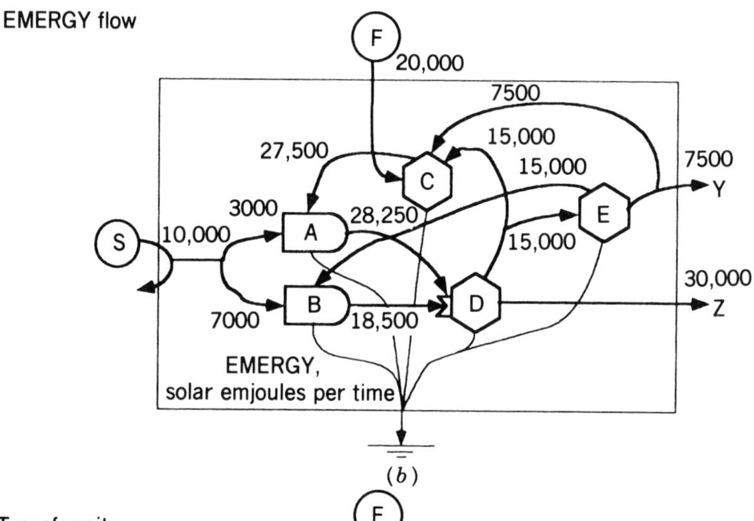

(b)

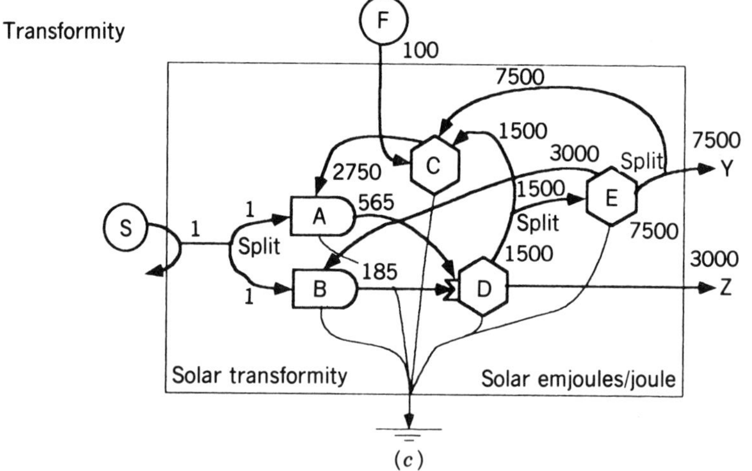

(c)

Figure 6.8

Figure 6.8. Example of evaluating EMERGY fluxes and transformities on a network by track-summing each input. The partially organized example includes splits, by-product branches, and energy flow data. **(a)** Network diagram with energy flows at steady state. **(b)** Flux of solar EMERGY evaluated at each point by summing all the contributing pathways leading to that point and which do not cross themselves; EMERGY splits in the same ratio as energy. Results of track summing:

EMERGY to A = 3000 + 3500 + 1750 + 20,000 = 28,250 sej/time unit:
10,000 from S splits to 3000 reaching A.
10,000 from S splits to 7000 through B and D, then splits to 3500 reaching C and A.
10,000 from S splits to 7000 through B and D, then splits to 3500 reaching E, then splits to 1750 to C, and finally reaches A.
20,000 from F passes through C to A.

EMERGY to B = 7000 + 1500 + 10,000 = 18,500 sej/time unit:
10,000 from S splits to 7000 reaching B.
10,000 from S splits to 3000 through A and D, then splits to 1500 through E reaching B.
20,000 from F passes through C, A, and D, then splits to 10,000 through E reaching B.

EMERGY to C = 1500 + 750 + 3500 + 1750 + 20,000 = 27,500 sej/time:
10,000 from S splits to 3000 through A and D, then splits to 1500 reaching C.
10,000 from S splits to 3000 through A and D, then splits to 1500 through E, then splits to 750 reaching C.
10,000 from S splits to 7000 passing through B and D, then splits to 3500 reaching C.
10,000 from S splits to 7000 passing through B and D, then splits to 3500 through E, then splits to 1750 reaching C.
20,000 from F goes directly to C.

EMERGY to D = 3000 + 7000 + 20,000 = 30,000:
10,000 from S splits to 3000 through A reaching D.
10,000 from S splits to 7000 through B reaching D.
20,000 from F passes through C and A reaching D.

EMERGY to E = 1500 + 3500 + 10,000 = 15,000:
10,000 from S splits to 3000 through A and D, then splits to 1500 reaching E.
10,000 from S splits to 7000 through B and D, then splits to 3500 reaching E.
20,000 from F passes through C, A, and D, then splits to 10,000 reaching E.

EMERGY to export Y = 750 + 1750 + 5000 = 7500:
10,000 from S splits to 3000 through A and D, then splits to 1500 through E, then splits to 750 reaching Y.
10,000 from S splits to 7000 through B and D, splits to 3500 through E, then splits to 1750 reaching Y.
20,000 from F passes through C, A, and D, then splits to 10,000 through E, then splits to 5000 reaching Y.

EMERGY to export Z = 3000 + 7000 + 20,000 = 30,000:
10,000 from S splits to 3000 through A and D reaching Z
10,000 from S splits to 7000 through B and D reaching Z
20,000 from S passes through C, A, and D to reach Z

EMERGY flows on other pathways in Figure 6.8b were evaluated similarly.

(c) Solar transformities for each pathway were obtained by dividing EMERGY flows in part b by energy flows in part a.

CALCULATING TRANSFORMITIES

Since by definition, transformities result from dividing the EMERGY by the energy they can be calculated from network diagrams that have energy flows marked on the pathways if the EMERGY flows can also be assigned to these pathways. Most of the diagrams in this chapter have three duplicate networks, each with different pathway numbers. The top one shows the energy flow, the middle one the EMERGY flow, and the bottom diagram the transformity (the values of the middle diagram divided by the values from the top one).

If a process with unknown transformity generates a known quantity of the same type of output energy as another process with known transformity, the unknown transformity can be obtained from the known process. The new transformity is the ratio of the known solar EMERGY flow to the energy flow of the unknown energy type. For example, in Figure 6.9, 1000 sej generates 1 J of type B energy that is also generated by 200 J of type A energy. The ratio of solar EMERGY to energy yields a solar transformity of type A energy of 5 sej/J.

EMERGY, MATERIALS, AND MONEY

Diagrams drawn to represent the systems window of attention often show the circulation of materials and money, sometimes alone and sometimes together with the energy system. Each has its own special principles and a relationship to EMERGY. Sometimes the systems are represented with network diagrams; at other times flows between sectors are given in tabular form (called inter-sector flow matrices). With network diagrams of energy, materials, and money, the EMERGY flows responsible for these cycles can be evaluated.

Figures 6.10, 6.11, and 6.12, show circulation of energy, material, and money in a three-sector example illustrating system relationships. In Figure 6.10a the

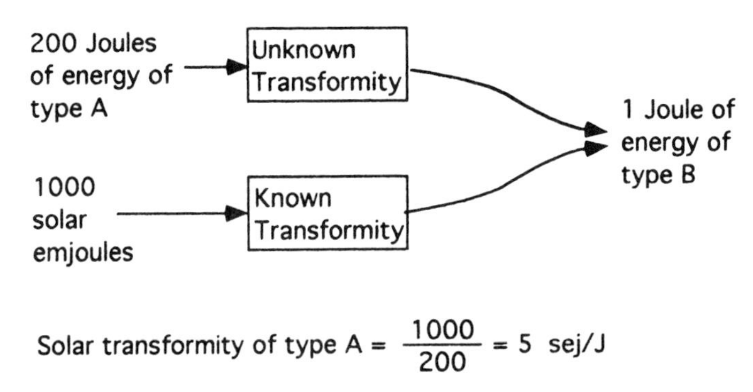

$$\text{Solar transformity of type A} = \frac{1000}{200} = 5 \text{ sej/J}$$

Figure 6.9. Calculating transformity where two different processes generate the same output product.

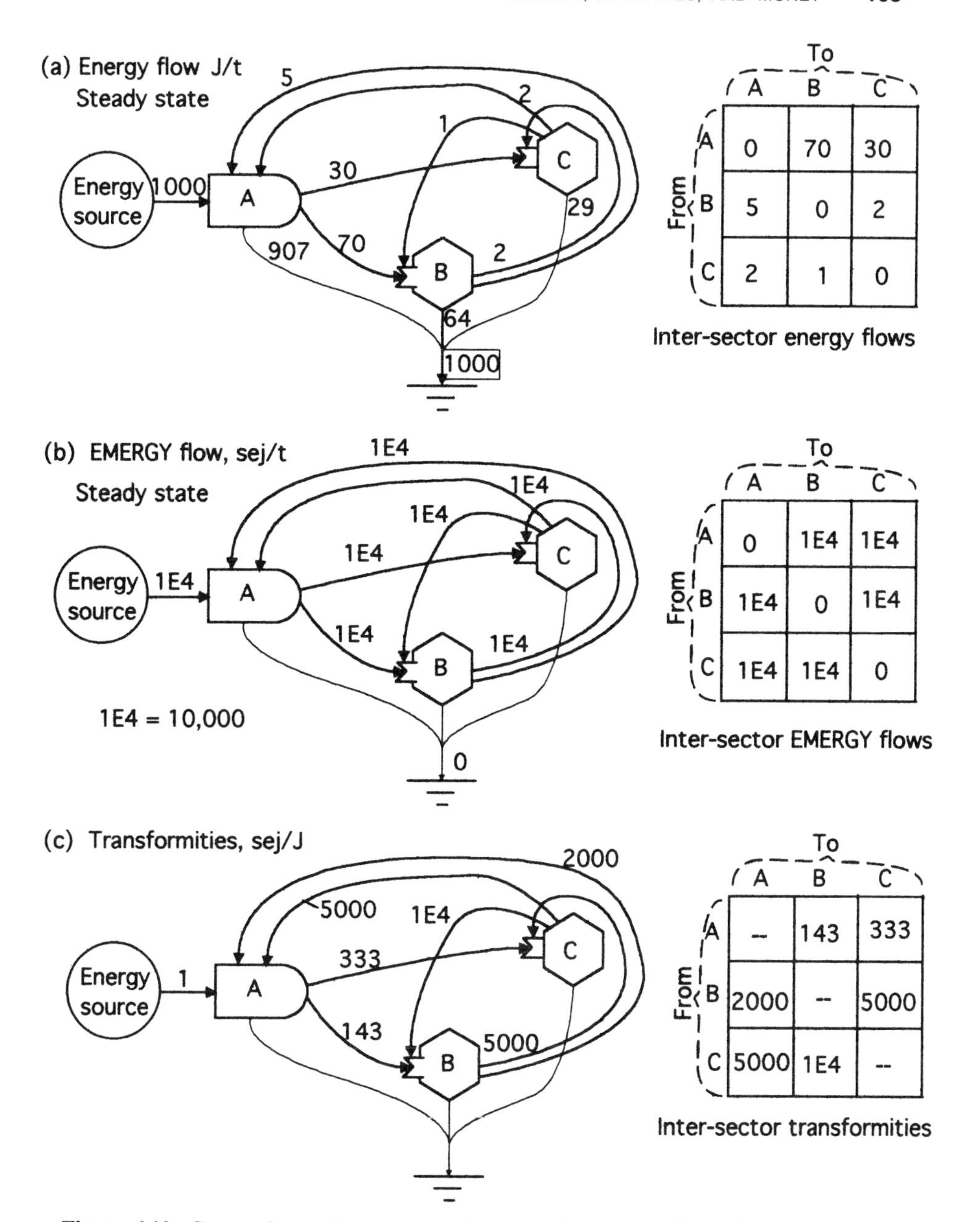

Figure 6.10. Comparison of network and matrix diagram methods for representing energy flows through a mutually essential three-sector system. **(a)** Energy flow. **(b)** EMERGY flow. **(c)** Transformities.

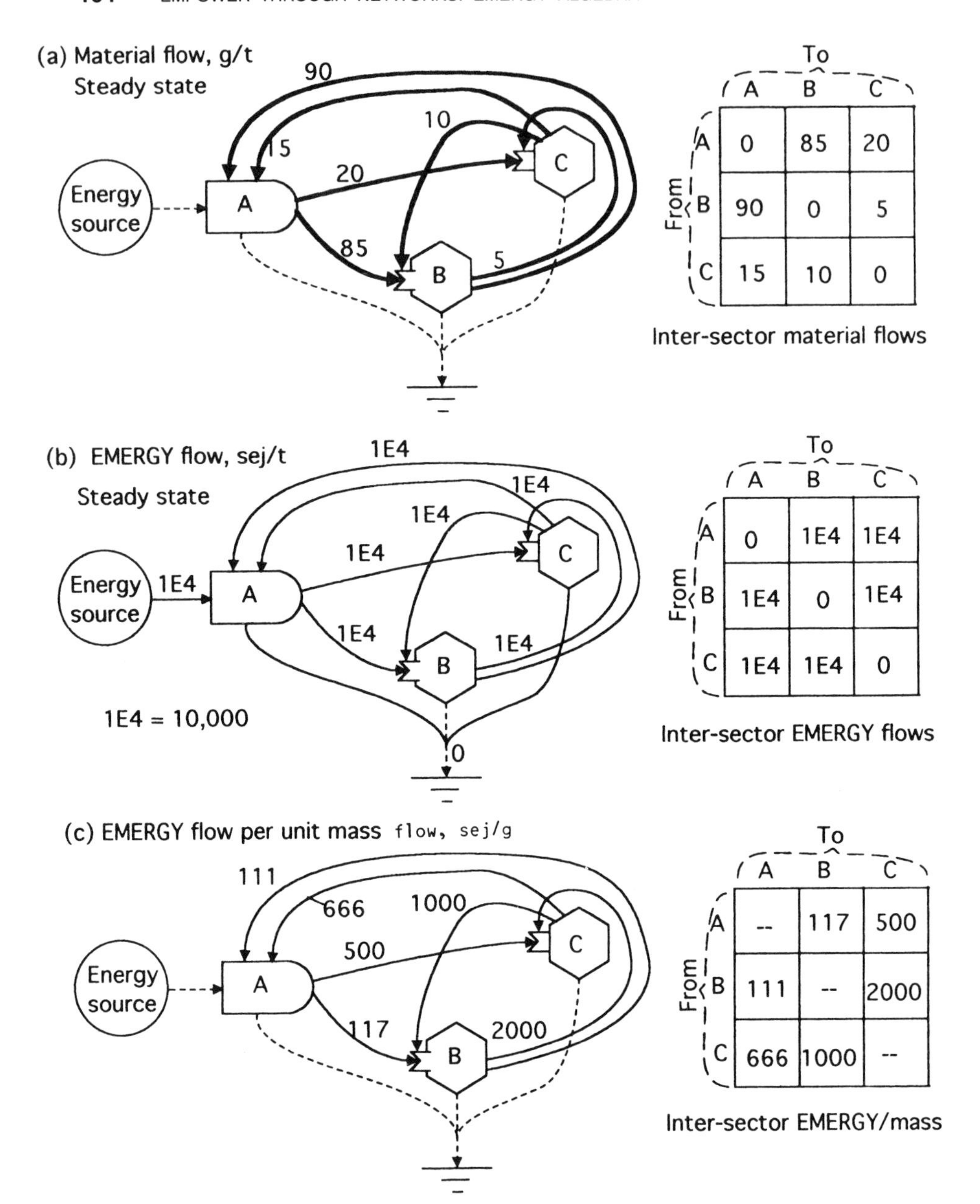

Figure 6.11. Comparison of network and matrix diagram methods of representing cyclic material flows for the system in Figure 6.10. **(a)** Materials flows. **(b)** EMERGY flow (same as in Figure 6.10). **(c)** EMERGY/unit mass.

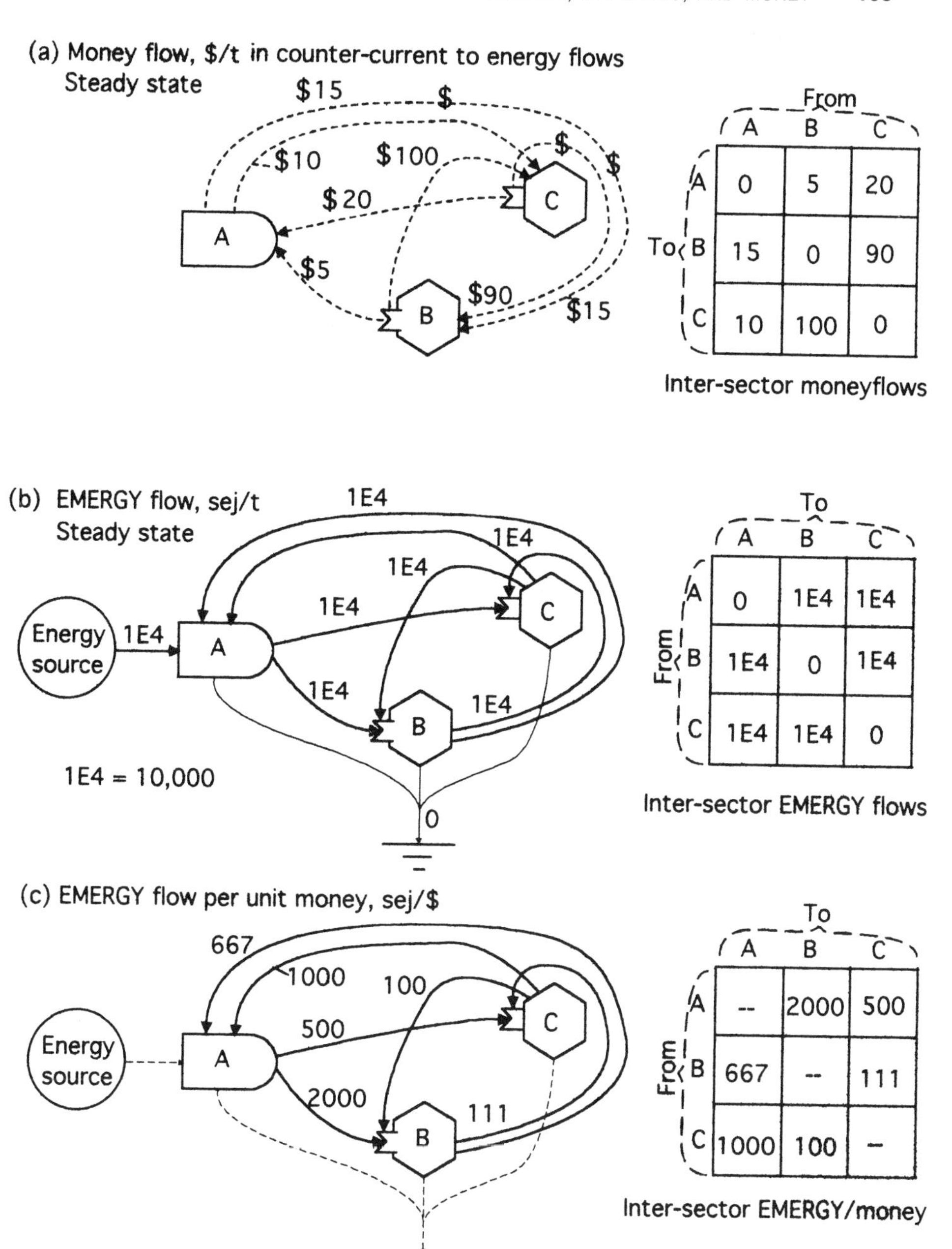

Figure 6.12. Comparison of network and matrix diagram methods of representing the circulation of money between people in the system of Figure 6.10. **(a)** Dollar circulation. **(b)** EMERGY flow (as in Figures 6.10 and 6.11). **(c)** EMERGY per unit money.

system is aggregated into three sectors, each necessary to the others. Therefore coproduct pathways of energy interact at the energy transformation units (A, B, and C) to which they contribute. EMERGY flows on all the pathways are the same (Figure 6.10b). From left to right, energy flows decrease and transformities increase, representing the hierarchy of energy transformations. The numerical values on the pathways are also given in matrix form. E4 is an abbreviation for 1×10^4.

Material Cycles

Figure 6.11a illustrates the materials cycle in a three-sector system such carbon or phosphorus circulating in a sealed aquarium. Materials are conserved, which means that materials flowing into a unit must be accounted for, either in storage or as an outflow. The EMERGY basis (Figure 6.11b) was related to the material cycle (Figure 6.11a) by dividing the former by the latter to obtain EMERGY per unit mass in Figure 6.11c. Materials that are concentrated by successive energy transformations may develop high EMERGY/material ratios. When materials are dispersed to the background environment again, the associated energy availability is lost, and with it the EMERGY. On the diagrams the lower concentrations of a dispersed material are to the left, whereas the materials to the right are more concentrated or are part of structures that required EMERGY to form. Georgescu-Roegen (1977) expressed fear for the modern economic system as unsustainable because of the dispersal of scarce materials and their entropy increase. However, as long as there are adequate EMERGY sources, the concentrating part of the material cycles can continue.

For example, a forest system growing on granite rock concentrates phosphorus from rainwaters as part of its operational structure. When the vegetation decomposes, the phosphorus may be dispersed in concentration again. The phosphorus-containing vegetation has the EMERGY content of its formation. The EMERGY per gram of phosphorus has increased in the process. The EMERGY per gram of vegetation has also increased. (Although both may be of interest, care may be needed to avoid confusing EMERGY per gram of structure and EMERGY per gram of a material contained *in* the structure.) The transformity of the materials in the vegetation is that of the vegetation.

Each kind of material cycle appears to occupy a characteristic zone in the energy-transformation hierarchy. C. Boggess (1994) related EMERGY flows to the phosphorus cycle. Note the distribution of quantities of several cycling materials on a graph of transformity on a logarithmic scale (Figure 6.13a). When the distribution of a material is plotted as a function of concentration on a linear scale, the result is usually a skewed pattern (Figure 6.13b), with most of the material at a lower concentration and decreasing amounts at higher concentrations. The material-cycle diagram (Figure 6.13c) shows the lesser concentrations passing to the end of energy-transformation chains that may be responsible for material distributions. The higher concentrations take more EMERGY, and therefore less material is processed.

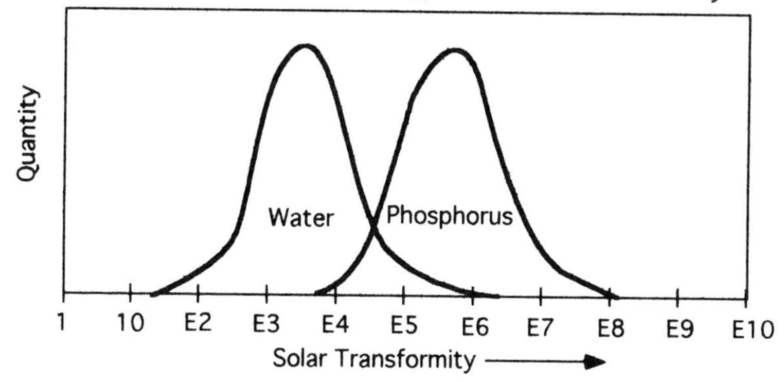

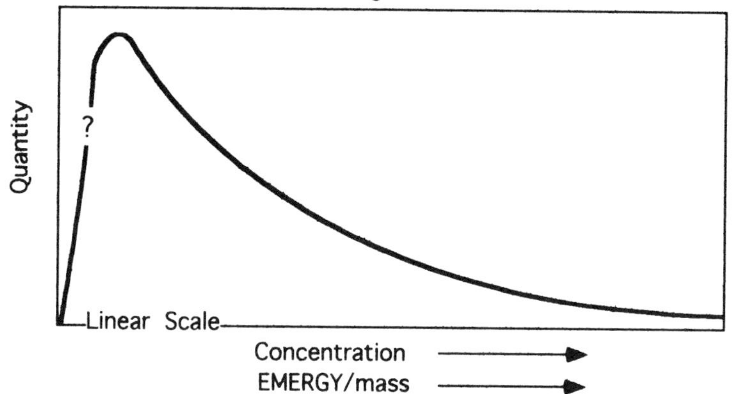

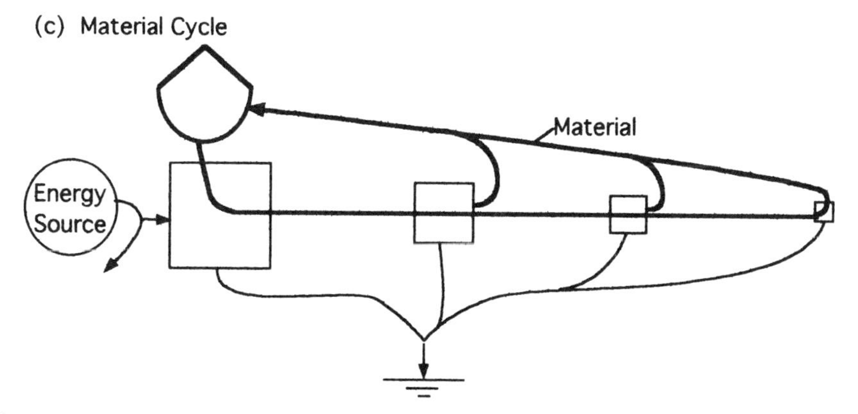

Figure 6.13. Distribution of materials in the energy-transformation hierarchy. **(a)** Quantity of material according to transformity on a logarithmic scale. **(b)** Typical material distribution with concentration on a linear scale. **(c)** Typical material circulation (heavy black lines) through several levels of the energy-transformation hierarchy.

In other words, each material cycle has a range of EMERGY/mass ratios depending on the structures in which the material is bound and the degree of concentration. Values appropriate to the place in the cycle, selected from previously tabulated EMERGY/mass quotients, may be used to evaluate EMERGY of materials.

Money

In a system of sectors involving people, money may circulate as a countercurrent, that is, a payment on each pathway for products circulating in the opposite direction (Figure 6.12a); notice the arrows pointing in the opposite direction from the flow of EMERGY (Figure 6.12b). Over the short range, money is conserved. (Money flowing into a unit is accounted for either as a storage or as an outflow.) In environmental-economic systems, money tends to be concentrated higher in the energy-transformation hierarchy. For example, A might represent farms, B industries, and C urban consumers. The EMERGY obtained for money paid is evaluated as the quotient of EMERGY per unit of money (Figure 6.12c). More EMERGY is obtained per dollar spent from the rural sector and less in the urban sector. As explained already in Chapter 4, money is only paid to people among whom it circulates. However, the real wealth that money buys depends on the EMERGY sources.

An average quotient of EMERGY per unit of money is a useful index for evaluating EMERGY where data on labor and human services are given in monetary units. The total annual EMERGY use of a state or nation is divided by the gross economic product. For example, see Figure 4.2, in which the U.S. value was calculated for 1992.

Double-Counting Cautions

If the EMERGY/money quotient for evaluating average services were based on the whole landscape energy of the country, there would be some double counting if that quotient were used to evaluate human services to a part of that landscape. For example, suppose the EMERGY contribution of agriculture to a nation was 10%; then 10% of the EMERGY of services calculated with the EMERGY/money quotient would be double counted if used to evaluate services to agriculture. If services were 20% of agricultural EMERGY and 10% too high, the error would be 2%. For small areas the double-counting error is negligible.

Some of the transformities calculated for geologic, climatic, and oceanic contributions from the earth are based on the total planetary EMERGY budget. To avoid counting the earth-contributed EMERGY twice (as already described with Figure 3.7), only the larger input should be used. (This procedure is equivalent to subtracting double counting from input sources and then summing all sources).

SUMMARY

The procedures given in this chapter calculate the EMERGY flows from systems diagrams. Principles were given for evaluating EMERGY flow through networks of various kinds including typical systems, different kinds of branches and junctions, kinds of sources, cycles of materials, and circulation of money. Materials and money are conserved as they circulate, whereas EMERGY is conserved only until the last available energy in a series is used up. Procedures were given for track-summing complex networks and determining transformities from network data. New pathway measures—EMERGY per unit mass and EMERGY per unit money—were introduced. Whether these EMERGY flows are correct depends on the aggregation of the network diagram and the assigned numbers being representative of the whole system's processes.

CHAPTER 7

EVALUATING ENVIRONMENTAL RESOURCES

Evaluating environmental production is required for management, since real wealth starts with environmental processes. Environmental resources include the aesthetic-informational and restorative contributions of natural ecosystems and parks as well as the more intensively harvested products of forestry, agriculture, fisheries, aquaculture, and mining. In this chapter examples of environmental EMERGY accounting are given with some of the ways of using EMERGY and emdollars to manage for maximum wealth.

AREA-TIME EVALUATION

Giampietro and Pimentel (1991) suggest a simple way of evaluating environmental works as the product of the area of the work times the duration of the work. For example, 1 ha of forest developing for 10 yr has the same value (10 ha-yr) as 10 ha of forest developing for 1 yr. This method assumes that all land areas of the earth and all periods of times contribute equal value to production.

An EMERGY equivalent to the land area–time units can be calculated using the annual EMERGY budget of the earth divided by the area of the land.

$$
\begin{aligned}
1 \text{ ha-yr} &= \frac{9.44 \times 10^{24} \text{ sej/yr}}{1.5 \times 10^{10} \text{ ha land}} \\
&= 6.29 \times 10^{14} \text{ sej/ha-yr}
\end{aligned}
$$

An annual emdollar value for the average hectare of land is obtained by dividing the above result by the EMERGY/money ratio. In 1993 emdollars this is:

$$\frac{6.29 \times 10^{14} \text{ sej/ha-yr}}{1.4 \times 10^{12} \text{ sej/\$}} = \$449/\text{ha-yr}$$

Area-time is an easy measure and easily adapted for courts and laws. However, it does not allow for the fact that the land surface of the earth is hierarchically organized in space. Rapidly rising mountains, river mouths, beaches, and cities are examples of concentrations of environmental work and value that would be underestimated using the area-time evaluation method that treats all land area the same. Nor does the area-time method recognize the pulsing surges of production that occur at different times (see Chapter 13).

EMERGY SIGNATURE

Every part of the earth has a set of environmental energy flows and storages on which its natural and economic processes depend. EMERGY evaluation of each of the inputs to an area characterizes the area's resources. In some places the wind is important, in other places the tides, and so forth. Water-wave EMERGY predominates beachfront coastal towns in Florida. Geologic mountain-building EMERGY predominates active mountain ranges. Tidal EMERGY predominates in many estuaries. River EMERGY dominates floodplains.

Procedures from Chapter 5 for creating an EMERGY evaluation table can be used to put all the available resources on a common basis. For example, Figure 7.1 shows the set of environmental resources used by the coastal counties of Texas. It plots the annual EMERGY flow of each arranged in order of increasing transformity. The pattern for each area is distinctive. We sometimes call this bar graph the *environmental EMERGY signature.*

WATER RESOURCES

In the organization of the geobiosphere, the hydrologic cycle has a special role as one of the main ways that the solar energy is coupled to the earth energies. In Chapter 3 we calculated an average EMERGY for the rain that the weather systems converge hierarchically to the land. The water carries chemical potential energy of its purity relative to the seawater from which sun and wind distilled it into vapor. Depending on the altitude, rain carries the geopotential energy of its elevation above the sea. In other words, the water in the cycle from sea to land is carrying two kinds of potential energy. Thus, there are two average solar transformities for rain, one for chemical potential and one for geopotential (see Table 3.2). In the self-organization of watersheds the geopotential energy carves the landscape, distributes sediments, and spreads the water out through vegetation in floodplains and coastal wetlands so that the chemical potential is better used by vegetation for organic productivity.

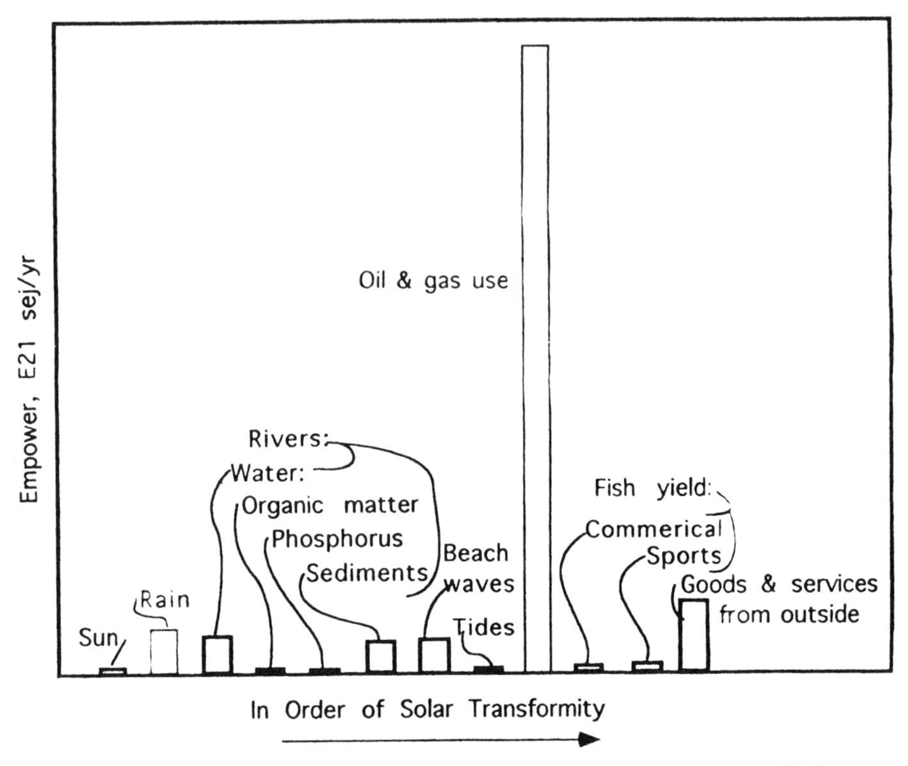

Figure 7.1. Environmental EMERGY signature of coastal zone of Texas (Odum et al., 1987a).

When rain falls on land, it splits into two flows: one component is transpired by the plants, and a remainder flow runs off downhill or into groundwater storages. To evaluate the EMERGY, an appropriate transformity is multiplied by the available energy used up in each of the two flow divisions (split). For the transpiration flow, the energy used is the Gibbs free energy for the concentration difference between the rainwater and the high-salinity water maintained within the leaves by sun and wind (for typical vegetation about 5 J/g of water transpired). For the runoff water, the energy being used up is the geopotential energy converted into frictional work by the flowing water. We calculate this physical energy use as the product of the water-level change, the water density, and the gravity (formulae are given in Appendix B). Water that leaves the boundary defined for evaluation carries its EMERGY to the next system. In some kinds of watersheds with converging streams, the transformities increase as the water collects and converges (e.g., see the Mississippi River example in Figure 2.8). Other transformities for water are given in Appendix C.

Like other environmental products, water contributes more EMERGY than is usually paid for it (Table 4.1; Odum et al., 1987a). Water is worth more to the public interest than people pay for it. In evaluating ecosystems, the water flows that contribute to transpiration and thus photosynthesis are a principal EMERGY source, for example, the salt-marsh evaluation (Figure 7.2 and Table 7.3). Much of the EMERGY in agricultural and forestry products comes from the EMERGY of the transpired water. As water is processed through purification processes for city use, there are additional EMERGY and dollar values added.

ECOSYSTEMS

The environmental landscape originally was covered with many kinds of ecosystems (forests, wetlands, grasslands, lakes, and so on). EMERGY accounting for an area often starts with a classification and mapping of ecosystem types and their contributions (see Chapter 15). EMERGY accounting may require evaluating ecosystems as a whole and its important components and processes. As in all EMERGY evaluations, a systems diagram is drawn and its inputs used to set up an EMERGY evaluation table. For example, a salt-marsh ecosystem is drawn and evaluated in Figure 7.2 and Tables 7.1–7.3.

A wide range of choice is available regarding details to include. For an overview, the energy systems diagram may be kept simple, but it must be complete in regard to EMERGY-contributing sources. The line items in the EMERGY evaluation table includes these main contributing sources plus a few of the main products stored and exported. Where interior details are of interest,

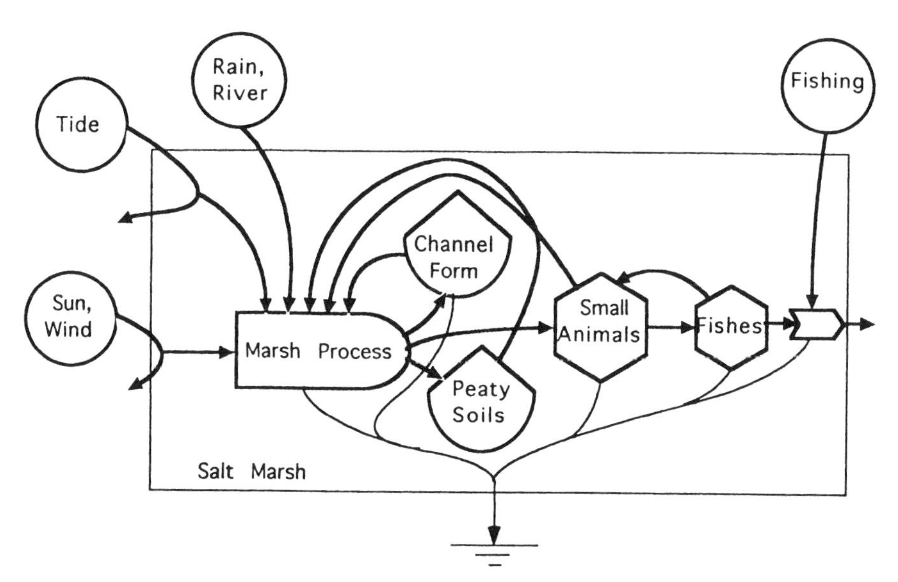

Figure 7.2. Energy systems diagram of a salt marsh aggregated according to the main necessary components and processes as a step in setting up EMERGY evaluation for Table 7.3.

TABLE 7.1. Annual EMERGY of Inputs to 1 ha of a Salt Marsh in Northwest Florida (Figure 7.2)

Note	Item, Units	Data	Solar Transformity (sej/unit)	Solar EMERGY ($\times 10^{14}$ sej/yr)	Em$ ($/yr)*
		Juncus			
1	Direct sun, J	5.95×10^{13}	1	0.6	30
2	Tidal absorption, J	0.7×10^{10}	23,564	1.65	83
3	Transpiration,† J	7.3×10^{10}	15,423	11.3	565
4	Total (Items 2 + 3)‡			12.95	648

* 1990 U.S.$ obtained by dividing annual EMERGY values by 2.0×10^{12} sej/1990 $

† Marsh transpiration, which is the fresh water used, includes that flowing in from land runoff and that flowing in from the seaside of the marsh. Transpiration integrates the combined energy inputs of insolation, wind energy absorbed, vapor pressure gradients in air, and freshwater activity in the estuarine waters.

‡ Here tide is added as a separate contribution, and is not included in the transformity for rain. In Table 3.2, tide is included in the global EMERGY budget making transformities slightly higher.

[1] Gainesville insolation with 10% albedo: $(1.58 \times 10^6$ kcal/m^2/yr$)(1-0.10)(1 \times 10^4$ m^2/ha$)(4186$ J/kcal$) = 5.95 \times 10^{13}$ J/ha/yr

[2] One-fourth of 80-cm tide absorbed:

$$(0.25)(1.0 \times 10^4 \text{ m}^2)(0.5)(0.8\text{m})(1025 \text{ kg/m}^3)(9.8 \text{ m/sec}^2)(705 \text{ tides/yr}) = 0.7 \times 10^{10} \text{ J/yr}$$

[3] Marsh transpiration, 4.0(?)mm/day for 365 days = 1460 mm:

$$(1.46 \text{ m})(1.0 \times 10^4 \text{ m}^2)(1 \times 10^6 \text{ g/m}^3)(5 \text{ J/g}) = 7.3 \times 10^{10} \text{ J/yr}$$

[4] Sum of items 2 and 3.

a more complex diagram can be made with more line items for the evaluation table. In either instance, the total EMERGY use must be included. In EMERGY accounting, the manner and detail of the aggregating should never be allowed to change the total EMERGY flows. Usually the initial diagram should be one without splits—in other words, with only parts and pathways necessary to the whole. See, for example, the EMERGY evaluation of the gross features of a salt marsh shown in Figure 7.2. The pathways of a more complex network with split-type pathways can be evaluated with track summing (see Chapter 6). For example, if the energy flows to clapper rails (a bird species) in salt marsh were known, their EMERGY could be estimated with track summing, even though clapper rails are not essential to the whole system but derive their energy from a splitting of pathways from some of the energy in small animals to larger consumers.

After making the diagram and identifying the EMERGY sources and main parts and pathways, a table evaluating these sources is prepared (Table 7.1). A sum of EMERGY use is included with whatever corrections are necessary to avoid double counting. Next a table for calculating transformities may be prepared (Table 7.2). The system's total EMERGY use is divided by each energy flow that is necessary to the whole to obtain transformities of these main divisions. Then a table of EMERGY storages is made (Table 7.3), where the flows into storage are multiplied by the transformities and the time of storing. The table includes emdollar equivalents.

Dominant Species

Sometimes we judge the integrity of complex ecosystems by their content of larger dominant wildlife species. A species that is important and controlling is sometimes called a *keystone species.* A larger species may occupy much of a hierarchical level and thus receive much of the system's EMERGY.

For example, baleen whales of southern-hemispheric oceans are high in a fertile food chain based on plankton growing in vigorously turbulent ocean waters. An evaluation of an average whale is given in Figure 7.3, attributing to them half of the EMERGY of their hierarchical level (Table 7.4). An emdollar value for each of these 15-year-old whales was found to be $2.6 million.

MITIGATION

As part of environmental management and development, people may want to swap one ecosystem area for another, and need a quantitative basis for stating how much of one type of environment is equivalent to another in terms of public value. Table 7.5 compares EMERGY flows and storages of several ecosystems on a hectare basis.

With mitigation banking, areas are purchased for protection and their natural values are restored. Mitigation credits are obtained that can be used by development projects in exchange for their displacement of protected areas. In other words, one environmental gain is substituted for a loss. Too often this has been done with opinions and politically affected committees. EMERGY evaluation can be used to define a mitigation credit in emdollar units. Then these credits can be assigned to the trade-off areas according to their emdollar production and/or storages.

NATURE PARKS

Among the less destructive uses of environmental resources are nature parks. Areas with ecosystems are combined with facilities for visitors to provide environmental experiences, aesthetics, recreation, psychological restoration, and so forth. Whereas the window of evaluation of the ecosystems alone (like Figure 7.2) may not involve the economy, parks are only supported and sustained if nature is used.

TABLE 7.2. Solar Transformities of Components and Processes of a Salt Marsh

Note	Item	Solar EMERGY (sej/ha/yr)	Energy (J/ha/yr)	Solar Transformity* (sej/J)
	Processes, replacement times less than 1 yr			
1	Solar insolation	—	—	1
	Juncus:	12.95×10^{14}		
2	Gross photosynthesis		6.1×10^{11}	2,119
3	Day net photosynthesis		3.9×10^{11}	3,320
4	24-hr net photosynthesis		1.91×10^{11}	6,780
5	Aboveground live biomass		1.86×10^{11}	6,962
6	Detritus production		1.17×10^{11}	11,068
7	Aboveground dead biomass		1.47×10^{11}	8,809
8	Underground organic matter		1.54×10^{11}	7,662
9	Peaty sediment		3.69×10^{11}	3,509
10	Fishes, crabs, shrimp	4.2×10^{11}	1.37×10^{4}	3.0×10^{7}
11	Tidal channel landform	3.3×10^{16}	19.6×10^{6}	1.6×10^{9}

* Quotient of two previous columns (solar EMERGY/energy). Solar transformities are based on Table 7.1.

[1] Unity by definition.

[2–9] EMERGY, 12.95×10^{14} sej/ha/yr from Table 7.1.

[2] Energy in gross production:

$$(5 \text{ g C/m}^2\text{/day})(8 \text{ kcal/g C})(4186 \text{ J/kcal})(365 \text{ days/yr})(1 \times 10^4 \text{ m}^2\text{/ha}) = 6.11 \times 10^{11} \text{ J/ha/yr}$$

[3] Energy in day net production:

$$(3.2 \text{ g C/m}^2\text{/day})(8 \text{ kcal/g C})(4186 \text{ J/kcal})(365 \text{ days/yr})(1 \times 10^4 \text{ m}^2\text{/ha}) = 3.9 \times 10^{11} \text{ J/ha/yr}$$

[4] Energy in 24-hr net photosynthesis:

$$(1.57 \text{ g C/m}^2\text{/day})(8 \text{ kcal/g C})(4186 \text{ J/kcal})(365 \text{ days/yr})(1 \times 10^4 \text{ m}^2\text{/ha}) = 1.91 \times 10^{11} \text{ J/ha/yr}$$

[5] Aboveground live biomass, 1111 dry g/m^2; replacement time, 1 yr:

$$(1111 \text{ g/m}^2)(4 \text{ kcal/g})(4186 \text{ J/kcal})(1/\text{yr})(1 \times 10^4 \text{ m}^2\text{/ha}) = 1.86 \times 10^{11} \text{ J/ha/yr}$$

[6] Energy in underground and detritus annual net production:

$$(0.7 \text{ g C/m}^2\text{/day})(11 \text{ kcal/g C})(4186 \text{ J/kcal})(365 \text{ days/yr})(1 \times 10^4 \text{ m}^2\text{/ha}) = 1.17 \times 10^{11} \text{ J/ha/yr}$$

[7] Aboveground dead biomass, 880 dry g/m^2; replacement time, 1 yr:

$$(880 \text{ g/m}^2)(4 \text{ kcal/g})(4186 \text{ J/kcal})(1/\text{yr})(1 \times 10^4 \text{ m}^2\text{/ha}) = 1.47 \times 10^{11} \text{ J/ha/yr}$$

[8] Underground biomass dry g/m^2 (Kruczynski et al., 1978); replacement time, 5(?) yr:

$$\frac{(4600 \text{ g/m}^2)(4 \text{ kcal/g})(4186 \text{ J/kcal})(1 \times 10^4 \text{ m}^2\text{/ha})}{5 \text{ yr}} = 1.54 \times 10^{11} \text{ J/ha/yr}$$

Everglades National Park was evaluated twice (DeBellevue et al., 1979; Gunderson, 1989; Table 7.6) Although the diagram first drawn to identify parts and processes was more complex, the main features of the park and its use are more simply aggregated for explanatory purposes in Figure 7.4, which also summarizes the annual EMERGY flows.

The investment ratio (ratio of inputs from the economy to inputs from nature) is less than 1, whereas most uses of environment for economic purpose have higher ratios, for example 7 on the average for the United States. Full economic development of an area (with an investment ratio of 7) leaves very little that is natural ecosystem. On an EMERGY or emdollar basis, the park contributes twice as much to the economy than might be inferred from using the dollar flows only. A preliminary evaluation of Yellowstone Park, with more visitors, was somewhat higher in investment ratio. The ratio for the much-visited Luquillo rainforest of Puerto Rico (Doherty et al., 1994) was only 0.15; this area appears to be only slightly impacted, not enough to lose much of its EMERGY contribution.

An important part of "natural" systems is the genetic information and biodiversity (to be considered in Chapter 12). Endangered species have very high transformity and EMERGY values, which are estimated from the environmental processes required for their replacement (see Chapter 12).

MINING AND MINERALS

In general, the scarce products from the earth are those that required more work for their formation and concentration. Therefore they tend to have

---

[9] Peaty sediment, 1 m thick: 10.5% organic matter, bulk density 0.7 (Coultas and Calhoun, 1976); replacement time, 100 (?) yrs

$$(0.105)(1 \times 10^4 \text{ m}^3/\text{ha})(0.7 \times 10^6 \text{ g/m}^3)(12 \text{ kcal/g})(4186 \text{ J/kcal})/100 \text{ yr} = 3.69 \times 10^{11} \text{ J/ha/yr}$$

[10] Fish, crabs, and shrimp, 8.2 g preserved weight per m^2; 2-yr replacement time; 5000 mg detritus/m^2 channels/day (Homer, 1977; Hall et al., 1986; Figure 2.6); transformity of detritus = 11,000 sej/J

EMERGY flow:

$$(5 \text{ g/m}^2\text{/day})(365 \text{ d/yr})(5 \text{ kcal/g})(4186 \text{ J/kcal})(11{,}000 \text{ sej/J}) = 4.2 \times 10^{11} \text{ sej/m}^2 \text{ channel/yr}$$

Energy flow: $(8.2 \text{ g/m}^2/2 \text{ yr})(0.2 \text{ dry})(4 \text{ kcal/g})(4186 \text{ J/kcal}) = 13{,}730 \text{ J/m}^2\text{/yr}$

[11] Tidal channel landform if generated in 100 yr. Half of tidal energy per marsh flooding-draining area—item 2, Table 7.1.

EMERGY: $(100 \text{ yr})(3.3 \times 10^{14} \text{ sej/ha/yr}) = 3.3 \times 10^{16} \text{ sej/ha}$

Energy of displacing mud: 2 m deep, 10% of area, 1000 kg/m^3 mud displaced; center of gravity, 1 m:

$$(0.1)(1 \times 10^4 \text{ m}^2/\text{ha})(2 \text{ m mean depth})(1000 \text{ kg/m}^3)(1 \text{ m})(9.8 \text{ m/sec}^2) = 19.6 \times 10^6 \text{ J/ha}$$

TABLE 7.3. Evaluation of EMERGY Stored in 1 ha of Typical Salt Marsh of Northwestern Florida

Note	Item, Units	Data	Solar Transformity† (sej/unit)	Solar EMERGY ($\times 10^{14}$ sej)	Em$ (1990 $)*
Juncus Marsh					
1	Aboveground live biomass	1.86×10^{11}	6,962	13	650
2	Underground marsh biomass	7.5×10^{11}	7,662	57	2,850
3	Aboveground dead biomass	1.17×10^{11}	8,809	10	500
4	Peaty sediment	3.69×10^{11}	3,509	13	650
	Total			100	4,800
5	Fishes, crabs, shrimp in channels	17.7×10^{6}	31×10^{6}	5.4	270
6	Tidal channel structure	19.6×10^{6}	1.6×10^{9}	313	15,650

* 1990 U.S.$ obtained by dividing annual EMERGY values by 2.0×10^{12} sej/1990 $.

† Table 7.2.

[1] Aboveground live biomass, 1111 dry g/m^2:

$$(1111\ \text{g/m}^2)(4\ \text{kcal/g})(4186\ \text{J/kcal})(1/\text{yr})(1 \times 10^4\ \text{m}^2/\text{ha}) = 1.86 \times 10^{11}\ \text{J/ha}$$

[2] Underground biomass, 4600 dry g/m^2 (Kruczynski et al., 1978);

$$\frac{(4600\ \text{g/m}^2)(4\ \text{kcal/g})(4186\ \text{J/kcal})(1 \times 10^4\ \text{m}^2/\text{ha})}{(5\ \text{yr})} = 7.5 \times 10^{11}\ \text{J/ha/yr}$$

[3] Aboveground dead biomass, 880 dry g/m^2:

$$(880\ \text{g/m}^2)(4\ \text{kcal/g})(4186\ \text{J/kcal})(1/\text{yr})(1 \times 10^4\ \text{m}^2/\text{ha}) = 1.17 \times 10^{11}\ \text{J/ha/yr}$$

[4] Peaty sediment, 1 m thick: 10.5% organic carbon (Coultas and Calhoun, 1976):

$$(0.105)(1 \times 10^4\ \text{m}^3/\text{ha})(0.7 \times 10^6\ \text{g/m}^3)(12\ \text{kcal/g})(4186\ \text{J/kcal})/100\ \text{yr} = 3.64 \times 10^{11}\ \text{J/ha/yr}$$

[5] Larger nekton in tidal channels; 5.3 g preserved wt/m^2 (Homer, 1977); using 10% of the area as tidal channels:

$$(5.3\ \text{g/m}^2)(0.1)(1 \times 10^4\ \text{m}^2/\text{ha})(0.2\ \text{dry})(4\ \text{kcal/g})(4186\ \text{J/kcal}) = 17.7 \times 10^6\ \text{J}$$

[6] Tidal channels covering 10% (?) of area; work of excavation:

$$(0.1)(1 \times 10^4\ \text{m}^2/\text{ha})(2\ \text{m})(1000\ \text{kg/m}^3)(1\ \text{m})(9.8\ \text{m/sec}^2) = 19.6 \times 10^6\ \text{J/ha}$$

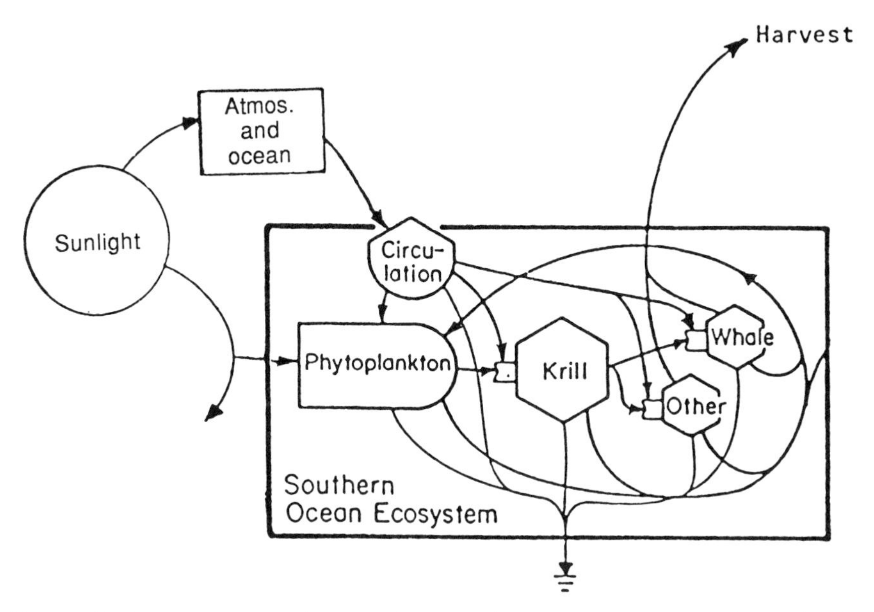

Figure 7.3. Energy systems diagram of the southern ocean ecosystem for evaluating a baleen whale.

TABLE 7.4. EMERGY Evaluation of One Southern-Hemispheric Whale

Note	Item	
1	Area of ocean per whale	3.8×10^7 m^2
2	Annual solar EMERGY supporting each whale	3.5×10^{17} sej/yr
3	EMERGY stored in average whale	52.5×10^{17} sej
4	Emdollars	\$$2.6 \times 10^6$

[1] Area of 3.8×10^{13} m^2 of area between Antarctic convergence and the south polar ice originally supported 1 million baleen whales (Knox, 1984).
[2] Half of the solar EMERGY per whale territory = annual solar EMERGY of the earth (9.44×10^{24} sej/yr) times the fraction of the earth's area per whale ($3.8 \times 10^7/5.1 \times 10^{14}$ m2).
[3] Average lifetime (15 years) times the annual EMERGY per whale.
[4] EMERGY per whale (52.5×10^{17} sej) divided by the 1990 EMERGY/$ ratio (2.0×10^{12} sej/$).
Source: Modified from Odum (1987c).

TABLE 7.5. EMERGY Evaluation of Ecosystems for Mitigation ($\times 10^{14}$ sej/ha)

	Plantation Forest*	Rain Forest*	Salt Marsh†
Annual contribution	3	6	13
Stored in ecosystem	15	1530	467

* From Table 12.1.
† Table 7.3 (Odum and Hornbeck, 1995).

TABLE 7.6. Principal EMERGY Inputs to Everglades National Park (Figure 7.4)*

Note	Item, Units	Raw Data (unit/yr)	Solar Transformity (sej/unit)	Solar EMERGY ($\times 10^{20}$ sej/yr)	Em$ Value ($\times 10^6$ 1985 $)
1	Federal Park budget, $	7.5×10^6	2.2×10^{12}	0.16	7.5
2	Visitors, $	242×10^6	2.2×10^{12}	5.32	242.
3	Fuels, J	1.71×10^{14}	6.6×10^4	0.11	5.1
4	Park concession, $	1.56×10^6	2.2×10^{12}	0.034	1.56
5	Rain, J	2.9×10^{16}	1.82×10^4	5.2	237
6	Freshwater inflow, J	2.7×10^{15}	4.8×10^4	1.3	59
7	Tide, J	2.15×10^{15}	1.6×10^4	0.34	17.2
8	Environmental total			6.84	313
9	Society total			5.62	256

[1] Federal Park budget from National Park files, $7,520,000.

[2] 780,000 people visited per year (Fla. Statistical Abstract, Florida Bureau of Economic and Business Research 1988):

$$(\$80/\text{day/person})(2\ \text{days/trip}) + (\$150/\text{trip}) = \$310/\text{person Dollar input}, \$2.42 \times 10^8$$

[3] Fuel use in boats and cars:
Boats: 6500 boat trips/yr; 20 gal/trip; 4.71×10^9 kcal/yr
Cars: 100,000 car trips/yr; 10 gal/trip; 4.09×10^{10} kcal/yr
Boats and cars: 4.09×10^{10} kcal/yr

[4] National Park Service files, $1,560,000.

[5] Park land area:

$$(3.24 \times 10^9\ \text{m}^2)(1.52\ \text{m/yr})(1000\ \text{kg/m}^3)(4.94 \times 10^3\ \text{J/kg}) = 1.46 \times 10^{16}\ \text{J/yr}$$

Florida bay:

$$(2.43 \times 10^9\ \text{m}^2)(1.2\ \text{m/yr})(1000\ \text{kg/m}^3)(4.94 \times 10^3\ \text{J/kg}) = 1.44 \times 10^{16}\ \text{J/yr}$$

[6] Chemical potential energy in freshwater inflow:

$$(450{,}000\ \text{acre-ft/yr})(1233\ \text{m}^3/\text{acre-ft})(1000\ \text{kg/m}^3)(4940\ \text{J/kg}) = 2.74 \times 10^{15}\ \text{J/yr}$$

[7] $(2.42 \times 10^9\ \text{m}^2$ bar area$)(0.5)(7.06 \times 10^2$ tides/yr$)(0.5\ \text{m})(0.5$ m range$)(9.8\ \text{m/sec}^2)$
$(1.03 \times 10^3\ \text{kg/m}^3) = 2.15 \times 10^{15}$ J/yr

[8] Sum of items 5–7.

[9] Sum of items 1–4.

Source: Simplified from DeBellevue et al. (1979) and Gunderson (1989), which include other small items.

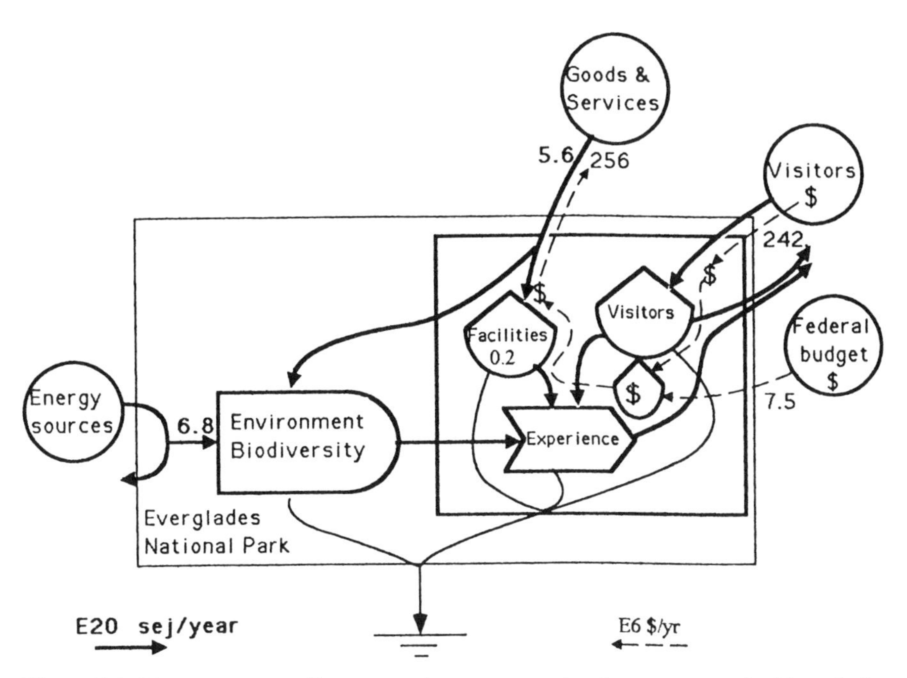

Figure 7.4. Energy systems diagram and EMERGY evaluation summary for Everglades National Park, Florida (Table 7.6).

higher EMERGY contents. Burnett (1981c) found that materials with more EMERGY contribution were less abundant. For example, phosphorus considered over the whole world is a scarce element, but phosphorous-concentrating processes acting over long periods of time can accumulate rich deposits. Figure 7.5 diagrams the phosphate concentrating that occurs when a system of acid swamp waters percolates through marine limestones over geologic time. Phosphate deposits were found to contain high EMERGY and transformities (Table 7.7).

EMERGY per unit mass (emjoules per gram) indicates the position a mineral has on the scale of earth scarcity and unit value. The EMERGY per mass of the main divisions of the earth's processes was estimated in Chapter 3. Where there is additional processing work by the earth, scarce materials may be concentrated further. Gold in South Africa was processed by more than one cycle of river erosion, deposition and uplift, each one accumulating grains and nuggets into the gravels at the bottom of the river. Gold is concentrated further by human mining and industrial works. EMERGY evaluation by Bhatt and Odum (H. T. Odum, 1991a) found a high transformity for gold.

When earth products are strip-mined, the environmental system is often destroyed as part of the earth is removed. After removal of the minerals, the earth may be returned to form a useful surface again as part of restoring the landscape. Some additional years are required for the ecosystems to be seeded and reestablished, for soils to develop, and for water conservation characteris-

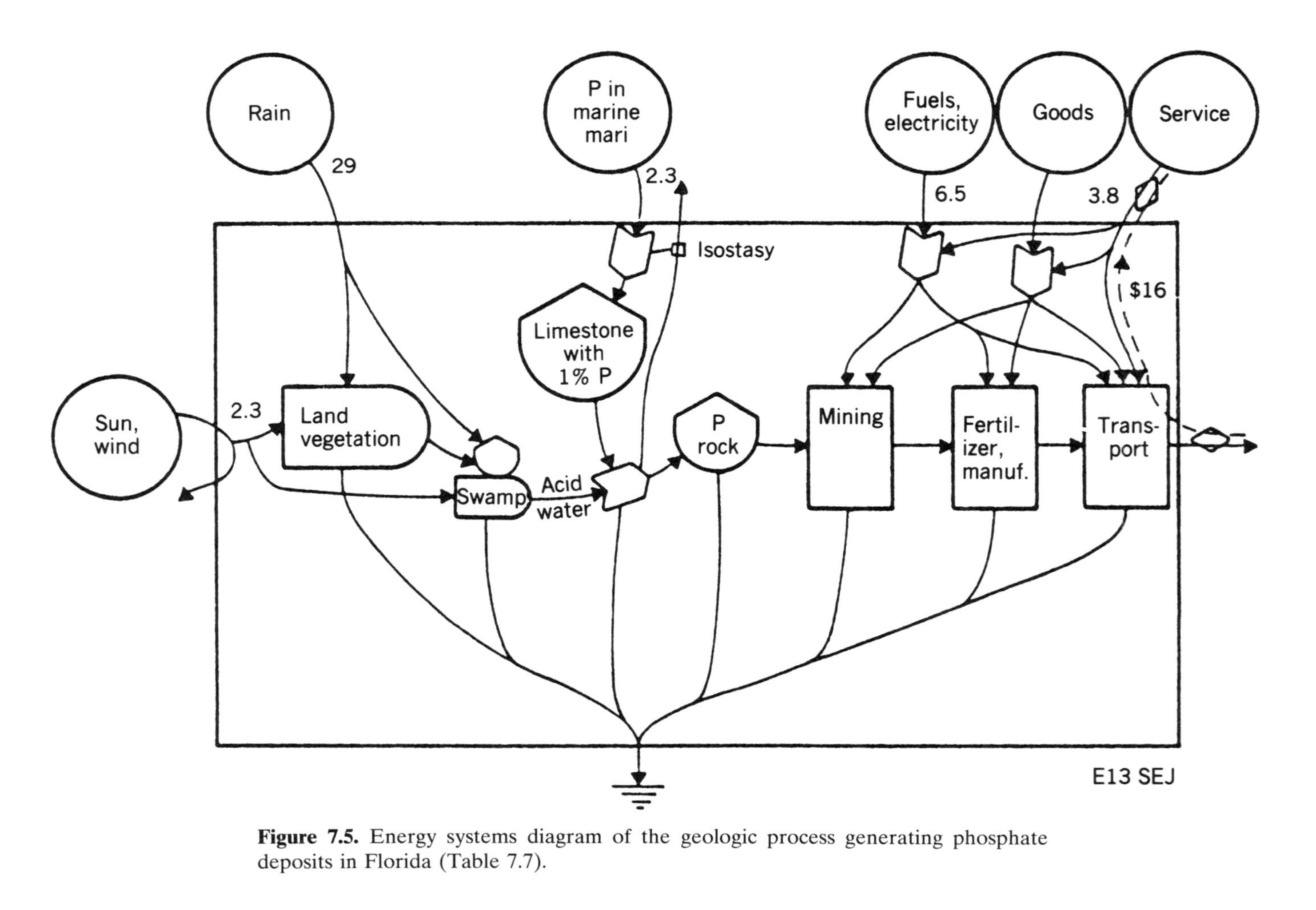

Figure 7.5. Energy systems diagram of the geologic process generating phosphate deposits in Florida (Table 7.7).

tics to be reestablished. The EMERGY contribution of the mined product should be compared with the interruptions and losses of EMERGY from ecosystems and water diverted during the mining and restoration period, and EMERGY diverted from other work for reclamation. Figure 7.6 shows the mining system, the yield of minerals, the inputs from the economy required for mining and restoration, and the several pathways of environmental contribution interrupted during the mining.

For the window of time of human society, the processes that develop mineral reserves is very slow (see Chapter 3). The decision to be made is whether mining is a net advantage and whether the post-mining use can be as productive. For phosphate mining in Florida there is a large net contribution (Table 7.8). However, EMERGY accounting found no net benefit for mining and processing oil shale (see Figure 8.2c). Kangas (1983), evaluating the EMERGY of landforms and their colonization by ecosystems, found that irregular, postmining lands, which had 20–40 yr of natural restoration through ecological succession and other processes, were more valuable than those lands that were bulldozed flat in well-intentioned restoration.

RENEWABLE RESOURCES

With agriculture, forestry, aquaculture, and fisheries, free environmental inputs (sun, winds, waters, rocks, tides, and so on) are combined with purchased inputs (labor, services, fuels, electric power, special equipment, materials, and so forth). Depending on the intensity of the economic use, free environmental inputs contribute 5–95% of the EMERGY of the products. Examples were given in previous chapters: silk agriculture (Figure 4.7), forest wood (Figures 1.5 and 5.1), and shrimp aquaculture (Figure 4.8). However, where agricultural and forest products are harvested from lands, bare ground and considerable soil erosion may result. Such using up of soils faster than they form is a nonrenewable source (see Figure 6.6). EMERGY of soil loss was important in Illinois corn agriculture (H. T. Odum, 1984b).

IMPACT

One of the first ways suggested for an inventory of impact interactions in a system was the impact table, a matrix with impact actions across the top, environmental properties down the side, and possible interactions indicated by the boxes within the table (Leopold et al., 1971). Systems diagrams and tables of interaction are alternate ways of representing the same network. Since a systems diagram of an environmental system has principal components, interrelationships and processes, it can be used to show direct and indirect effects of an impact (H. T. Odum, 1972b; Wang et al., 1980). Examinations

TABLE 7.7. EMERGY Evaluation of Phosphate Formation and Mining Florida: Forming and Mining 200 kg 10% P Rock after Its Concentration from 2000 kg Original Limestone Marl by Percolation of Organic-Rich Swamp Waters*

Note	Item	Data, Units	Solar Transformity (sej/unit)	Solar EMERGY ($\times 10^{12}$ sej)
	Phosphate Rock Formed, 200 kg			
1	Direct sun, 10 m^2 of land	2.31×10^{14} J	1	23.
2	Limestone marl uplift	2×10^6 g	1.0×10^9 J/g	200.
3	Rain and runoff	1.87×10^{10} J	1.82×10^4	340
4	Total solar EMERGY into rock formation (2 + 3)			540
	Mined Rock Phosphate, 180 kg			
5	Service	$2.84	4.1×10^{12}	11.6
6	Fuel	1.58×10^8 J	6.6×10^4	10.4
7	Electricity	7.03×10^6 J	2.0×10^5	141.
8	Total solar EMERGY in mined phosphate rock (d +e + f + g)			703

Solar EMERGY per Unit Calculated by Dividing Total Solar EMERGY Previously Used by the Energy or Weight Respectively

Note	Item	Solar Transformity (sej/J)	Solar EMERGY per Mass (sej/g)
9	Phosphate rock in the ground	7.7×10^6	2.7×10^9
10	Mined rock	10.1×10^6	3.9×10^9/g phosphate 1.78×10^{10}/g P

* Time for development of phosphate deposit estimated from rate of percolation through Florida swamp: 10% of rain over 10 m^2 runs into swamp area of 1 m^2; 2.5 cm/wk (130 cm/yr) percolated down carrying 100 mg/l organic matter that oxidizes, generating acid as it percolates, dissolving limestone.

$$\frac{(1.3\ m^3/m^2/yr)(100\ g\ org/m^3)(72/180\ g\ C/g\ org)}{12\ g\ C/mole} = 4.33\ mole/m^2/yr$$

To dissolve 90% of limestone of density 1.9 at 4.33 moles/m^2/yr:

$$\frac{(1 \times 10^6\ cm^3/m^3)(1.9\ g/cm^3)(0.9)}{(100\ g/mole\ CaCO_3)} = 1.71 \times 10^4\ moles\ CaCO_3$$

$$\frac{(1.71 \times 10^4\ moles\ CaCO_3/m^3\ rock)}{(4.33\ moles\ acid/m^3/yr)} = 3949\ yrs$$

[1] Direct sun on 10 m^2 area from which waters converge to 1 m^2:

$$(10\ m^2)(1.4 \times 10^6\ kcal/m^2/yr)(4186\ J/kcal)(3949\ yr) = 2.31 \times 10^{13}\ sej/10\ m^2$$

[2] Direct solar energy on estuary operating ecosystem accumulating marl from shells, 1 m^3 in volume per 10 m^2. Same as in note 1, 2.31×10^{14} sej/10 m^2.

[3] Rain runoff, 10% of the rainwater on 10 m^2 converging to swamp and operating its vegetation and leaching period:

$$(1.08 \text{ m/yr rain used})(10 \text{ m}^2)(1 \times 10^6 \text{ g/m}^3)(4.4 \text{ J/g})(0.10) = 4.75 \times 10^6 \text{ J/yr}$$

$$(4.75 \times 10^6 \text{ J/yr})(3949 \text{ yr}) = 1.87 \times 10^{10} \text{ J rainwater/10 m}^2$$

[4] Solar EMERGY-total of 2 and 3. Gibbs free energy in the phosphorus concentration relative to prevailing solutions from which phosphorus was first concentrated by shellfish system to 1% and further by swamp solution to 10%. Solution equilibrium with solid phosphate taken as 1 ppm:

$$\text{Free energy/g} = \frac{(8.33 \text{ J/mole/deg})(300°)}{33 \text{ g/mole}} \ln\left(\frac{1.0}{0.01}\right) = 348 \text{ J/g}$$

Phosphorus content concentrated initially at 1%, with a density of 2 g/cm^3:

$$(0.01)(1 \times 10^6 \text{ cm}^3/\text{m}^3)(2 \text{ g/cm}^3)(348 \text{ J/g}) = 6.96 \times 10^6 \text{ J/10 m}^2$$

[5] Services, phosphate rock, $15.8/Tn in 1977 when solar EMERGY/$ was 4.1×10^{12} sej/$:

$$(180 \text{ kg rock})(\$15.8 \times 10^{-3}/\text{kg}) = \$2.84$$

[6] Fuel mining, 1 ton:

$$(318 \times 10^{10} \text{ kcal/15.1} \times 10^6 \text{ tons}) = 2.1 \times 10^5 \text{ kcal/Ton (Rushton, 1981)}$$

$$(180 \text{ kg/1000 kg/t})(2.1 \times 10^5 \text{ kcal/t})(4186 \text{ J/kcal}) = 1.58 \times 10^8 \text{ J}$$

[7] Electric power:

$$1400 \times 10^{10} \text{ kcal/15.1} \times 10^6 \text{ tons} = 9327 \text{ kcal/ton}$$

$$(180 \text{ kg/1000 kg/ton})(9327 \text{ kcal/ton})(4186 \text{ J/kcal}) = 7.03 \times 10^6 \text{ J}$$

[8] Sum of d, e, f, and g.

[9] Solar transformity of unmined phosphate rock, 200 kg rock:

$$\text{Gibbs energy:} \quad (200 \text{ kg})(1000 \text{ g/kg})(348 \text{ J/g}) = 6.96 \times 10^7$$

$$\frac{540 \times 10^{12} \text{ sej}}{6.96 \times 10^7 \text{ J phosphate}} = 7.75 \text{ E6 sej/J unmined phosphate rock}$$

$$\text{EMERGY/g:} \quad \frac{540 \times 10^{12} \text{ sej}}{2 \times 10^5 \text{ g phosphate}} = 2.7 \times 10^9 \text{ sej/g unmined phosphate rock}$$

[10] Solar transformity of mined phosphate rock:

$$\frac{703 \times 10^{12} \text{ sej}}{6.96 \times 10^7 \text{ J phosphate}} = 1.01 \times 10^7 \text{ sej/J mined phosphate}$$

Solar EMERGY per mass:

$$\text{Per gram phosphate:} \quad \frac{703 \times 10^{12} \text{ sej}}{1.8 \times 10^5 \text{ g phosphate}} = 3.9 \times 10^9 \text{ sej/g mined phosphate}$$

$$\text{per gram P, 22\% of Calcium phosphate:} \quad \frac{703 \times 10^{12} \text{ sej}}{(0.22)(1.85 \times 10^5 \text{ g phosphate})} = 1.78 \times 10^{10} \text{ sej/g P}$$

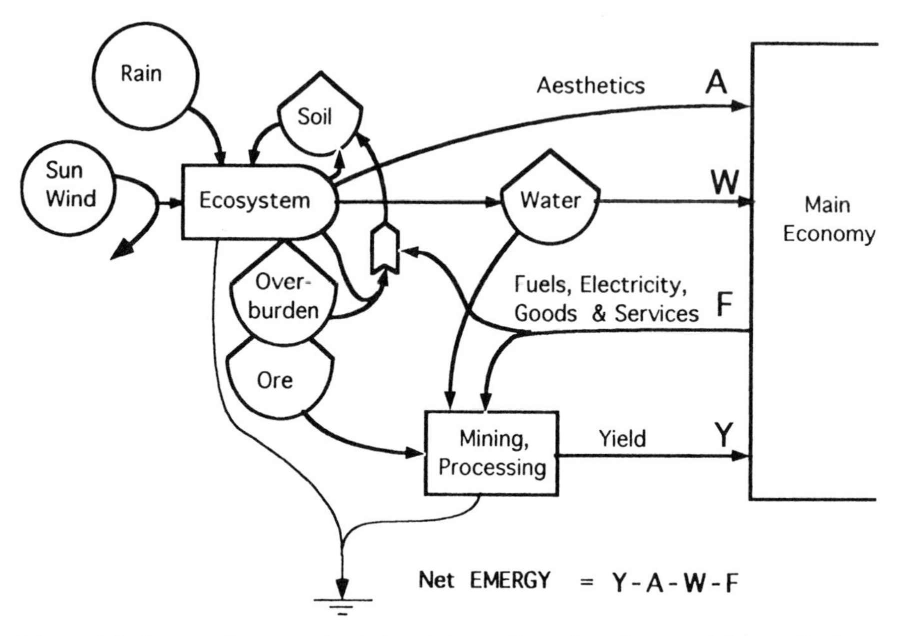

Figure 7.6. System diagram of mining, interruption of environmental contributions, and reclamation. Net EMERGY evaluation equals the yield minus environmental losses.

of the diagram and its changes allows a person used to the systems language to select those parts most affected by the change.

For example, Figure 7.7a is a systems diagram of a wetland ecosystem, which was impacted by acid wastewaters containing lead from a battery recovery operation (Sapp Superfund site, Jackson County, Fla.). Studies on ecology, biogeochemistry, EMERGY, and economic analysis were made by Ton (1990, 1993) and Pritchard (1992). EMERGY evaluations in Figure 7.7b,c compare an all-technology system of battery lead recovery and waste treatment with acid washing of batteries followed by wetland filtration of lead wastewaters. Both systems contributed high value by reusing lead, a high-transformity product. The technological system uses more resources, but the wetland may require long-term preservation because of its lead storage.

An energy systems diagram is also a mathematical model for the system. For example, Figure 7.7a is an impact model showing effect of the Sapp Swamp ecosystem on lead processing, and vice versa—the effect of lead on the ecosystem. One of the normal ways of studying the properties of a model is to write the equations equivalent to the diagrams, write a computer program, and simulate the effect of a change (such as an impact) on the various parts and processes over time. With a somewhat simplified version of Figure 7.7, Ton (1990, 1993) wrote equations and simulated the uptake of lead, mortalities of the organisms, and the deposit of much of the lead in the permanent peaty detritus. The systematic exploration of model responses is sometimes called

TABLE 7.8. EMERGY of Land Reclamation and Restoration after Phosphate Mining in Florida

Note	Item	Data	Solar EMERGY/Unit (sej/unit)	Solar ($\times 10^{20}$ sej)	Em$* ($\times 10^9$ 1992 $)
			Mining		
1	Motor fuels	1.32×10^{16} J	6.6×10^4/J	8.7	0.5
2	Electric power	5.9×10^{16} J	2.0×10^5/J	118.	8.4
3	Service	1.39×10^9	4.4×10^{12}/1977 $	62	4.4
4	Phosphate yield	9.56×10^{13} g	3.9×10^9 sej/g	3720	264
			Environment		
5	Reclamation cost	0.11×10^9 $	4.4×10^{12}/1977 $	4.8	0.34
6	Water diverted	3.8×10^{14} J	4.8×10^4/J	0.18	0.012
7	Diverted land production	1.89×10^{17} J	1×10^4/J	18.9	1.34

* EMERGY flow divided by 1.4×10^{12} sej/1992 $.

[1–6] From impact plan of Mississippi Chemical Corporation.

[6] Gibbs free energy of fresh water relative to transpiring plant leaves.

[3,5] 1977 Estimated costs:

$$(\$15.8/\text{ton})(9.56 \times 10^7 \text{ ton}) = \$1.51 \times 10^9$$

Source: Modified from an evaluation of mining in Hardee County by Rushton (1981).

a *sensitivity analysis.* An impact study and a quantitative sensitivity analysis have much in common.

One approach for using EMERGY accounting to assess impact is to make an *EMERGY change table.* This is an EMERGY evaluation table that contains only line items that have been changed. For example, impacts of lead on the Sapp Swamp are given in Table 7.9.

Information Amplified Impacts—Valdez Oil Spill

Figure 7.8 is an approximate EMERGY evaluation of the way television networks may have caused an inappropriate response to the *Valdez* oil spill in Prince William Sound, Alaska. With support from the Cousteau Society, EMERGY evaluations were made of the oil spill, pipeline interruptions, ecological damage, social disruption, the Alaska pipeline operation, alternative shipping alternatives, and the overall economy of Alaska (M. T. Brown et al., 1993; Woithe, 1992). Then the EMERGY flows of the television industry were evaluated for an assumed hour of *Valdez* coverage over the summer of 1989 (Table 7.10).

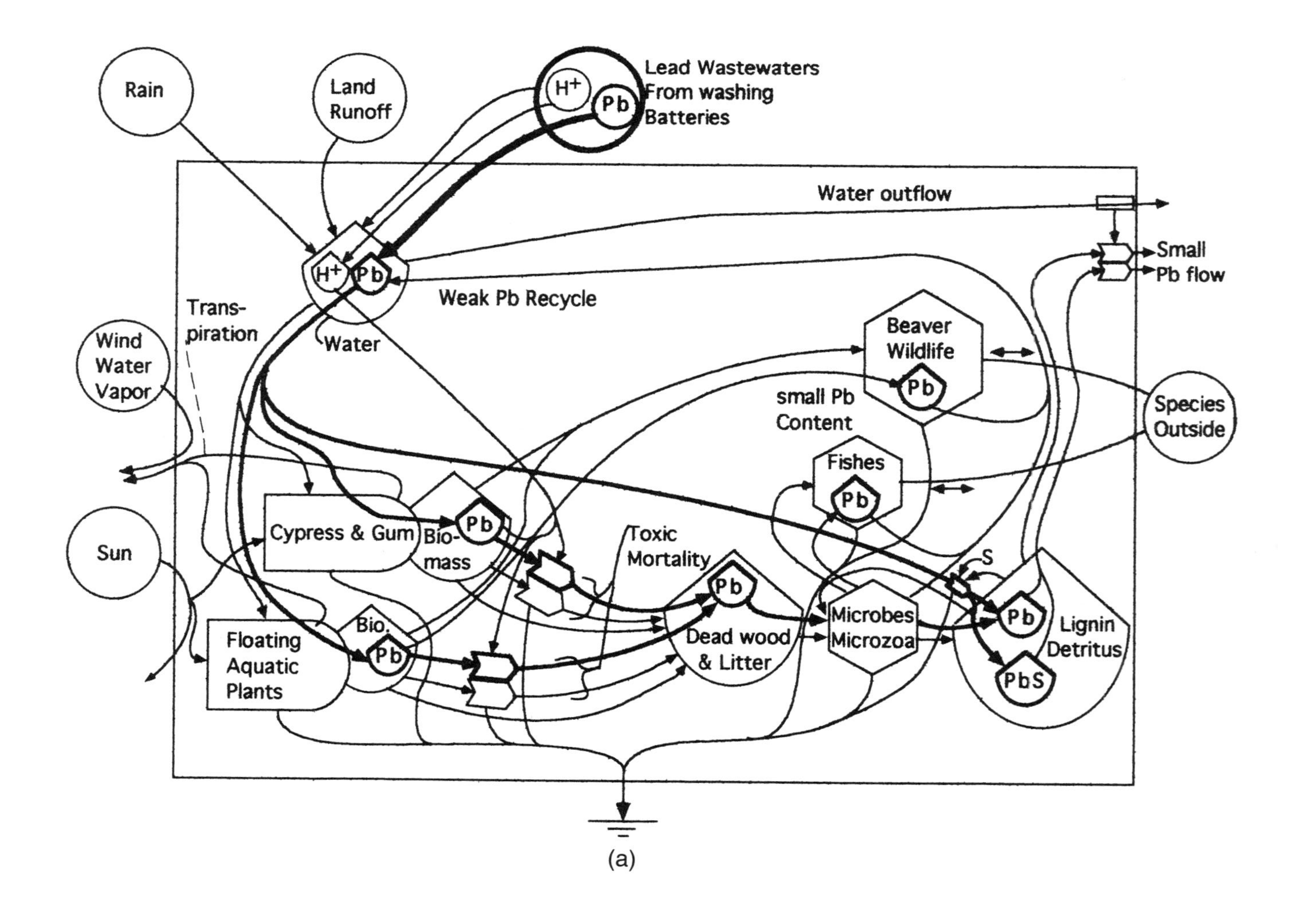

Rain
Land Runoff
Lead Wastewaters From washing Batteries
H+
Pb
Water outflow
Small Pb flow
Weak Pb Recycle
Trans-piration
Water
Wind Water Vapor
Beaver Wildlife
small Pb Content
Species Outside
Fishes
Cypress & Gum
Bio-mass
Toxic Mortality
S
Sun
Floating Aquatic Plants
Bio.
Dead wood & Litter
Microbes Microzoa
Lignin Detritus
PbS

(a)

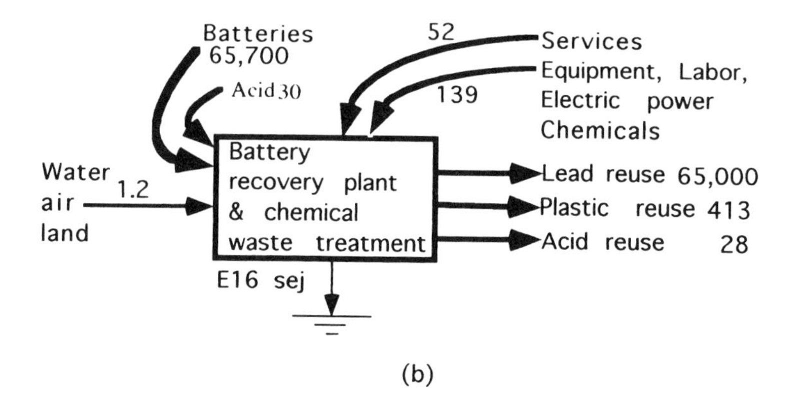

(b)

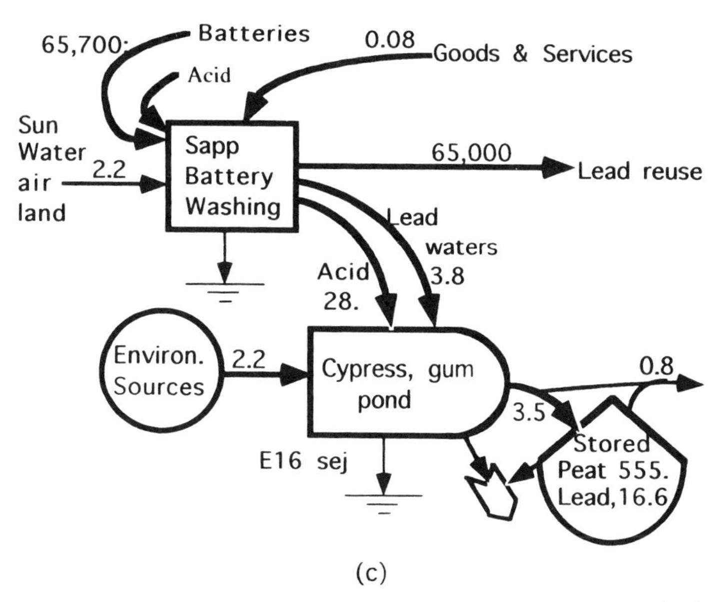

(c)

Figure 7.7. Evaluation of alternatives for recovery of lead in automobile batteries; Pritchard (1992). Project of D. T. Sendzimir Foundation with the University of Florida. **(a)** Energy systems diagram of Sapp Swamp, a cypress wetland that filtered acid battery wash waters in Jackson County, Fla. Thick lines indicate main pathways of lead, its filtration, and impact. **(b)** EMERGY evaluation of technological methods. **(c)** EMERGY evaluation of acid washing with wetlands receiving wastewaters.

TABLE 7.9. EMERGY Evaluation of Lead Impact on Sapp Swamp, Florida*

Note	Item	$\times 10^{16}$ sej/yr	Em$/yr†
1	Annual production before impact	7.10	47,333
2	Lead inflow	2.47	16,467
3	Annual production after impact	2.20	14,666

* 29.2-ha Superfund site, Jackson County, Fla.

† Solar EMERGY divided by 1.5×10^{12} sej/1991 $.

[1] Based on gross primary production and a solar transformity of 1317 sej/J.

[2] Based on solar EMERGY per mass of 7.3×10^{10} sej/g.

[3] Based on diurnal curve analysis of oxygen in waters and leaf-area index estimation of tree contributions.

Source: Data from Pritchard (1992).

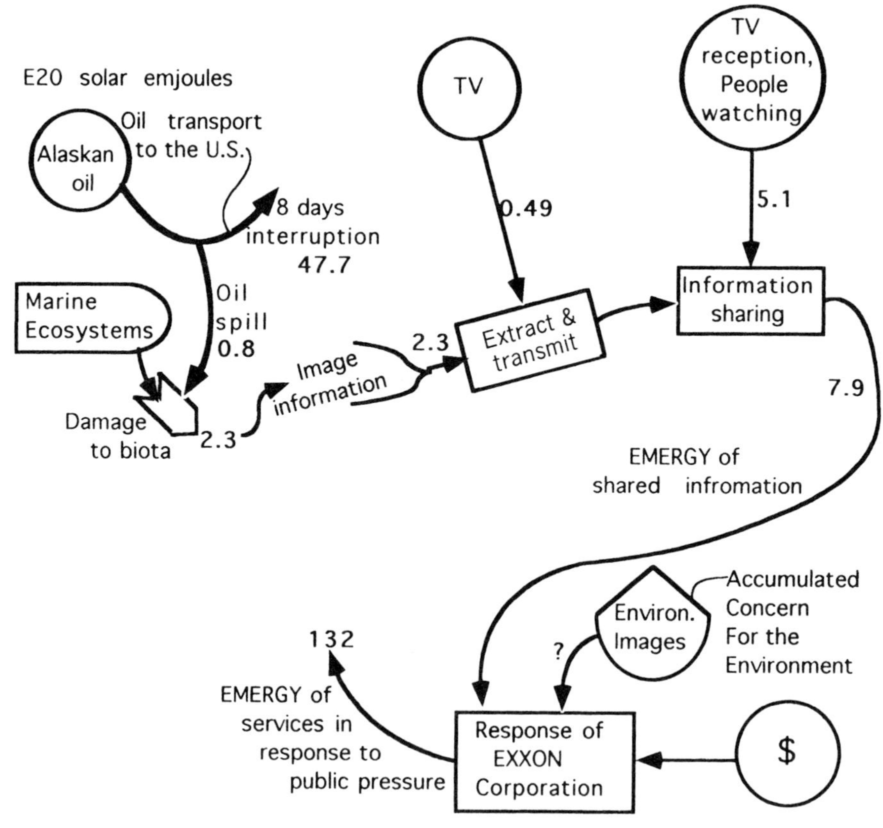

Figure 7.8. EMERGY of *Valdez* oil spill and its information role based on 1 hr of television transmission and 0.5 hr of reception per person (see Table 7.10).

TABLE 7.10. EMERGY of *Valdez* Oil Spill and Its Information Role: Based on 1 hr of Television Transmission and 0.5 hr of Watching per Person

Note	Process	EMERGY ($\times 10^{20}$ sej)	EM$* (million $)
1	EMERGY of phenomenon (spill)	4.0	250
2	EMERGY of damage	2.3	143
3	EMERGY of information of that damage	2.3	143
4	EMERGY of copying and transmitting	0.49	30.6
5	EMERGY of receiving	0.55	34.4
6	EMERGY of watching and sharing, USA	5.1	318
7	Cumulative total	7.9	494
8	Response by Exxon and government	132.	8250.
9	Oil flow interrupted	47.7	2981.

* Expressed in 1991 U.S.$ using 1.6×10^{12} sej/$.

[1] Oil spill evaluation: 4.0×10^{20} sej in oil loss.

[2] From Woithe (1992).

[3] Assume EMERGY of the information transmitted was what was required to create the image, (the EMERGY of the damage).

[4] Data for TV copying and transmission taken from Table 12.5: 42.7×10^{22} sej/8760 hr/yr = 0.49 sej/hr.

[5] Receiving in USA from Table 12.5, 14.0×10^{22} sej/yr; 7 hr/day)(365 days) = 2555 hr/person watching TV:

$$(14.0 \times 10^{22} \text{ sej/yr})/(2555 \text{ hr}) = 0.55 \times 10^{20} \text{ sej/hr watching}$$

[6] EMERGY added for people watching ½ hr each:
EMERGY per person per hour from total annual EMERGY of U.S.:

$$(365)(24) = 8760 \text{ hr/yr}$$

$$(900 \times 10^{22} \text{ sej/USA/yr})(0.5/8760 \text{ hr/yr}) = 5.1 \times 10^{20} \text{ sej}$$

[7] Total of items 4–7.

[8] $2.5 billion expenditure by Exxon and $0.1 billion by federal government; half spent in Alaska and half in mainland USA:

$$(0.5)(2.6 \times 10^{9} \text{ \$/yr})(8.6 \times 10^{12} \text{ sej/\$}) + (0.5)$$

$$(2.6 \times 10^{9} \text{ \$/yr})(1.6 \times 10^{12} \text{ \$/yr}) = 1.32 \times 10^{22}$$

[9] Eight-day interruption of oil shipments: 0.477×10^{22} sej.

As a result of the world's public reaction to the spill, Exxon Oil Company, by 1994, had paid $3.5 billion into the Alaskan system, taking it away from other normal pathways in the world economy. In 1994 another jury awarded $5 billion more in damages (still under appeal). As the diagram suggests, by the time the circle had been completed of extracting, transmitting, amplifying, and showing information to millions of people—causing reactions that sent billions of dollars to Prince William Sound–there had been an EMERGY amplification 100-fold. The damage to the community was partly due to the large payments of money causing so much empower injection into a small area in a short time without many ways of joining it to useful work.

EMERGY that was diverted was far out of proportion to any concept of repair or replacement. Was the information overamplified and inappropriate to the fundamental concept of maximizing good EMERGY production and use? Or, did the information system serve to generate a useful new symbol of worldwide significance to galvanize people around the need to protect the work contributions of the renewable environment everywhere? Since half the world's empower is from the renewable environment, the amplification might have been appropriate, but was the response to add more EMERGY to the disaster correct? The *Valdez* example provides some indication of the way EMERGY evaluation of information can help put television determinism in proportion. The management of world television information may be the most important means for adapting the global system to the future. At issue is the question of whether shared information can and should control economic power or vice versa.

CUMULATIVE IMPACT ASSESSMENT

A concept popular in some government agencies is cumulative impact assessment. The concern is that important environmental properties may accumulate the effects of many small impacts until the collective effect becomes large enough to exceed thresholds for sustaining the desirable features of the environmental system. For example, many activities of humans may contribute to lowered groundwater tables, such as pumping out water, substituting ornamental vegetation with higher transpiration rates, and draining and ditching—all of which reduce recharge. The cumulative impact on groundwater would be given in units of water level; cumulative impact on water quality would be given in units of water purity; and so on.

The various impacts can also be evaluated in EMERGY and emdollar units. As a general common denominator for different kinds of units, evaluating EMERGY units allows the cumulative-impact-assessment concept to be applied to the total energy and health of the system. In other words, the cumulative impacts of each of several parts of a system can be expressed as a single number, the *percentage of total system function.*

The immediate effect of a change tends to be a negative one, since it may change the conditions for which the system was adapted to perform at maximum. Environmental systems may reorganize so that what was an impact to the previous system is used as a stimulus to the replacement system. For example, a toxic impact becomes an aid in competition to those species resistant to the impacts that prevail thereafter. A high EMERGY value of a change may be regarded as a temporary drain, and potentially a resource for a reorganized system. In other words, EMERGY accounting can measure the immediate impact, the losses during the time of reorganization, and finally compare the newly adapting pattern with the old one.

RISK ASSESSMENT

Risk is a name for evaluating the environmental hazards that have been important in recent years in government agencies such as the Environmental Protection Agency (EPA). As sometimes defined in this context:

Risk is the *probability* of an *undesired* environmental *impact.*

It is the chance of something bad happening to something else. Risk has four dimensions: (1) the quantity of items at risk; (2) the unit impact of the hazard, (3) the frequency of the impact, and (4) the societal value given to the impact. After enough education and experience, the human perception of importance may eventually correspond with the real importance. In the short run, the assessment may have to consider current public perceptions, but for the long run, the perception may correspond to the longer-range real impact. Figure 7.9 suggests the relationships of the components of risk assessment. The impacting influence interacts with the environmental component to produce the impact, which interacts with human perceptions to produce the risk assessment.

For example, we multiply the quantities of a population or an ecosystem by an impact index, such as mortality, times the probability (frequency of past occurrence), times some measure of desirability. Data for each of the inputs to the risk assessment may be a distribution showing the range of properties. For example, there are many waters of low copper concentration but few with high concentration. Thus, there is a distribution in the impacting influence (e.g., concentration of a toxic element), a distribution in the impacted entity (e.g., distribution of numbers and sizes in a fish population), and distribution

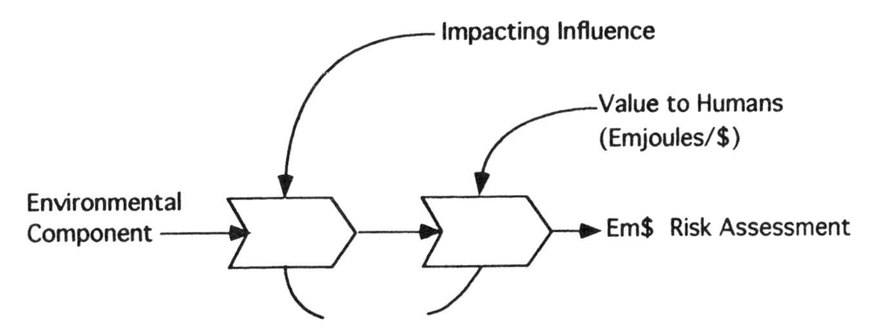

Figure 7.9. Systems diagram of risk assessment.

in human perceptions of importance. O'Neill et al. (1982) suggested that risk be evaluated with dynamic models of interaction between a distribution of environmental entities and the distribution of impacting influence along with statistical parameters of each input to provide probabilities. Examples were assembled in a "risk assessment forum" (Eastern Research Group, 1993).

These methods are highly empirical, are hard to generalize, and are converted to costs and values with difficulty. EMERGY-transformity may be helpful in the future. Using the interactions of Figure 7.9, risk may be simulated and/or evaluated in EMERGY flow units. The empower of each of the inputs in Figure 7.9 may be evaluated as the product of the flow times its EMERGY per unit flow (transformity, EMERGY per mass, EMERGY per bit, EMERGY per monetary unit). The EMERGY values of the interacting flows are added. The final empower (EMERGY flow per unit time) may be expressed in emdollars of public value by dividing by the EMERGY/money quotient. In other words, the emdollar values of risk are high when there is a large quantity impacted, a large EMERGY of impacting input, and a large EMERGY in the information of perception.

An EMERGY distribution may be made from each input distribution. A set of data may be arranged hierarchically by plotting the quantity of each item

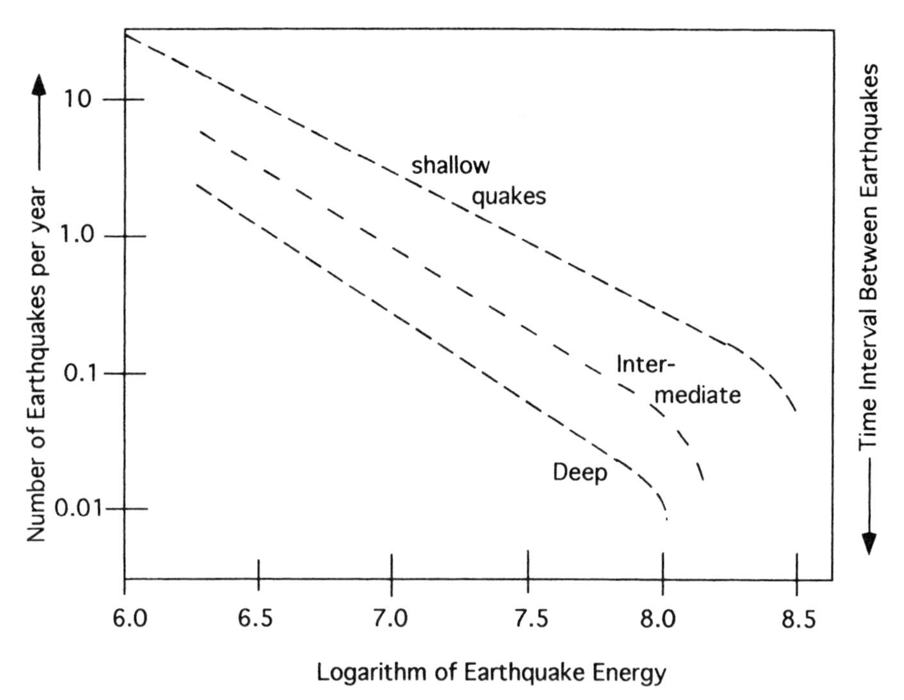

Figure 7.10. Energy intensity of earthquakes increasing with time interval between earthquakes. Number of earthquakes is plotted per 1/10 unit of energy (Gutenberg and Richter, 1949). Transformity may be assumed to be proportional to the time for accumulating strain between quakes. Reprinted with permission of University of Chicago Press.

as a function of solar transformity. Next, we multiple the quantity (stored or flowing) times its transformity to get the EMERGY or empower, which is also plotted as a function of solar transformity, thus making an EMERGY signature graph. Next we multiply each EMERGY value by its expected frequency (probability), and plot the product as a function of solar transformity. Then the EMERGY of the environmental entity and that of the impact may be combined. Annual dollar values to the public welfare (emdollars) are obtained by dividing the empower values by the EMERGY/money quotient for the year.

Transformity evaluations *alone* may help in evaluating risk. As previously described (Chapter 2), the higher the transformity, the higher the unit impact. The higher the transformity of the environmental component being impacted, the more important it is to ecosystems and to humans. Higher emdollar values go with higher transformities. The higher the transformity, the longer the time between impacts (Figure 7.10). People tend to recognize transformities that are numerically within the range of human affairs as important to them. For example, heavy metals have high impacts, impact perceptions and risk, and have large economic effects and high transformities.

SUMMARY

Methods and examples were given of EMERGY accounting of environmental contributions and impacts including lands, waters, wilderness nature, products harvested for economic use, and indices of environmental impact used in public environmental management. Normally undervalued when market values are used, the environmental values expressed in EMERGY or emdollars were compared with human economic endeavors on an equivalent basis.

CHAPTER 8

NET EMERGY OF FUELS AND ELECTRICITY

Urban civilizations primarily run on fuels and electricity. If the sources of fuels and electricity deliver much more EMERGY than is required to obtain them, then the economy is abundant and the standards of living are high. In other words, prosperity depends on the net EMERGY of its primary sources. Although fuels operate basic industries and most transportation, electric power with higher transformity operates at the higher levels of the energy hierarchy of civilization (Figure 8.1). In this chapter we evaluate the net EMERGY yield of fuels, electric power, and energy sources proposed as substitutes.

FUELS

A fuel is a resource that generates concentrated heat to operate processes and machines of society. Fuels are required for the heat engines of industry and transportation, for cooking and to heat people, homes, and for conversion into electricity and other forms of energy. Good policy on fuels in the short run first uses the sources with the highest EMERGY yield ratio (Figure 8.2, previously defined in Chapter 4). The higher the yield relative to what is used to process it, the more EMERGY goes to support the rest of the economy. Coal production in New Zealand is an example of a good EMERGY yield ratio of a fuel (12.3 in Figure 8.2b).

Unusable Fuels

Shale oil is a famous example of a fuel that apparently has no net EMERGY yield. The evaluation in Figure 8.2c was used in testimony to the U.S. Congress advising against attempts to develop the oil shale. However, a joint project of the federal government and private industry was authorized despite the evidence, and several billion dollars were invested in joint synfuels projects before shale oil was proven to be uneconomical and large-scale developments abandoned in the 1980s.

Fuel Substitution

Marchetti and Nakicenovic (1979) document the substitution of fuels since the start of the Industrial Revolution. Economic development first uses those sources of heat with the highest net EMERGY yield because they are cheapest, shifting to those of lower ratio later. Since technology for substitution of fuels exists, market prices help select the sources with the largest net contributions of heat to be used first. Evaluating the EMERGY yield ratio of a fuel source predicts when it will become economical. A main fuel that provides the highest EMERGY yield ratio to an economy can be called its *primary source.*

Fuel Comparisons

Enough net EMERGY evaluations of different kinds of fuels have been done to set priorities. Values in Figure 8.3 are typical. Coal, oil, and natural gas in the richer geologic reserves generated 6–60 times more EMERGY than was used to process them. Their importance to an economy was 6–60 times their effort, justifying much larger economic and military efforts than might be inferred from their dollar costs. The EMERGY yield ratios in mining and processing in Figure 8.3 may be compared with the EMERGY yield ratio from buying fuel (Figure 4.4b). Since the 1973 energy crisis, the EMERGY yield ratio has

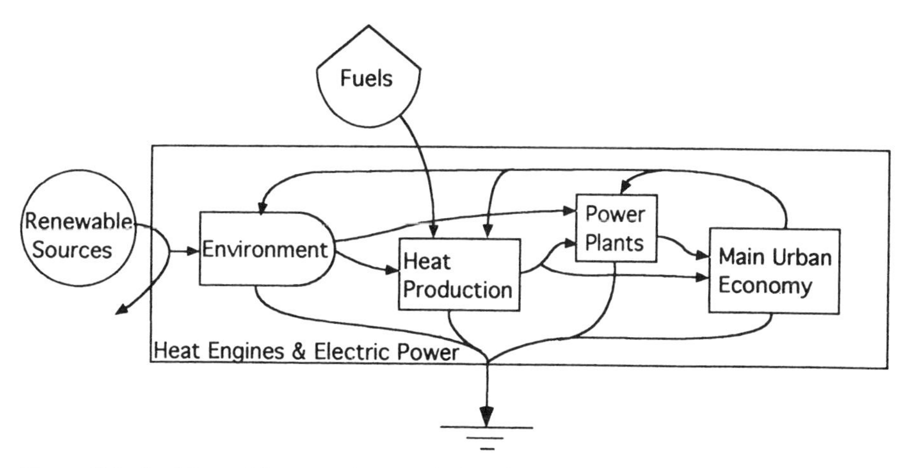

Figure 8.1. Position of heat engines and electric power in the hierarchy of the human economy.

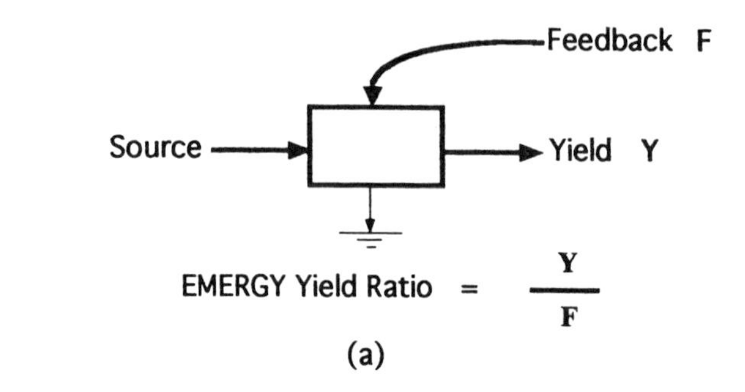

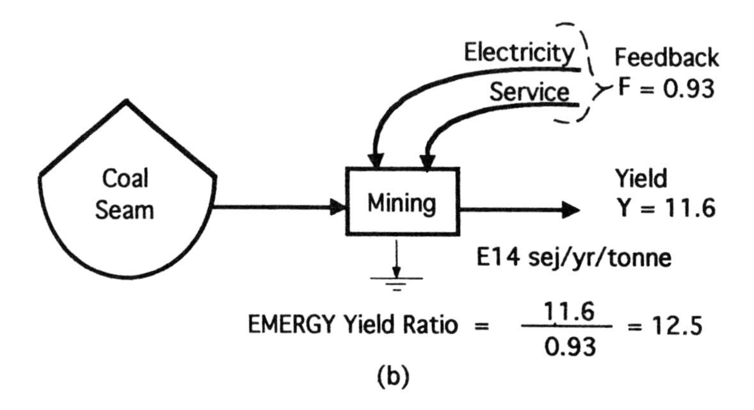

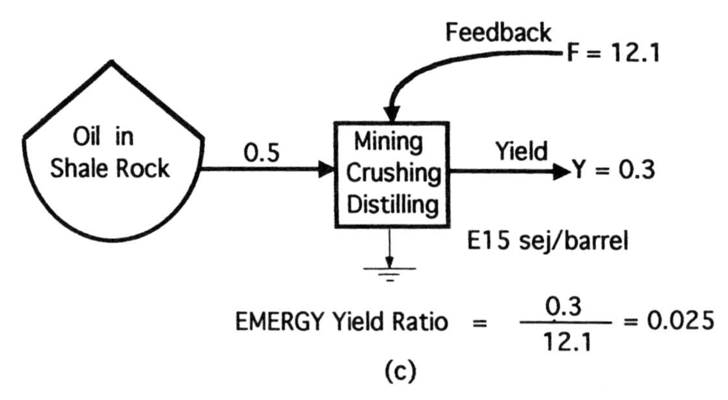

Figure 8.2. Diagrams of EMERGY yield ratio for evaluation of a primary resource. **(a)** Definitions of yield Y, feedback F, and EMERGY yield ratio. **(b)** Coal deposit with net EMERGY yield in New Zealand. **(c)** Oil shale without net EMERGY yield. For details see tables on page 139.

(b) EMERGY Analysis of Coal Mining in New Zealand, 1 metric tonne (tn)

Note	Item	Raw Units	Solar EMERGY per unit (sej/unit)	Solar EMERGY ($\times 10^{14}$ sej/yr)	Em$* (1984 U.S.$)
1	Mining services	11 $/Tn	3×10^{12}/$	0.33	15.0
2	Energy used mining	6.22×10^7 J	4×10^4/J	0.03	1.4
3	Coal mined	2.9×10^{10} J	4×10^4/J	11.6	527.3
4	Transport service	15 $/Tn	3×10^{12}/$	0.45	20.5
5	Transport energy	2.27×10^8 J	5.3×10^4/J	0.12	5.5

* EMERGY flow in column 5 divided by 2.2×10^{12} sej/$ for U.S. in 1984.
[1] Mining costs in 1980 $; EMERGY/$ ratio, 3×10^{12} sej/$ based on 1980 EMERGY analysis of New Zealand (H. T. Odum and Odum, 1983).
[2] Rate per Tn from a Texas strip mine.
[3] (1 Tn)(1×10^6 g/Tn)(7 kcal/g)(4186 J/kcal) = 2.9×10^{10} J
[4] Transport costs to markets in New Zealand, 1980 $.
[5] (560 btu/ton-mile)(400 miles)(1013 J/btu) = 2.27×10^8 J.

(c) EMERGY Analysis of One Barrel of Shale Oil from Pilot Plant, Anvil Points, Colo., 1944. Liquid Fuel from Oil Shale Where Yield of Liquid Fuel was Calculated from Separately Determined Transformity.

Note	Item	Data, Units	Transformity (sej/unit)	Solar EMERGY ($\times 10^{15}$ sej)	Em$ Value* (1984 U.S.$)
1	Shale oil yield	5.9×10^9 J	5×10^4/J	0.30	136.
2	Processing service	5513 $	2.2×10^{12}/$	12.1	5513.
3	Oil from rock	1.0×10^{10} J	5×10^4/J	0.5	227.
4	Vegetation interrupted	3×10^7 J	1.8×10^4/J	0.0005	0.22
5	Reclamation service	$0.031	2.2×10^{12}/$	0.00007	0.03

* EMERGY flow in column 5 divided by 2.2×10^{12} sej/$ for U.S. in 1984.
[1] (1 bbl)(5.8×10^6 btu/bbl)(1013 J/btu) = 5.9×10^9 J/bbl.
[2] Appropriations, $\$53.5 \times 10^6$ (1970 $) for 22,903 bbl of shale oil:

$$\frac{(53.5 \times 10^6 \text{ 1970 \$})(2.36 \text{ 1983 \$/1970 \$})}{(22{,}903 \text{ bbl})} = 5513 \text{ 1983 \$/bbl}$$

[3] 1.7 bbl of oil extracted to generate 1 bbl of final yield.

varied from 3 to 30 as the price of world crude oil has varied from \$35 to \$12 per barrel. In general, the economy turned to more growth when the yield ratio rose and vice versa—declined when the price rose.

Global Decline in Net EMERGY

In the developed countries the very high net yields (60/1) early in this century have declined as the rich sources close to the surface have been used up, leaving mainly smaller sources and those deeper in the earth. Hall et al. (1986) show graphs of the energy yield ratios of fuels in the United States declining sharply in this century, especially after the oil embargo of 1973. (Their name for this ratio is "energy return on investment," abbreviated EROI.) The fuel that was economical in the period 1980–1994 had an EMERGY yield ratio that ranged from 3 to 12.

The long-range trend over the earth on the average is a decline in the EMERGY yield ratios of fuels. Here and there are found new gas, oil, and coal deposits that have high EMERGY yield ratios, but on the average the ratio is declining, perhaps 1–2% per year. This means that more and more of the economy and human service becomes involved in getting the fuels, and fewer other activities are possible.

---

[4] Vegetation interruption estimated for transpiration, 100 yrs.
Area disturbed per barrel:

$$\frac{(35 \text{ gal oil/ton})}{42 \text{ gal/barrel}} = 0.83 \text{ bbl/ton}$$

$$(1 \text{ bbl})/(0.83 \text{ bbl/ton}) = 1.2 \text{ ton/bbl}$$

$$\frac{(1.2 \text{ ton/bbl})(0.5 \text{ m}^3\text{/ton})}{10 \text{ m deep}} = 0.06 \text{ m}^2\text{/bbl}$$

Productivity lost estimated from displaced transpiration:

$$(0.06 \text{ m}^2)(100 \text{ yr})(0.5 \times 10^6 \text{ g/m}^2\text{/yr})(10 \text{ J Gibbs free energy/g}) = 3 \times 10^7 \text{ J}$$

[5] Reclamation cost:

$$\frac{(\$2050\text{/acre})}{(4.05 \times 10^3 \text{ m}^2\text{/acre})} = \$0.51\text{/m}^2$$

$$(\$0.51\text{/m}^2)(0.06 \text{ m}^2\text{/bbl}) = \$0.031\text{/bbl}$$

Source: Data assembled by Gardner (1977).

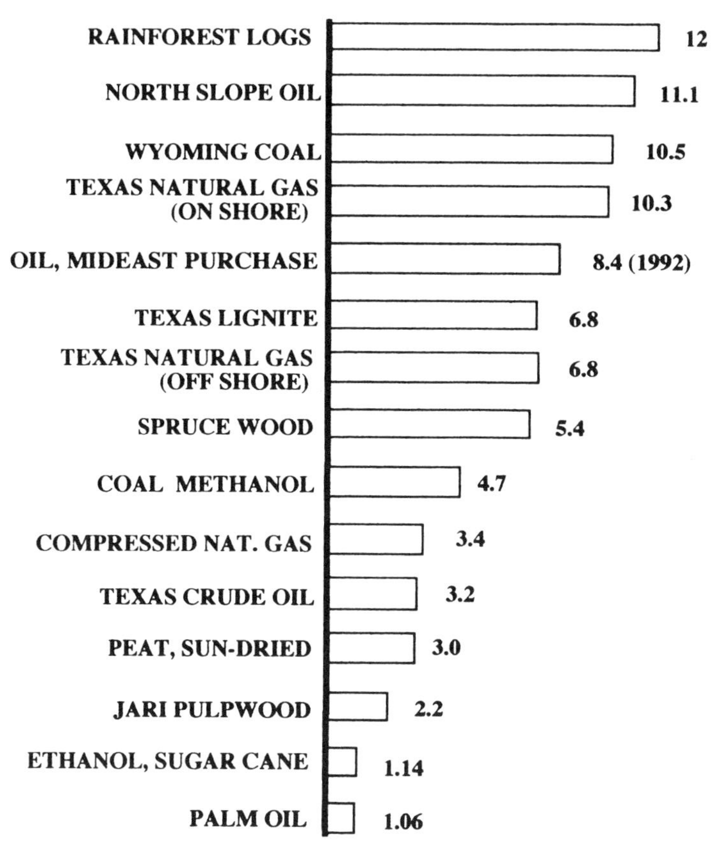

Figure 8.3. EMERGY yield ratio of fuels: Examples of rainforest logs from Odum and Odum (1983); North Slope oil from Brown et al., (1993); Wyoming coal from Ballentine (1976); Texas natural gas—shore and offshore—coal methanol, compressed natural gas, and Texas crude oil from King and Schmandt, (1991); Mideast oil purchase and Texas lignite from Odum et al. (1987a); Jari pulpwood and palm oil from Odum et al. (1986); spruce wood from Doherty et al. (1995); sugarcane, Figure 8.5; Peat, Table 5.2.

Peat

When organic matter is deposited under water or in wetlands where soils are often low in oxygen, peat deposits may develop. Peat has been a regular fuel for fireplaces and more recently for peat-fueled power plants, as in Ireland. Wet peat has no net heat unless the vaporized water is recondensed in special boilers to release its heat of vaporization. In other words, it won't burn.

However, after drying in the sun, peat becomes a good fuel with a good net EMERGY (see Table 5.4).

Like virgin forest wood, long periods of time were used in building organic deposits, which are nonrenewable within the time span of the present economy. The transformities of peat are higher than those of many fuels, which means that peat should be used for high-quality purposes. Many peat deposits are part of the necessary platform of wetland ecosystems that are important in hydrologic cycles and water quality. Wetlands of the elevated type have low transpiration and conserve water. Mining such deposits for fuel may leave a lake-filled hole afterwards, with higher water losses to evaporation. Peat is an excellent filter, removing many inorganic and organic toxic substances from percolating waters (see Figure 7.7c). Peaty wetland filters are one of nature's means of maintaining water quality.

Coal

As the oil and gas run out, large reserves of coal of many varieties remain in many countries (for example, New Zealand, Figure 8.2b). A thick lens, close to the surface in Wyoming where it can be strip-mined at little cost or disturbance, may have an EMERGY yield ratio of 40. As analyzed by Ballentine (1976), the yield ratio drops to 6 after 1000 miles of railroad transport. Even as far as Chicago there was an EMERGY-yield-ratio equivalent to other fossil fuels typically available. If the use of coal was for conversion to electricity, greater net EMERGY yields were obtained by converting to electricity before transport.

Lignite

Lignite is peat that has been transformed into coal but still retains a high moisture content. Lignite can be burned without further drying. Lignite located near the surface and used nearby has a good net EMERGY value. A detailed evaluation made of the Big Brown lignite mine (Figure 8.6) found EMERGY yield ratios similar to those of world fuel sales (Odum et al., 1987a). The mines were economical, as predicted by the EMERGY yield ratio.

Oil

Oil drilling may find oil near the surface with little effort, or it may encounter nothing, becoming a "dry hole." Thus, the EMERGY yield ratio may vary from very high values of 100 or more down to zero. Whereas intensive drilling in the United States has exhausted larger pools of oil close to the surface, drilling in many places such as Saudi Arabia still provides oil with little effort. Thus, different regions of the world are characterized by different EMERGY yield ratios. About 10% of oil's energy is required for transoceanic shipment and another 10% of oil's energy is used in refineries, which separate raw petroleum into fractions (i.e., gasoline, kerosene, heating oil, and so forth).

In spite of the large EMERGY required from the economy in steel and services in building and maintaining a pipeline, the oil at its destination may yield a high net EMERGY. M. T. Brown et al. (1993) calculated that, for the oil delivered through the Alaska pipeline since its inception the net EMERGY ratio was 11.

Natural Gas

Natural gas is the most concentrated natural fuel, because it contains fuel gases that are high in energy content per gram: methane and hydrogen. Natural gas can burn at higher temperatures than other fuels. Where pipelines are available, transmission costs are small. Because gases under pressure move through the earth easily, natural gas may be drawn from locations deeper in the ground.

Like coal and oil, natural gas originates from sedimentary deposits of organic matter from primary photosynthetic production in past ages. The idea was considered briefly that methane might exist in limitless quantity deep in the earth as a result of original processes of the solar system, but exploratory drilling 6 km deep in the ancient rocks of Sweden revealed no evidence for this (Science, 1990).

NET EMERGY YIELD AND FREQUENCY OF HARVEST

Although developing knowledge of renewable systems of lower energy use is good policy for the future, renewable fuel systems have lower EMERGY yield ratios than the "nonrenewable" (slowly renewable) sources and are used first. The reason for this concerns the time of replacement. It takes time to accumulate reserves from environmental processes. The longer the time for accumulation, the more EMERGY is available for harvesting. The EMERGY yield ratio depends on the time of accumulation. Coal and oil deposits required long periods of time for their creation, virgin forests less, and biomass plantations only a few years. As the frequency of harvest is increased, the net yield relative to what has been put back into the system by human work is less. The more of the work that is left to nature, the higher is the net EMERGY, but the longer the time required (Figure 8.4).

Nonrenewable reserves require only the small EMERGY of harvesting and transport, and consequently a very high net EMERGY yield may be obtained for reserves at the surface of the earth. Examples are virgin forests, thick coal deposits, oil near the surface, and natural gas. Many researchers have confirmed the limited yields of biomass production per unit time (Leach, 1976; Heichel, 1976; Slesser and Lewis, 1979; Smil, 1983a; Spreng, 1988). The slope of the line in Figure 8.4 is the EMERGY *yield ratio per unit time,* an index of solar energy transformation to biomass. It may describe the time-based, thermodynamic limit to solar energy conversion.

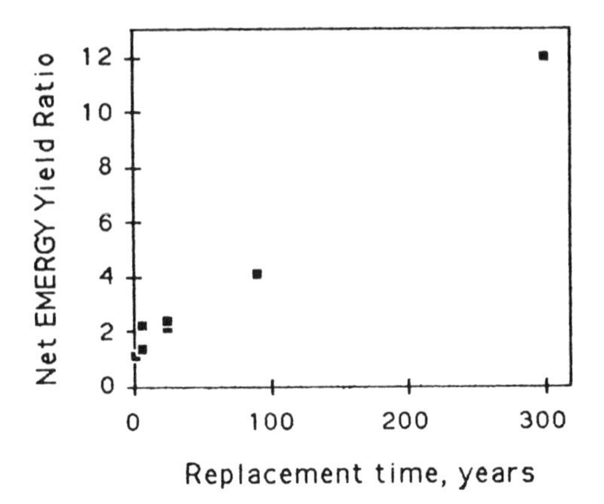

Figure 8.4. EMERGY yield ratio of environmental production as a function of the frequency of harvest. Harvest frequencies and yield ratios for several sources are: New Zealand Monterrey pine, 24 yr, 2.1 (Table 5.2); corn, 1 yr, 1.1 (Odum, 1984b); palm oil, 1 yr, 1.06; surgarcane alcohol, 1 yr, 1.10 (Fig. 8.5); rainforest wood, 300 yr, 12.0; Jari plantations, 7.? yr, 2.2; plantation willow in Sweden, 6 yr, 1.34; spruce, 90 yr, 4.1; slash pine, Florida, 25 yr, 2.4.

NET EMERGY CHANGE WITH PROCESSING

With each step that a fuel is transformed, its net EMERGY decreases. More resources are fed into each transformation, and there is usually less yield for the next stage. For example, sugarcane in Bahia, Brazil (Figure 8.5), when cut, initially has an EMERGY yield ratio of 2.1. From left to right in the figure, transformations eventually produce the liquid fuel ethanol, and in the process, the net EMERGY decreases until what is fed back is almost what is yielded. The net energy yield ratio of sugarcane alcohol calculated in the same way for Louisiana is similar (da Silva et al., 1978; Hopkinson and Day, 1980; H. T. Odum and Odum, 1983; E. C. Odum and Odum, 1984). Such a fuel is neutral as far as its effect on the economy that uses the fuel. It contributes little more than it uses.

Upgrading Fuels to Higher Transformity

For many advanced processes, higher temperatures are needed than can be achieved with ordinary fuels because they are not concentrated enough or contain too much water. Fuels of higher quality and higher transformity are made by partially burning ordinary fuels so that the low-temperature burning part of the fuel is used to concentrate the rest. The transformed fuel product can be used later to support a higher temperature combustion.

One of the oldest fuel upgrades is the conversion of wood to charcoal. A 17th-century Swedish example has been evaluated (Sundberg et al., 1991).

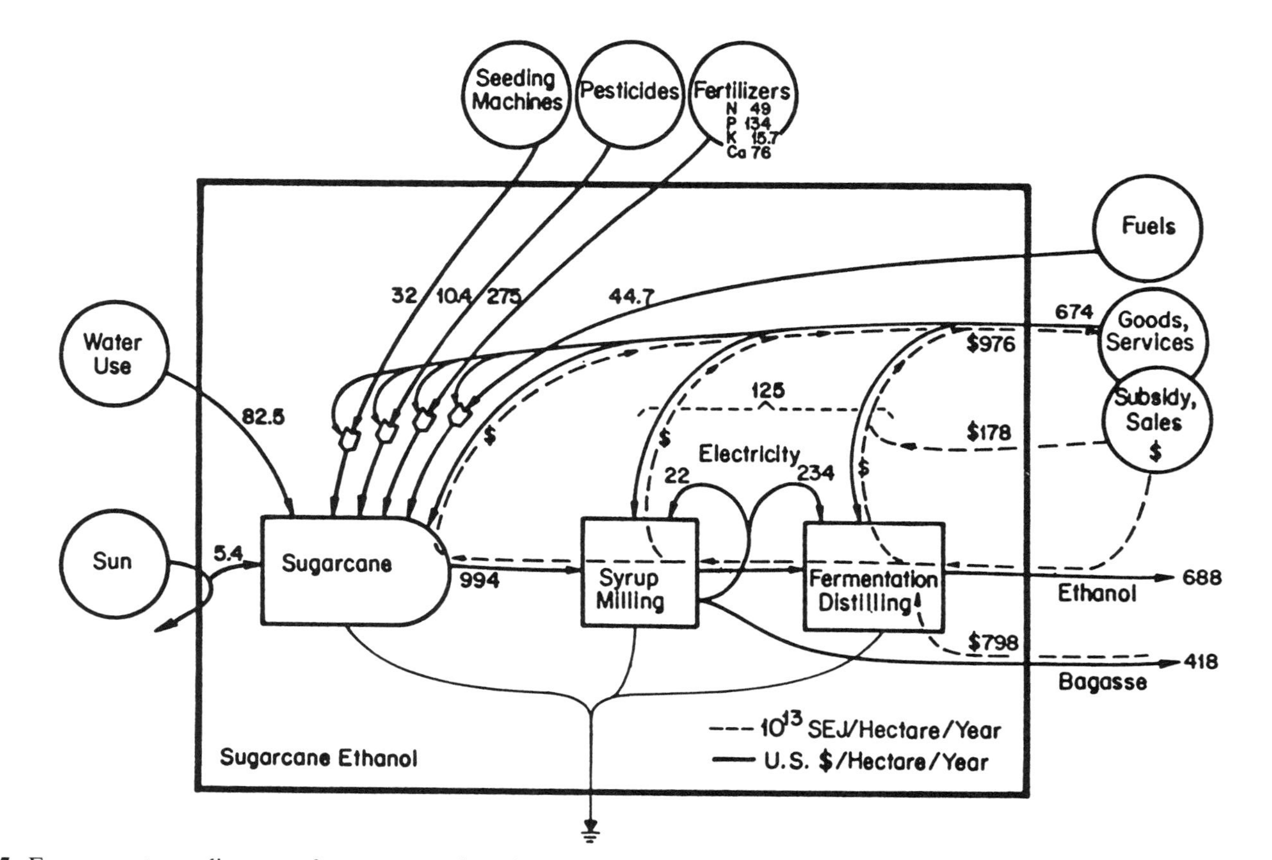

Figure 8.5. Energy systems diagram of sugarcane-ethanol system in Bahia Brazil, with solar EMERGY and dollar costs per hectare per year marked on the pathways (E. C. Odum and H. T. Odum, 1984; Odum, Brown, and Christianson, 1986).

Coal is upgraded in a similar way to coke. Both charcoal and coke have been used to get the high temperatures for steel making.

Liquid fuels for transportation require some processing with increases in transformity over their initial states in nature. Only about 10% of the energy is required to distill motor fuels from petroleum, a liquid mixture. Forty percent or more is required to convert coal to liquid motor fuel. For biomass crops on a renewable basis, most of the energy yielded is required for the production and conversion to liquid form.

Net EMERGY and Transformity

Each time an environmental product is further transformed into a more highly developed product in the economy, additional EMERGY is added and the transformity is increased. If the higher transformity is developed by collecting and concentrating dilute energy, using more EMERGY from the free environment, the EMERGY yield ratio of the product is increased. If, however, the EMERGY for the transformations is being supplied by the economy, the net EMERGY yield decreases. For example, sun is successively transformed in the biologial food chain to form photosynthate, leaves, and then wood. As more solar EMERGY from the environment goes into the concentrating process, the solar transformity increases. With higher concentrations of organic matter accumulating the EMERGY yield ratio is greater.

On the other hand, if the wood is further processed by economic transformations (e.g., into building materials and paper), and if the EMERGY for these transformations comes from the economy, the EMERGY yield ratio decreases. In other words, transformity and EMERGY yield ratio are correlated up to the point where products enter the economy, after which they are inversely correlated.

BIOMASS AND RENEWABLE FUELS

We conclude from Figure 8.4 that there is a thermodynamic time factor in net EMERGY yields. For the window of time and space of our society, net EMERGY yields from biomass and other short-period conversions cannot compete with the yields of nonrenewable fuels until they are gone. It means that the civilization-supporting potential of renewable energy is much less than that of fossil fuels.

Wood

In historical times wood was the main fuel (Sundberg et al., 1994). However, now global fuel use is too great to be supported on wood. The EMERGY in standing wood now remaining would replace only a small part of the fuel uses if fossil fuel sources were interrupted.

When old growth forests are exploited without replanting to make them renewable, very high EMERGY yield ratios result, producing a very large temptation for short-range profit. Old-growth forests are dense, valuable chemical depositions with high transformity, whereas the wood from fast-grown plantations is low-density, with low calorie contents, not even good for fireplace burning. The quantity of virgin-forest storages now remaining on the earth are small compared with the fossil fuels, and may be needed for the more important purpose of preserving biodiversity.

Because fresh wood has high water content, its use as a fuel in simple fireplaces depends on drying. The heat of vaporization is so high that the water in green wood absorbs much of the heat that is released into changing the state of water from liquid to gas. If space, sun, dry wind, and roof shelters are available for drying first, the energy content can be markedly increased. The EMERGY of the additional inputs lowers the EMERGY yield ratio. Drying is one very good way to use solar energy. Burning in special furnaces so that the water vaporized is recondensed in special chambers, releasing the heat of vaporization again, requires greater EMERGY inputs in technology of the furnace.

EMERGY yield ratios are pertinent to considerations of wood for fuels as fossil fuels decrease in availability. Naturally grown wood has EMERGY yield ratios as high or higher than coal, oil, and gas, but because of the length of time required to grow, it is really a nonrenewable resource.

Forestry plantations are planted, fertilized, weeded, thinned, and cut at an early age of 20–90 yr. With many more paid inputs, plantation wood has lower EMERGY yield ratios than the natural wood does, but the yields per year are greater. As with agriculture, plantation wood is a means for combining the fuel-based inputs of the economy with the environmental inputs of the land to generate increased productivity.

That wood is a net EMERGY yielder means we can, in the future as in the past, operate a prosperous economy on a renewable, solar energy basis, but not as much or as intensively as now. The EMERGY yield of economically intensive biomass plantations cannot alone substitute for fossil fuels and native woods until these two fuel sources are gone and the economy has been reorganized again to do less (less population, smaller cities, fewer autos).

ELECTRIC POWER

Electricity is a higher-transformity energy than most fuels. Electric power is very flexible and versatile for supporting higher-quality functions of the economy. In 1983 a solar transformity for electricity was temporarily standardized as 2×10^5 sej/J, pending more research. Table C.1 (Appendix C) contains calculations of the transformity of electric power with different methods and examples. The average there is 1.74×10^5 sej/J. If further work establishes lower values or some other more accurate value, then the 1983–1994 studies can be recalculated. However, the effect of a 17% change in electric-power transformity on most other EMERGY evaluations would be only a few percent except where electric power is a major part of the input sources.

Net EMERGY of Electric Power

Generating electricity from fuels uses up inputs from the economy so that electricity has a lower EMERGY yield ratio, about 2.5. For example, strip mining of lignite in Texas (Figure 8.6) has an EMERGY yield ratio of 6.8, but after the lignite is transformed into electricity in a nearby power plant, the EMERGY yield ratio is 2.2, which is much less.

For heat-generating uses where direct use of fuel is adequate, electricity (EMERGY yield ratio of 2.2) does not compete with typical fuels (ratio of 3–12). High-transformity energy should not be used for low-quality purposes, because it wastes energy.

Alternate Sources for Electric Power

Electric power can be generated from many other kinds of energy including solar voltaic cells, windmills, wave-absorbing machines, tidal power plants, hydroelectric plants, and nuclear power plants. Whether there are alternate sources of electric power as good as the electric power from fuels has been highly controversial in this century, affecting national and international policies for energy. EMERGY evaluation helps to clarify conflicting claims. The

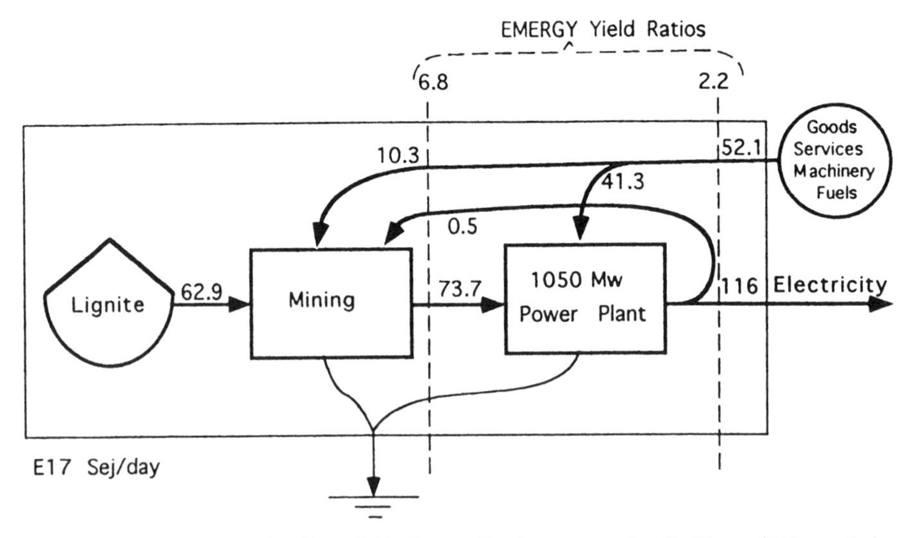

Figure 8.6. EMERGY evaluation of Big Brown lignite power plant in Texas (Odum et al., 1987a). Reprinted by permission of L. B. J. School of Public Affairs, The Univ. of Texas.

EMERGY yield ratios of several processes for generating electric power are compared in Figure 8.7 and discussed as in the following sections:

Tidal Power Plant at La Rance, France

At La Rance in northern France tidal seawaters surging in and out of an estuary were dammed and diverted through turbines in order to generate electricity. Complex computer programs fit the mechanical loading to the tidal amplitude and period as it varied under lunar and solar influence (Figure 8.8). In Table 8.1 the EMERGY contribution was given a tentative value by difference. An important contribution from the environment was the geologic basin, which made possible the tide-harnessing dam.

When net EMERGY was evaluated (Table 8.1) without the lost contribution of the estuarine ecosystem, a high EMERGY yield ratio (15) was found. However, determining the net benefit of operating this power plant instead of letting the tide serve the estuary would require subtracting the contributions of the estuarine ecosystem and fisheries that existed before construction.

Geothermal Electricity

Traditional power plants use fuels to generate heat concentrations in boilers, making steam to turn turbines to generate electricity by spinning coils of wire

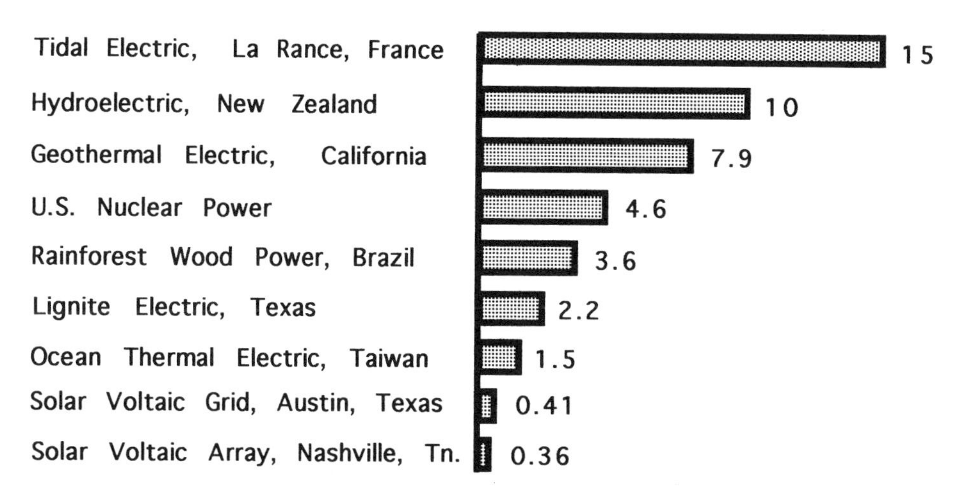

Figure 8.7. EMERGY yield ratios for electric power sources. Data for tidal electric from Figure 8.8 and Table 8.1; geothermal electric, Figure 8.9; hydroelectric, New Zealand (H. T. Odum and Odum, 1979) Brazil, Table C.1; U.S. nuclear electric, Figure 8.10; wood power, Brazil, Table C.1 (H. T. Odum et al., 1986); lignite-electric, Figure 8.6 (H. T. Odum et al., 1987a); solar-voltaic power grid, Austin, Tex. (King and Schmandt, 1991), Figure 8.11 and Table 8.2; solar-voltaic array, Nashville, Tenn. (unpublished); unpublished evaluation of ocean thermal electric plan of Tseng et al., (1991).

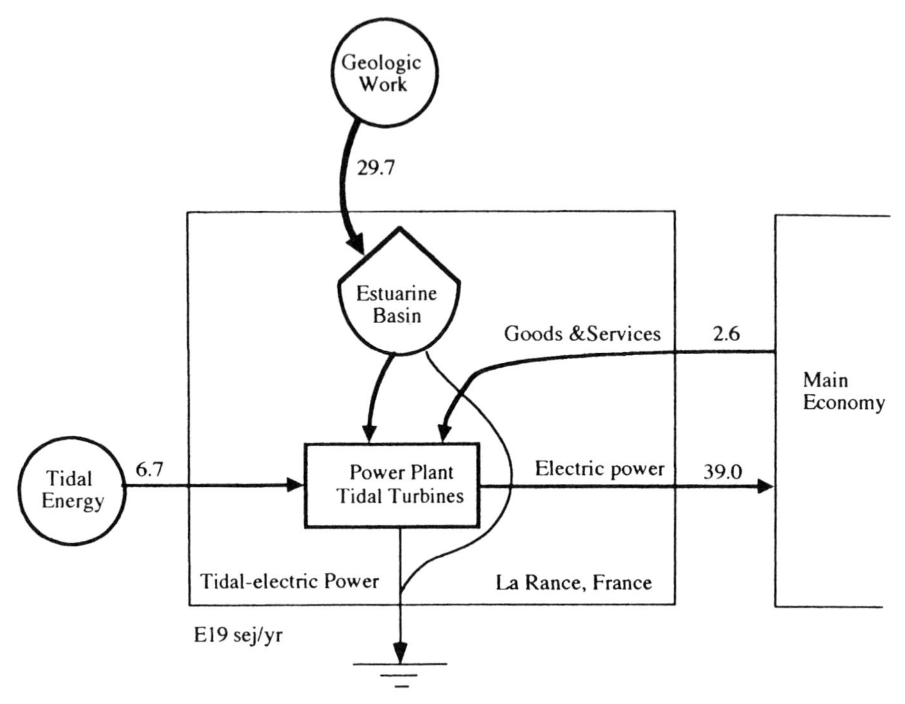

Figure 8.8. EMERGY evaluation of tidal power plant at La Rance, France (data from Table 8.1).

in magnetic fields. In volcanic regions, high temperatures in the ground are close to the surface and can be used as the heat source to operate power plants. Operations in New Zealand and California have been generating electric power for decades. As shown in Figure 8.9, waters percolating in the hot rocks are converted to steam that is collected in pipes and carried into power plants. If heat is removed faster than it is being supplied from below, the rocks will cool after a few years and new holes will have to be drilled in order to draw energy from different rocks.

Gilliland (1975) analyzed a California source of geothermal energy in a volcanic area and found a good net EMERGY yield (Figure 8.9: 5.0/0.68 = 7.9). Her article started the long controversy over whether to evaluate energy alternatives in energy units, money units, or EMERGY units (Slesser, 1978). The potential for geothermal electric plants is limited to volcanic areas. In most areas of the earth, temperatures near the surface are not high enough to generate much net EMERGY. The same is true of the temperature gradient in the ocean, called the ocean thermal energy conversion (OTEC). The geothermal energy of hot springs in Iceland is piped many miles for general heating of the city of Reykjavik.

TABLE 8.1. EMERGY Evaluation of Tidal Power Plant at La Rance, France

Note	Item	Data & Units (J,g or $/yr)	Solar Transformity (sej/unit)	Solar EMERGY ($\times 10^{19}$ sej/yr)	EM$ ($\times 10^6$ Em$/yr*)
1	Tidal energy used	4.0×10^{15} J	16,842/J	6.7	48
2	Goods and Services	4.7×10^6 $	5.5×10^{12}/$	2.6	18
3	Estuarine basin	By difference		29.7	210
	Sum of inputs			39.0	276
4	Electrical output	1.95×10^{15} J	2.0×10^5/J	39.0	276

EMERGY indices:

EMERGY yield ratio: $\frac{39.0}{2.6} = 15.$

EMERGY investment ratio: $\frac{2.6}{6.7 + 29.7} = 0.07.$

* Solar EMERGY divided by 1.41×10^{12} sej/1991 U.S.$

[1] Tidal height, 7 m; tidal pool area, 2.2×10^7 m^2; 705 tides/yr:

$$(0.5)(7\text{ m})(2.27 \times 10^7\text{ m}^2)(705/\text{yr})(9.8\text{ m/sec}^2)(1.02 \times 10^3\text{ kg/m}^3) = 4.0 \times 10^{15}\text{ J/yr}$$

[2] ($4.7 million 1974 $)($5.5 \times 10^{12}$ sej/1974 $) = 2.58×10^{19} sej/yr.

[3] Basin EMERGY = output EMERGY minus EMERGY of items 1 and 2:

$$(39.0 \times 10^{19} - 11.1 \times 10^{19} - 2.6 \times 10^{19}) = 25.3 \times 10^{19}\text{ sej/yr}$$

[4] (544×10^6 kW-hr/yr)(860 kcal/kW-hr)(4186 J/kcal) = 1.95×10^{15} J/yr.

Source: Data from Lawton (1972).

Hydroelectric Power

When the potential energy of elevated water is controlled by a dam and the falling water used to turn turbines, electric power is generated, often cheaply. Hydroelectric plants utilize the extensive earth and planetary energies that transformed sunlight into rain, generated the elevated lands, and carved the river basin required to converge the hydrologic energies. The accumulated prior work of nature in developing suitable hydroelectric sites is a very high EMERGY contribution to the natural capital required for the reservoir. Just imagine the EMERGY requirement if bulldozers were required to build a mountain to catch the rain at a high elevation and then carve out a basin suitable for a dam.

Where water flows are steady, only very small reservoirs with a small dam are needed to maintain a steady output of electricity. Where waters are sharply

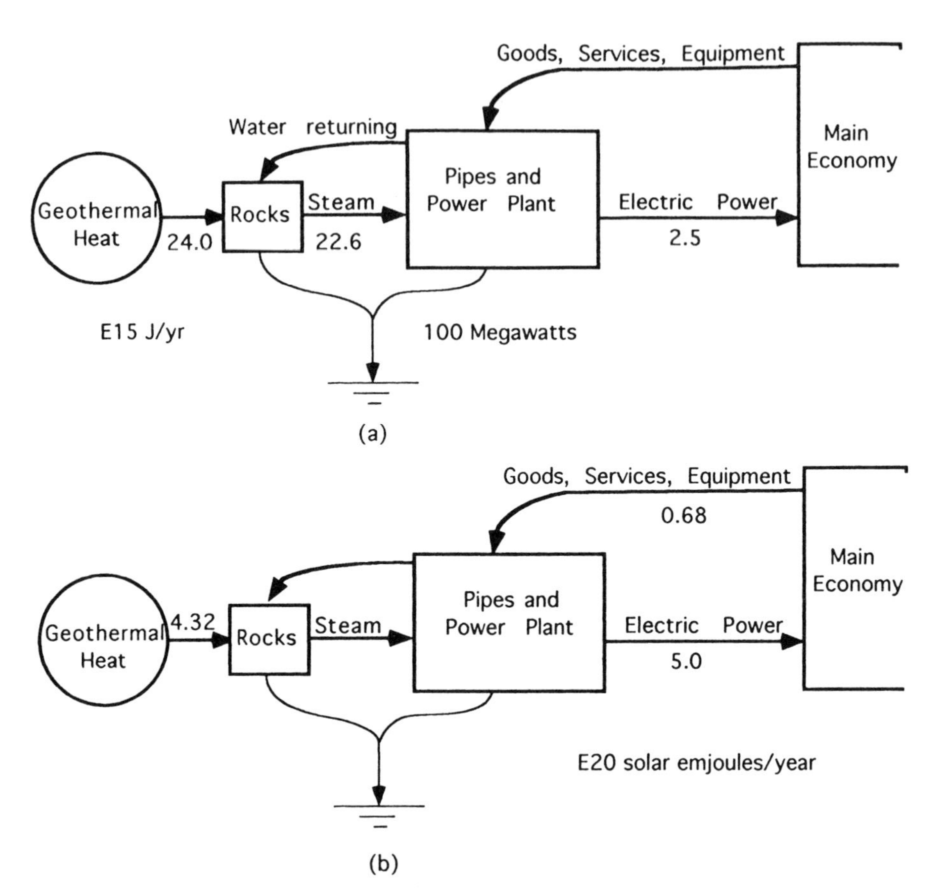

Figure 8.9. EMERGY evaluation of 100-MW geothermal electric power plant, modified from Gilliland (1975) using revised EMERGY/money ratios and solar transformity of geothermal heat:

Note	Item	Data per yr	Solar EMERGY/unit (sej/unit)	Solar EMPOWER ($\times 10^{20}$ sej/ yr)	Em$ 1991* ($\times 10^{6}$ U.S.$)
1	Geothermal heat	24×10^{15} J	18,000	4.32	306
2	Goods and services	9.7×10^{6} $	7.0×10^{12}/$	0.68	48
3	Electric power	2.5×10^{15} J	2.0×10^{5}/J	5.0	354

* 1991 U.S. Em$ obtained by dividing solar empower by 1.41×10^{12} sej/1991 U.S. Em$.

[1] Includes 6% lost in piping; solar EMERGY from electric power yield minus goods and services:

$$\text{Solar transformity} = \frac{(5.0 - 0.68) \times 10^{20}\ \text{sej/yr}}{24 \times 10^{15}\ \text{J/yr}} = 18{,}000\ \text{sej/J}$$

[2] Economic feedbacks in 1972 U.S.$; 7.0×10^{12} sej/1972 $ from Table D.1.

[3] Efficiency of conversion 11%.

seasonal or irregular, large reservoirs are required; these involve high costs, large areas, large evaporation losses, and short lifetimes due to silting-in. A full EMERGY-net-benefit evaluation requires subtracting the contributions the river made before it was dammed.

M.T. Brown (1986b), studying dams on the Amazon, compared the EMERGY yielded in electric power with the EMERGY of the forest systems that were reduced by being covered with a less productive aquatic system. He estimated the break-even point as 2 ha/kW of capacity. Low, broad reservoirs with less power per area covered should not be built.

Nuclear Electric Power

Nuclear power begins with the exploration and mining of nonrenewable uranium ore. Large amounts of electric power were used to concentrate the fissionable isotopes and make fuel elements. Large EMERGY inputs were in the form of capital structures and in the extensive high-quality services of engineers and government regulators. Large EMERGY losses result from accidents, decommissioning, and storage of radioactive wastes.

Attempts to put into practice the dream of deriving unlimited energy from nuclear fission were somewhat disappointing at first. Nuclear fission power is too hot in its core, 5000°C, to be used easily for a human economy that runs at 300°C. Much of the heat has to be dumped into the atmosphere through cooling systems to get the temperature down to a level (1000°C) at which machinery can be operated. Although nuclear reaction energy is enormous and intense, and works well within stars where gravity is large and temperatures normally high, on earth nuclear fission either blows itself apart (atomic bombs) or requires much EMERGY to contain it, cool it down, and finally use it to operate thermal electric turbines. However, recent EMERGY analysis (Figure 8.10) suggests there has been a gradual increase in engineering efficiencies so that for making electricity, nuclear plants are now more yielding than before.

When a net EMERGY evaluation was attempted in the early period of nuclear power construction, there was no net EMERGY (Lem, 1973); since most plants were under construction, interest payments on capital were large, more fuel was being prepared than used, and there were large contributions from the government through the Atomic Energy Commission.

Later, the system necessary to operate a nuclear power plant through a complete life cycle was analyzed by Kylstra and Han (1975) and others, although some studies omitted some inputs. The EMERGY yield ratio for nuclear power ranged from 1.5 to 3.0, depending on what was included. As a general energy source for the heat engines of society, the EMERGY yield ratio between 1 and 3 did not compete with that of fossil fuel in 1980s Ranging from 3 to 9. In other words, nuclear power plants were not economical as a general energy basis for the economy. As long as fossil fuels were cheap, nuclear power plants provided a way to use the fossil fuels to get additional net EMERGY, but as soon as prices of fossil fuels rose sharply in 1973, some of

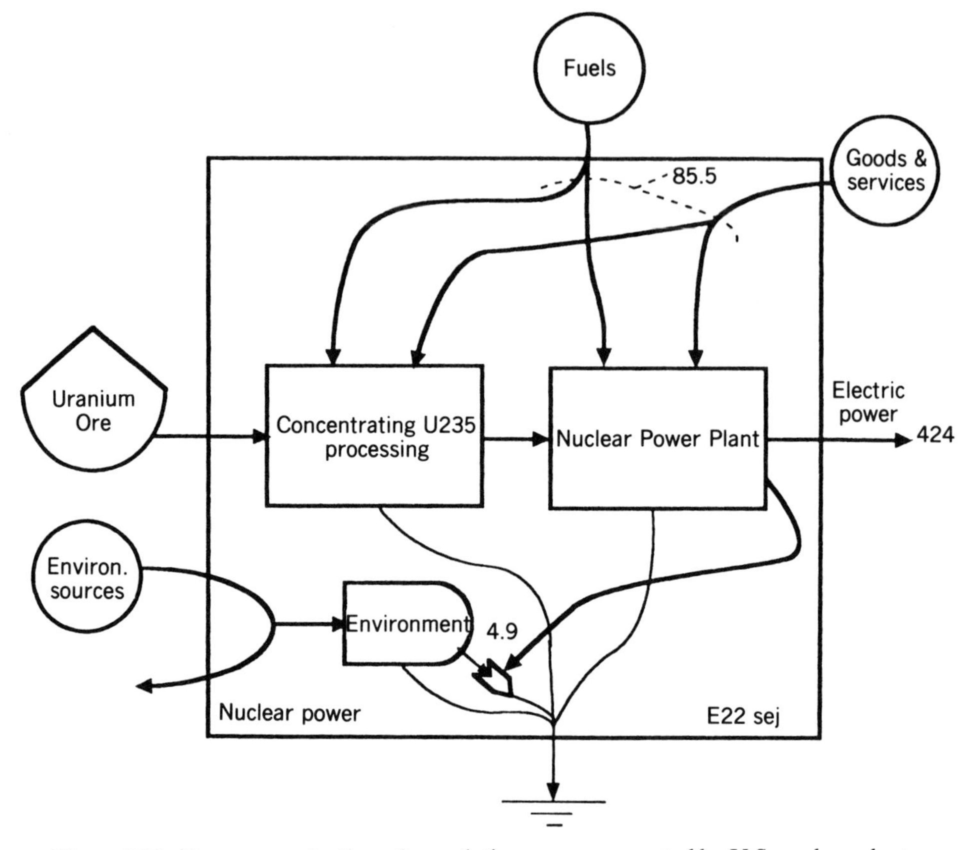

Figure 8.10. EMERGY evaluation of cumulative power generated by U.S. nuclear plants using data from Lapp (1991):

Note	Item	Solar EMERGY ($\times 10^{22}$ sej/yr)
1	Administration	15.1
2	Construction	31.1
3	Cancellation	3.8
4	Operation and maintenance	6.8
5	Fuel cycle	26.3
6	Decommission	3.6
7	Three Mile Island	0.46
8	Steel, concrete, wood, copper and aluminum	1.49
9	Processing energy and materials	2.0
10	Energy required for construction	0.63
	Total feedbacks from economy:	91.28
11	Yield of electric power (5.8×10^9 MWh*)	424
	EMERGY yield ratio = 424/91.3	4.6

* MWh = megawatt-hr electrical

the economy's EMERGY that previously went into the nuclear plants had to go for the fossil fuel. This made all the inputs to nuclear plants more expensive and thus made them less economical. Those plants constructed early on were subsidized indirectly with cheap fossil fuel, which made everything purchased in construction cheap.

Since electricity generated from fossil fuel had similar yield ratios, there was little advantage to use nuclear power then, especially with the large financial costs associated with long periods of construction, permitting, and safety precautions. Also, there were uncertainties about probabilities of accidents, decommissioning costs, and long-term waste storage costs. After the Three Mile Island accident, construction of new plants decreased, especially in the United States.

Lapp (1991) prepared an EMERGY evaluation of the U.S. nuclear power requirements and contributions from its inception through 1990 (summarized with Figure 8.10). The new analysis showed a higher EMERGY yield ratio, about 4.5, which included a part of the Chernobyl impact and some efforts to evaluate waste storage and decommissioning. In this later period, without the high EMERGY requirements of new construction, the U.S. plants have been operating with fewer interruptions and better efficiencies. The results of the analysis suggest that present nuclear plants are better than fossil fuel at generating electricity. There may be a period of time now in which nuclear power will again be in favor, although the uranium fuel available at reasonable costs to support fission plants may be limited to part of a century.

Breeder Power Plants

A breeder reactor system collects the plutonium by-products of fission and uses them in further nuclear reactions. Thus, more energy is obtained from the original uranium than with normal fission alone. The breeder plants require that the highly radioactive wastes of fission be processed through chemical separations so as to concentrate the plutonium and form it into fuel elements for further nuclear power operations. There is doubt that a high EMERGY yield ratio can be obtained because the intense radioactivity and human toxicity of the materials to be processed require robot operations. Since plutonium can be used to make atomic bombs easily, the risks from loss of plutonium to terrorists are great and the security costs high. Pilot operations in the United States were canceled in the Carter administration, but operations were developed in France and other countries. An EMERGY evaluation of the breeder process has not been made.

Fusion Electricity

The net EMERGY yield of fission reactors with a 5000°C core was only moderate, partly because reactions were too hot for operations on earth. Since fusion reactions operate at greater than 50 million° C, the EMERGY inputs to control, contain, cool down, and harness such intense energy on earth are likely to be much larger than with nuclear fission. The EMERGY inputs required in the experimental operations achieved so far are enormous compared to those that were involved in fission pilot plants in a similar stage of development. For these reasons it seems unlikely that fusion can ever be a net EMERGY yielder on earth. Fusion is a net EMERGY yielder within the sun and other stars where the gravity is sufficient to hold hot reactions together.

In 1989 there was a diversion of research money, and public discussion of some experiments suggesting a small amount of fusion had occurred in solutions at ordinary laboratory temperatures. Whether fusion occurred or not, it should have been immediately obvious that the proposed ingredients—heavy water (water containing the isotope deuterium) and the scarce element palladium—contained far too much EMERGY for the heat formed to be a net yield.

Solar Voltaic Power

For generating electricity from sunlight, solar voltaic cells have been intensely studied for most of this century, gradually increasing efficiencies while also increasing EMERGY requirements and costs. EMERGY evaluations of solar vol-

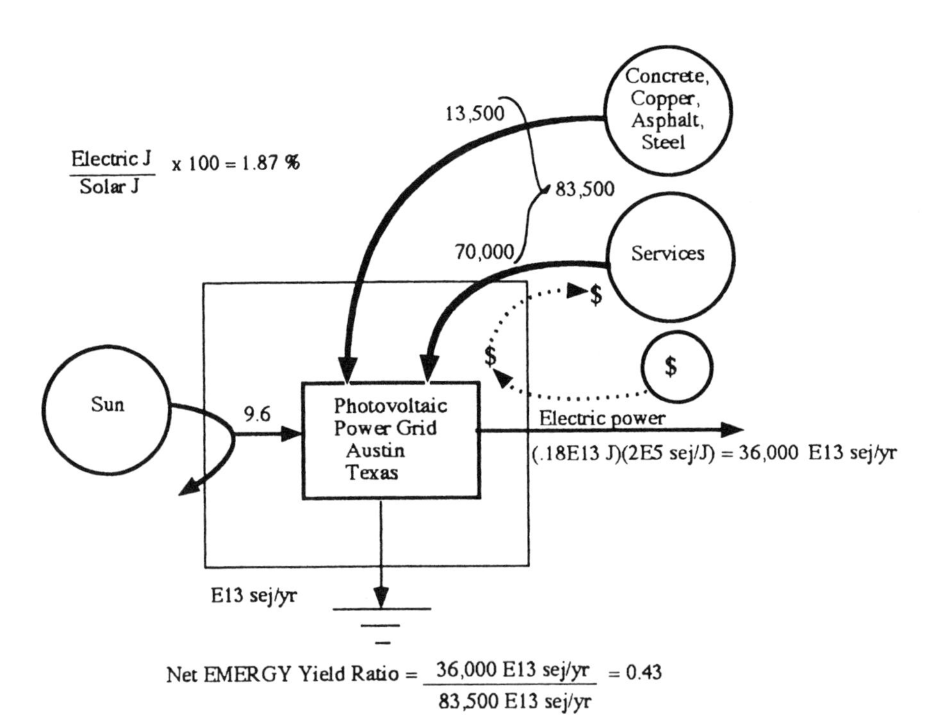

Figure 8.11. EMERGY analysis of a solar photovoltaic power grid in Austin, Tex., using data in Table 8.2 from King and Schmandt (1991).

TABLE 8.2. EMERGY Analysis of Solar Voltaic Power Installation in Austin, Texas. Annual Contributions and Requirements

Note	Item	Data and Units (J,g or $)	Solar Transformity (sej/unit)	Solar EMERGY ($\times 10^{15}$ sej/yr)	Em$ ($/yr*)
1	Solar energy	2.67×10^7 kWh	1	0.0962	60
2	Design installation	1.94×10^4 $	2.8×10^{12}	54.3	27,160
3	Materials cost	1.63×10^4 $	2.8×10^{12}	45.6	22,820
4	Collector cost	5.68×10^4 $	2.8×10^{12}	159.	79,520
5	Administrative cost	1.2×10^5 $	2.8×10^{12}	336.	168,000
6	Operation and maintenance	4.0×10^4	2.8×10^{12}	112	56,000
7	Concrete	2.23×10^3 lb	1.5×10^{13}	33.4	16,725
8	Structural steel	1.16×10^4 lb	4.16×10^{11}	4.8	2,413
9	Steel building	1.47×10^2 lb	4.16×10^{11}	0.06	30
10	Rebar	7.8×10^3 lb	4.16×10^{11}	3.24	1,622
11	Copper wire	26.7 lb	4.49×10^{13}	1.2	600
12	Total from economy (items 2–11)			749.7	374,951
13	Electric power output	1.8×10^{12} J	2.0×10^5	360	180,000

$$\text{EMERGY yield ratio: } \frac{360 \times 10^{15}\ \text{sej/yr}}{749.7 \times 10^{15}\ \text{sej/yr}} = 0.48$$

$$\text{EMERGY investment ratio: } \frac{749.7 \times 10^{15}\ \text{sej/yr}}{0.096 \times 10^{15}\ \text{sej/yr}} = 7809$$

* Solar EMERGY in column 5 divided by 1.6×10^{12} sej/1991 $.

[1–12] Based on an interview with the site engineer for the city of Austin.

[1] $(2.67 \times 10^7$ kWh/yr$)(3.6 \times 10^6$ J/kWh$) = 9.61 \times 10^{13}$ J/yr

[13] $(5.0 \times 10^5$ kWh/yr$)(3.6 \times 10^6$ J/kWh$) = 1.8 \times 10^{12}$ J/yr

Source: R. King, personal communications, 1991. Data from King and Schmandt (1991).

taic arrays have yet to show any net EMERGY yield. The power grid at Austin, Tex., evaluated by King and Schmandt (1991) is taking more EMERGY out of society than it is generating (Figure 8.11; Table 8.2). The main EMERGY inputs are in the materials, and especially in the human services involved. If the analysis is made using energy (instead of EMERGY), then most of the high transformity inputs such as human services are greatly underestimated and net yields are obtained, which is why there is so much confusion in this field.

With research and manufacturing experience, the goods and services required to make the solar cells have been decreasing a little each year. Costs per unit power have been decreasing slowly. However, when this process of improving efficiency has gone as far as is thermodynamic, the efficiency of these cells may approach that of the green plant chloroplast, which is nature's photovoltaic cell. Studies in biophysics providing curves of efficiency as a function of light intensity for isolated chloroplasts show them to be more efficient than hardware cells. It may be the natural conversion of sunlight to electric charge that occurs in all green-plant photosynthesis after 1 billion years of natural selection may already be the highest net EMERGY possible. A century of plant physiological research has not succeeded in increasing the basic efficiency of gross photosynthesis. Misleading comparisons have been made by comparing the net photosynthesis of plants, after they have used much of the transformed energy for maintenance and reproduction, with the gross photoelectric conversion of solar cells (without taking into account their maintenance, cleaning, replacement, and so on).

LOCATING INDUSTRIAL PLANTS AND THEIR COOLING

Whether to construct an industrial plant depends on evaluation of alternatives within a larger systems window of time and space. However, after a firm decision has been made to construct the plant, EMERGY accounting can determine the best location for optimum interaction with the environment. Figure 8.12 is a generalized diagram for citing power plants and considering alternatives. Where the EMERGY output is set (for example, 1000 MW of electric power), the best system is one that uses other inputs of goods and services, fuels, materials, environment, and so on, the least.

The heat engines that run much of industry depend on a large temperature difference between heat sources and the environment and cannot derive net EMERGY from the last few degrees of temperature difference (e.g., 7°C). Consequently, they are designed to release large volumes of this lukewarm heat to the environment. Nuclear energy has more heat release than fuel plants because of its high temperature in the core of its nuclear reactions (e.g., 5000°C), which would melt the steel if they were not cooled some first. In other words, usually not all of the available energy from a nuclear plant is harnessed.

To carry this heat away into the environment, while at the same time maintaining a cool side of the heat engine, requires some cooling method. This may be accomplished by circulating waters from lakes and estuaries, circulating waters in a specially constructed reservoir, or exchanging the heat with air in a cooling tower. The large volumes of low-temperature fluid discharged to the environment still contain enough available energy to drive work processes in the environment, causing currents and other processes. Nature's heat engines can use the low-temperature gradients because they use their own fluid structures for the transformation process (not having

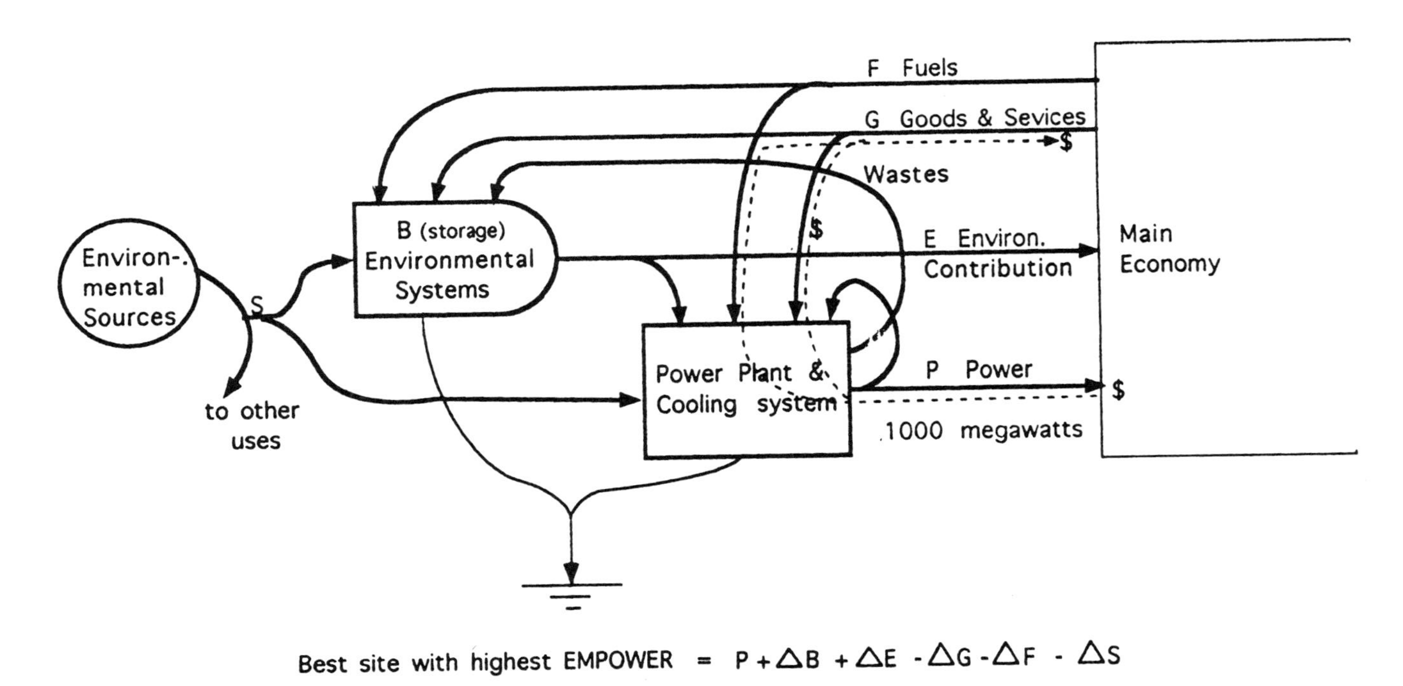

Best site with highest EMPOWER $= P + \Delta B + \Delta E - \Delta G - \Delta F - \Delta S$

Figure 8.12. Generalized diagram for evaluating power-plant alternatives (modified from Odum et al., 1983a).

to divert EMERGY for steel equipment). The method of cooling affects the environmental impact, operational costs, and location of the plant.

EMERGY evaluation has been used to select the cooling method for maximum net benefit overall. Odum et al. (1977a) compared estuarine once-through cooling with cooling towers for power plants at Crystal River, Fla., and Kemp et al. (1977b) evaluated the plant at Anclote, Fla. As simplified in Figure 8.13, release of a new condition tends to negatively impact the ecosystem which was there before, but may stimulate one that is adapted to the new energy source.

Slightly heated waters are a potentially useful contribution to environment (e.g., by improving offshore circulation, or by increasing the metabolism for species adapted to higher temperatures), but because the power plants at Crystal River were frequently on and off, releases were sporadic; only small organisms could adapt to the alternating temperature regimes, and the measured gross plant productivity in the grass flats near discharge was reduced by about half. However, substitution of the cooling tower caused EMERGY resources in its manufacture to be diverted from uses elsewhere in the economy. The environmental impact of that much manufacturing and maintenance (of the cooling tower) was estimated using the average investment ratio of the United States. Building the cooling towers, when considered on the larger scale of estuary and economy, was a net EMERGY loss. On the larger scale of the plant and the affected economy elsewhere, there was more environmental impact with the tower than without. This study delayed the tower construction for a decade, but eventually the national policy to build cooling towers without regard for the overall benefit prevailed again.

Three alternatives for cooling a nuclear power plant near Chicago were evaluated for the Nuclear Regulatory Commission (Odum et al., 1983a).

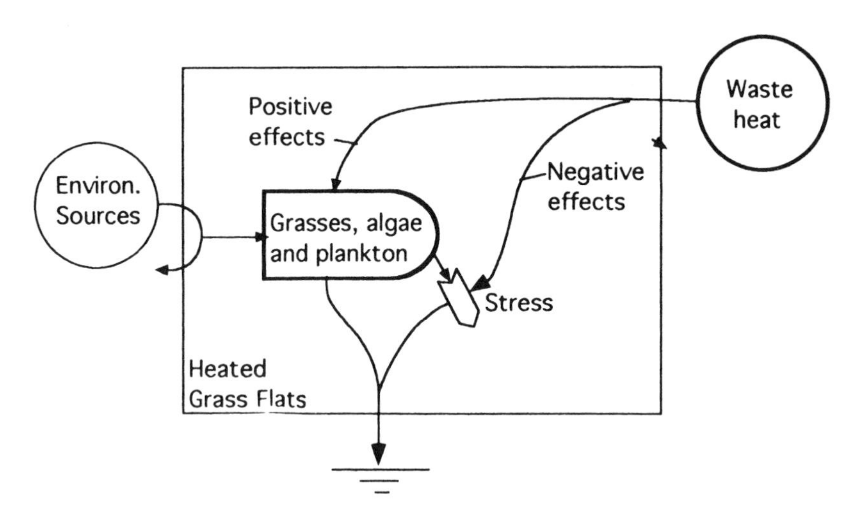

Figure 8.13. Plus-minus nature of environmental impact: the example of hot-water release to estuarine ecosystems at Crystal River, Fla.

EMERGY analysis showed that the productive agricultural land lost when a cooling lake was constructed was greater than the EMERGY of aquatic ecosystems lost if cooling had been done in existing lakes. Because it used so much additional energy and manufacturing, the cooling-tower alternative diverted more EMERGY than necessary and was a poor use of conservation dollars.

EVALUATING ENERGY TRANSFORMATION SERIES

Fuels and electricity may be converted to other forms with a series of energy transformations. There are alternative pathways and combinations of processes. For example, to deliver transportation, electricity can be used to make hydrogen, which is then used to power an automobile, or electricity can be used to drive an electric motorized car. Or wood can be used to make electricity, wood can be used to make a liquid motor fuel, and so on. In order to evaluate alternative series, we arrange the choices to be considered into a series in a systems diagram and evaluate the EMERGY yield ratio for the combined processes. An example in Figure 8.14 compares five alternative pathways for production of hydrogen. Among the series shown, the one with the highest net EMERGY ratio is the preferred process (e.g., hydropower in Figure 8.14c). The chart was prepared by selecting the same arbitrary output of electrical power and/or hydrogen gas for each series, calculating other numbers from EMERGY yield ratios of processes involved.

EMERGY EVALUATION OF ENERGY CONSERVATION MEASURES

Many authors have pointed out the savings to be had with more-efficient use of fuels and electricity by using more-efficient equipment and procedures. Usually, a conservation opportunity is measured as the money saved by using less fuel or electricity for the same function. Although an increased cost for goods and services is required to make a process efficient, there is often a net savings in money. See, for example, papers on appliances by Capehart and Blackburn (1989) and the summary by Smil (1991). In preliminary evaluation of some of these examples using EMERGY, the net savings in emdollars was found to be much larger than when evaluated in dollar costs only. This is because of the very large net EMERGY of fuels and electicity, which causes any savings in these resources to contribute more than the EMERGY increases in equipment and service costs. Since the free market does not maximize benefits to society as a whole in these situations, some special incentives are needed.

Whereas energy conservation in the sense of increasing efficiency of use has a net benefit, an economy that conserves, in the sense of restricting fuel use, tends to reduce its net EMERGY and thus its ability to compete economically. Taxing fuels is sometimes offered as an incentive for energy conservation,

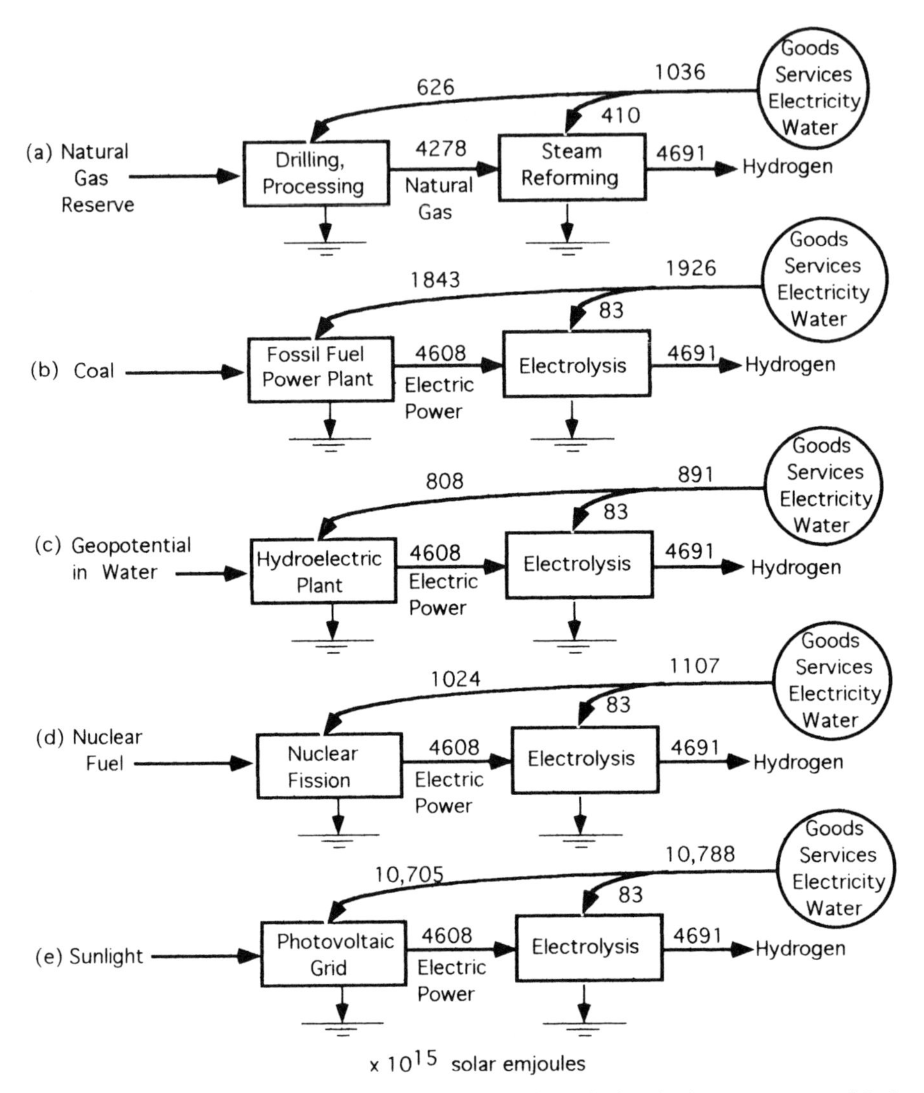

Figure 8.14. EMERGY flows for alternate ways of producing hydrogen gas, modified from Barbir (1992). **(a)** Steam forming from natural gas. The EMERGY yield ratios are: natural gas, 6.8; the conversion, 11.4; and the combined process, 4.5. The solar transformity is 76,300 sej/J. **(b)** Hydrogen production from electric power: 160,000 sej/J of coal. The EMERGY yield ratio of electricity from coal, 2.5; combined ratio, 2.4. The solar transformity of hydrogen is 204,000 sej/J. **(c)** Hydroelectric power with a solar transformity of 85,437 sej/J and an EMERGY yield ratio of 5.7; combined, 5.3. The solar transformity of hydrogen is 110,563 sej/J. **(d)** Requirements for conversion as in part b combined with EMERGY analysis of U.S. nuclear fission power (Lapp, 1991); electric power transformity, 160,000 sej/J, with an EMERGY yield ratio of 4.5; combined, 4.2. The solar transformity of hydrogen is 203,956 sej/J. **(e)** Hydrogen from solar-driven photovoltaic cells. Combined EMERGY yield ratio, 0.43. The solar transformity of hydrogen is 69,000 sej/J.

but reducing fuel use has a negative amplifier effect on the economy that may be greater than the increases in efficiency. If the tax reduces luxury and waste, the effect is beneficial. However, if the tax reduces fuel use for productive processes, then there is harm to the economy. There is a negative effect on the economy that is proportional to the net EMERGY of the fuel being conserved. For example, if the fuel source has an EMERGY yield ratio of 6, then any reduction in use of that fuel causes a six fold reduction in empower production.

NET EMERGY AND GLOBAL CARBON DIOXIDE

Because global consumption of fuels is occurring faster than their production by the environment, carbon dioxide has been increasing, affecting the climate. By using the fuels with the highest EMERGY yield ratio, we generate the most EMERGY contribution with the least release of carbon dioxide (Burnett, 1981b). Although biomass is more renewable, its EMERGY yield ratio is less than that of fossil fuels, and substitution would not reduce carbon dioxide release.

SUMMARY

Procedures and examples were introduced for evaluating the net EMERGY of fuels and electricity, the primary sources of economic vitality and urban development. Where available energy is transformed in a series of processes, EMERGY is wasted if high-transformity energy is used when energy of lower transformity will suffice. The net EMERGY of sustainable biomass was found to be proportionate to the time of productive energy transforming. Therefore, renewable systems with short production cycles are not able to support the energy concentration of the present civilization. Net EMERGY evaluations were made comparing present fuels and electric power plants with proposed alternative energy sources. U.S. Nuclear fission has increased its efficiency and net EMERGY. Evaluations of solar voltaic cells continue to show no net EMERGY yield.

CHAPTER 9

EVALUATING ALTERNATIVES FOR DEVELOPMENT

For several centuries human society has expanded by using the high net EMERGY of "nonrenewable" resources. Urban growth was centered where EMERGY flows supporting the economy converged spatially. Economic development in technological cities was accelerated by cheap fuels and electric power interacting with life-supporting resources (water, air, and land) from the rural environment. Since economic development is apparently empower-dependent, EMERGY accounting can be used to choose development plans that can be sustained. This chapter provides principles and examples of EMERGY accounting to evaluate alternatives for economic development.

For the time-and-space window of regional economic development, productivity depends on the interaction of the high-transformity products of the cities interacting with the environmental inputs in a matching process. As explained in Chapter 4, economic production is based on the interaction of purchased inputs (commodities, raw materials, and so on) with free inputs from environmental production and reserves (Figure 9.1a). An EMERGY investment ratio may be calculated for determining whether the economic-environmental interface has the matching that contributes most to systems productivity. As diagrammed in Figure 9.1b, the investment ratio used in this chapter is the quotient of purchased EMERGY divided by the free EMERGY. (Several related indices were given in Figure 5.3 and Table 5.3 as part of EMERGY accounting procedures). The term "EMERGY investment" may be appropriate since the index measures the extent to which the economic system has invested EMERGY from sources elsewhere.

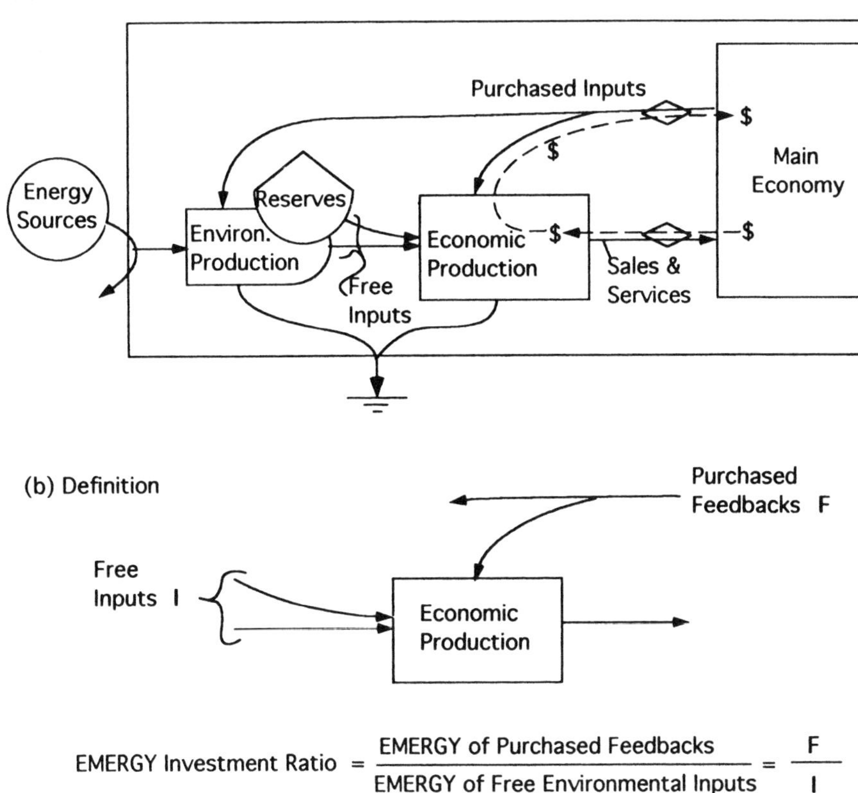

Figure 9.1. Energy systems window of typical resource production. **(a)** Environmental-economic interface. **(b)** Diagram defining the EMERGY investment ratio.

EMERGY-MATCHING CONCEPTS

Theory given here relates EMERGY matching to the traditional concepts of limiting factors in production processes. As already explained in Chapters 1 and 2, energy systems designs that prevail consist of a web of energy transformation units (see Figure 2.2). Each energy transformation uses two or more different kinds of energy to interact in a production process. Each type of energy generates more production when it interacts in a mutually amplifying process with another input of either higher or lower transformity (see Figure 2.1).

According to well-known concepts of limiting factors, any input to a productive process can become limiting when the other factors are available in excess. However, self-organizational processes, either in nature or by humans, tend to develop systems designs around the production process so that the inputs become equally limiting (having a similar derivative of production as a function of the input factor). This is done by accumulating and recycling the potentially limiting factors.

Self-organization tends to retain systems where the effect of a flow is commensurate with EMERGY requirements that contributed to that flow. Therefore, similar EMERGY flows mean similar effects. The best use of EMERGY flow for maximizing production comes when the EMERGY fed back to production matches the input EMERGY. Matching EMERGY inflows to production is equivalent to balancing potentially limiting factors. For example, in Figure 9.2a, the EMERGY of feedback matches that of the source inflow, so that neither input to production is more limiting than the other. EMERGY is efficiently used when applied equally to both inputs. The EMERGY investment ratio is 1.0, and the system is then limited only by the external source.

Next suppose, as in Figure 9.2b, that there is a second, high-transformity input to production, bringing a larger input (source 2). The larger EMERGY now makes the other input (source 1) more limiting. Because of the increased inputs, production is increased, but at a lesser rate. Production is far to the right on the curve of diminishing returns. The EMERGY investment ratio is larger. The uneven EMERGY matching increases production, but EMERGY is less efficiently used.

The condition of uneven EMERGY matching is prevalent in environmental-economic developments of this century. As nonrenewable resources have enriched the economy, more feedbacks have passed to environmental interfaces, thus increasing the production, increasing the EMERGY investment ratio, and producing with less efficiency. The feedbacks from the economy to the environmental interface are usually purchased—accompanied by a countercurrent of money (Figure 9.2c). The higher the investment ratio, the more intensive is the environmental use, and the more money circulates. What intensity of development prevails?

Each economic producer (Figure 9.2c) may be in economic competition with similar users elsewhere. The users with lower EMERGY investment ratios get more of their EMERGY free from environment and thus have to purchase less. With lower costs they can sell their products for less, and thus they can capture the markets. In other words, evaluating the EMERGY investment ratio provides an *a priori* method of determining if some economic-environmental use will be economical. It will tend to be economical if its ratio is less than or equal to the one prevailing in the region. Many undeveloped areas of the world where regional EMERGY investment ratios are small may compete better for markets in competition with developed areas like the United States with EMERGY investment ratios averaging 7 or more. Table 9.1 compares the average EMERGY investment ratios within countries.

An area with economic uses at a low EMERGY investment ratio may not be able to maintain this pattern because the potential for economic growth is large, and more intensive developments can displace the less intensive by

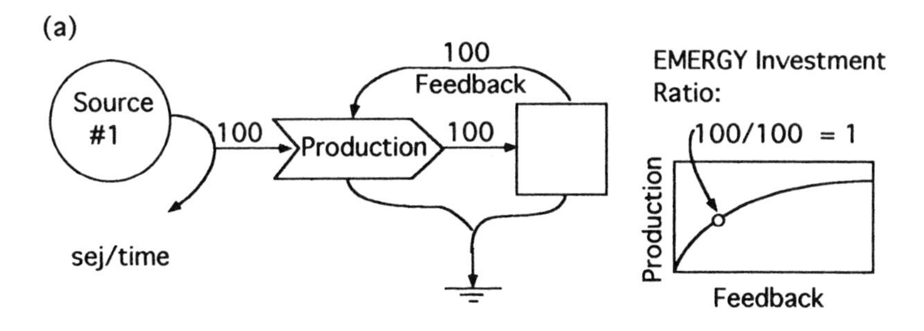

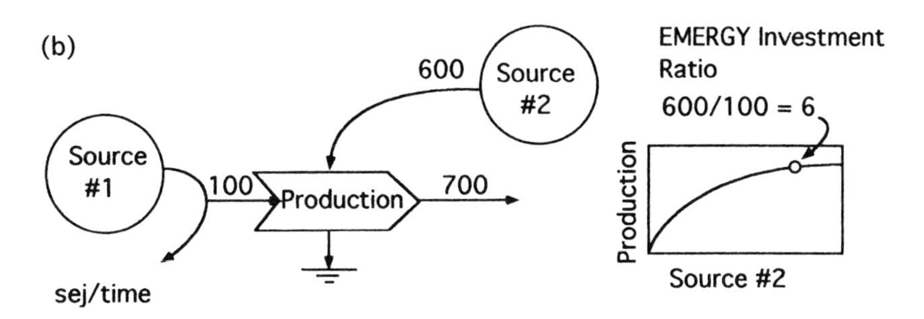

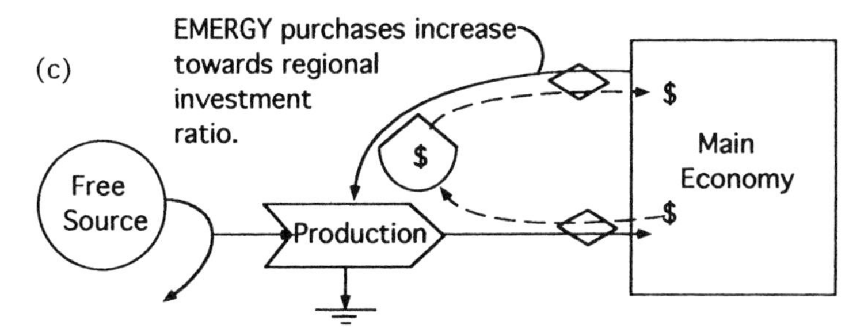

Figure 9.2. EMERGY matching in production processes. **(a)** A system with equal EMERGY matching. **(b)** A high EMERGY investment ratio with EMERGY-rich feedbacks. **(c)** Market-controlled development, which tends to develop the prevailing EMERGY investment ratio of the region.

drawing in more EMERGY and money. Thus, the EMERGY investment ratio that is sustainable tends to be that of the region. Because developed countries are rich with nonrenewable energy, the matching that prevails there is high and production-inefficient.

As global free-trade systems develop, what is economical use of the environ-

TABLE 9.1. Regional Investment Ratios within States and Nations (Environmental Loading Ratio)

Note	Location	Environmental Renewable Component ($\times 10^{20}$ sej/yr)	Economic Component* ($\times 10^{20}$ sej/yr)	Economic/ Environment Ratio†
1	West Germany	193	17,300	90.0
2	Taiwan	73	1,787	24.5
3	Texas	410	6,330	15.4
4	Italy	1,208	11,442	9.5
5	Switzerland	87	646	7.4
6	Spain	255	1,835	7.2
7	U.S.A	8,240	58,160	7.1
8	Sweden	511	3,597	7.0
9	China	9,450	62,460	6.6
10	Mexico	1,390	4,750	3.4
11	Soviet Union	9,110	29,140	3.2
12	Dominica	1.8	4.8	2.7
13	Scotland	389	679	1.8
14	Japan	8,100	13,480	1.66
15	WORLD	94,000	138,000	1.47
16	Australia	4,590	3,960	1.10
17	Thailand	779	811	1.10
18	India	3,340	3,410	1.0
19	New Zealand	438	353	0.81
20	Brazil	10,100	7,600	0.70
21	Ecuador	891	483	0.54
22	Papua New Guinea	1,052	163	0.15
23	Liberia	427	38	0.09
24	Alaska	4,500	360	0.08

* Total annual EMERGY use minus environmental renewable component from the column 3.

† Ratio of annual solar EMERGY brought into an area by purchase and that contributed free by the environment. An average for a country's internal regions was estimated by dividing the nonrenewable and imported empower use (economic component) by the nation's free renewable environmental empower use. It was reasoned that most local areas within a country have to buy their nonrenewable empower. Years of evaluation vary from 1980 to 1993.

Sources: 1,6,11,12,16,18,23, H. T. Odum and Odum (1983); 9, Lan (1993); 2, Huang and Odum (1991); 4, Ulgiati et al. (1994); 13,14,20, modified from student project reports, "Energy Analysis" Class, Univ. of Florida, 1988–1993; 5, Pillet and Odum (1984); 10, Brown et al. (1992); 17, Brown and McClanahan (1992); 19, Odum et al. (1988); 7,3, Odum et al. (1987); 8, Doherty et al. (1995); 21, Odum and Arding (1991); 24, Woithe (1992); 22, Doherty (1990), Doherty et al. (1987a) (1993); 15, Figure 10.6.

ment may be affected more and more by the world EMERGY investment ratio. The ratio of the purchasable and transportable nonrenewable resource use of the world to the world's renewable global EMERGY use is about 1.0. In other words, less developed areas will tend to compete with developed areas with high investment ratios.

EMERGY Attraction Value

After the environmental EMERGY production of an area is evaluated (see Chapter 7), the total environmental EMERGY may be multiplied by the regional EMERGY investment ratio to obtain the area's *EMERGY attraction value.* It is the potential for EMERGY flow (or its emdollar equivalent) if economic development reaches the regional average. In other words, the potential for development is estimable from the environmental resource evaluation. In a sense there are two environmental values: (1) the environmental EMERGY contribution by itself and (2) the total EMERGY flow after the environmental flow is matched with purchased EMERGY (EMERGY attraction value).

However, let's not forget that economic use of the environment, to be sustainable, requires a reinforcing feedback to the environmental production system (see Figure 4.6). For example, feeding back land care may be required to sustain agriculture. On theoretical grounds this reinforcing feedback might return as much EMERGY to the environment as the economic uses draw from it. When the environment is supplying a net EMERGY, the feedback is less (i.e., the investment ratio is low, as with many nonrenewable uses), and the arrangement may not be sustainable. When the EMERGY feedback is greater than that from the production (i.e., high investment ratio), the efficiency is lower, and some of the attracted, purchased EMERGY would be better used elsewhere.

EVALUATING ALTERNATIVE DEVELOPMENTS

Among existing and planned developments, EMERGY accounting can compare productivity and potentials. If those systems prevail that produce more EMERGY and utilize it more efficiently, then systems with greater empower (useful EMERGY flow) may be better and more likely to continue. (See the maximum empower principle defined in Table 2.2 and further summarized in Chapter 15). In recent times of rich nonrenewable energies (cheap fuels), systems previously based on environmental energy sources alone have generally been displaced by alternative systems under economic use because purchased EMERGY is added to that of the original environmental production. By interfacing the economy, purchased EMERGY flow (empower) over that of the primitive system. The displacement of primitive systems by economic ones is consistent with the maximum empower principle as long as there are available sources of cheap EMERGY to purchase.

For comparing alternative systems, we use the procedures of Chapter 5 to make an EMERGY evaluation table for each system that includes all the incoming sources used in the system. Then the existing and proposed alternatives are compared with the regional potential based on the regional EMERGY investment ratio. Figure 9.3 diagrams four alternatives often compared in evaluating developments.

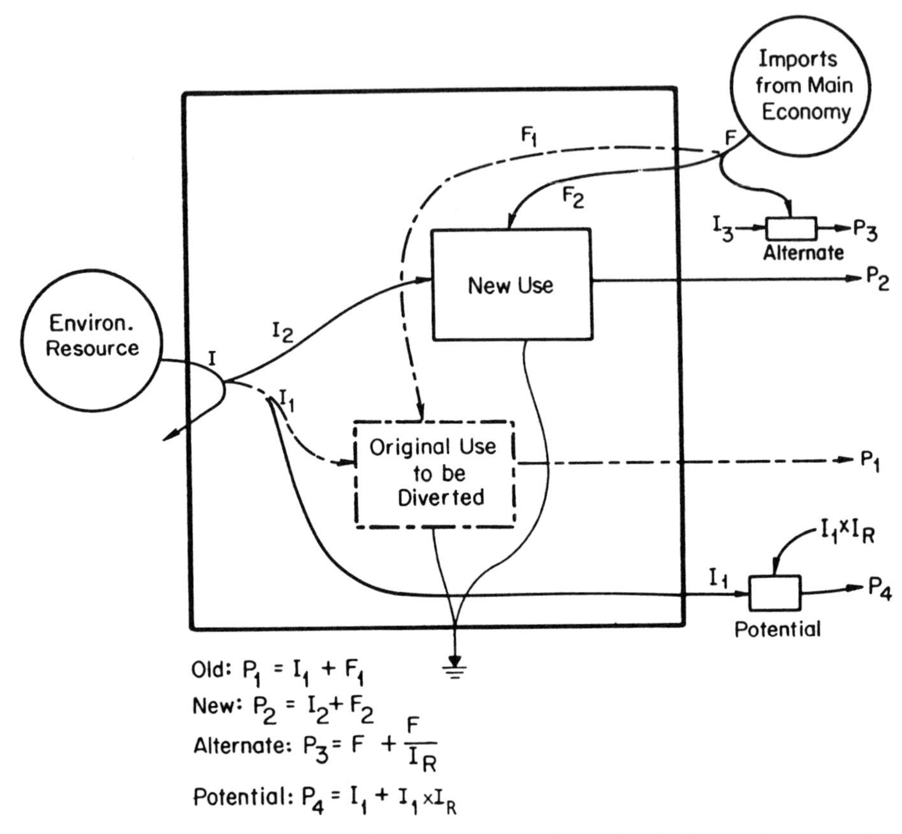

Figure 9.3. Procedure for comparing two systems (P_1 and P_2). For each system the contribution (P) is the sum of the EMERGY of its environmental input (I) plus the EMERGY of its economic input (F). P_3 is the output of the alternative use of the economic input if it is matched with other environmental resources (I_3) according to the prevailing EMERGY investment ratio, I_R. P_4 is the output of the potential full use of the environmental input if it is matched with economic input according to the regional EMERGY investment ratio I_R.

EMERGY of the Primitive System

The original system before economic development receives only the free environmental EMERGY flow (input I_1) plus a small inflow from the surroundings (feedback F_1). In Figure 9.3 primitive production is P_1.

EMERGY of the Developed Alternative

After economic uses develop, the new system usually still receives some environmental EMERGY (I_2), but then it receives more EMERGY flow (in feedback F_2) from the larger economy. The new use is shown in Figure 9.3 with an increased EMERGY production P_2.

EMERGY of Alternative Investment

A particular development might be compared with the intensity of development that is predicted assuming the same feedback investment F had been matched with input of environmental resources according to the regional investment ratio (I_R). In Figure 9.3, EMERGY flow P_3 is the sum of the EMERGY of F and the environmental EMERGY flow I_3 (equal to F/I_R).

EMERGY of Potential Matching

A development should also be compared with the economic matching that could result if all the original environmental EMERGY flow I_1 was retained and matched with feedback EMERGY according to the regional investment ratio I_R. In Figure 9.3, this alternative EMERGY flow P_4 is the sum of the original environmental EMERGY flow I_1 plus the feedback from the larger economy ($I_1 \times I_R$).

For example, one of the summary diagrams from our study of development alternatives for the Mississippi River (Odum et al., 1987c) is Figure 9.4 with details in Table 9.2. The development that took place with channelization, levees, shipping, floodplain agriculture, and so forth (Figure 9.4b) increased the total EMERGY flow from 46 billion to 76.2 billion Em$/yr, but it diminished the environmental EMERGY from 46 billion to 15.7 billion Em$. Empower was increased.

This development may be compared with the alternative that matches the investment of 60.5 million Em$ with environment according to the investment ratio ($I_R = 7$): ($60.5 + 60.5/7 = 69.1 \times 10^9$ Em$/yr). The observed development is close to that which is economical according to the regional matching principle.

Figure 9.4c shows the potential (368 billion Em$/yr) if development could be revised to restore the original environmental contributions while also combining these with compatible economic development attracted according to the EMERGY investment ratio ($I_R = 7$). This comparison shows that the economic vitality of the Mississippi system might be increased six fold by a more holistic pattern of development.

The evaluation in Figure 9.4 omits the very large EMERGY flows of sediment erosion, now being diverted to the Gulf of Mexico but which formerly enriched the lowlands and expanded the Louisiana marshes. Inclusion of the sediments would increase the estimate for long-run economic benefit by many times (Odum et al., 1987c).

EVALUATING AREAL EMERGY DISTRIBUTION

Theory is given here about the way EMERGY flows are distributed over landscapes and how EMERGY accounting and mapping can evaluate existing structures and the suitability of areas for appropriate developments. As introduced

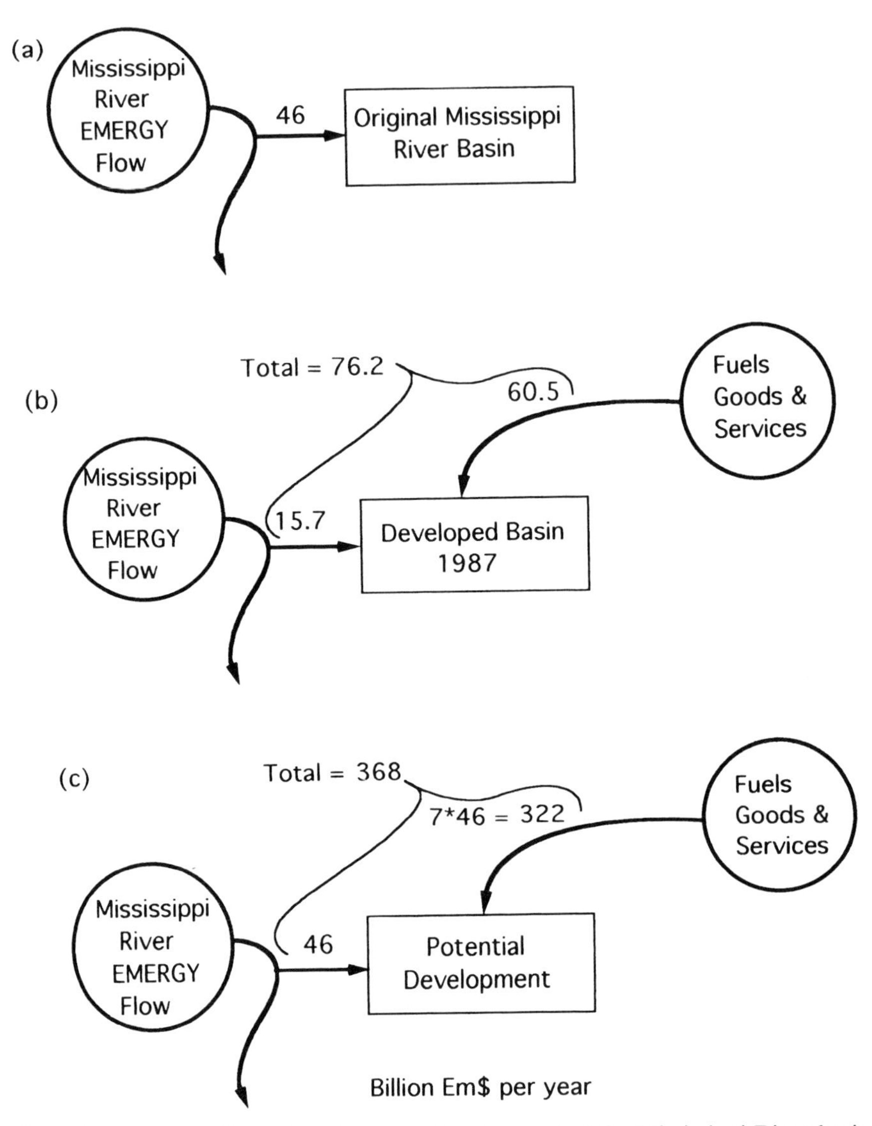

Figure 9.4. EMERGY flows for alternative development of the Mississippi River basin (Odum et al., 1987c). **(a)** Before development. **(b)** Developed basin. **(c)** Potential development.

in Chapter 2 (see Figure 2.5), self-organizational processes organize the energy-transformation hierarchy into spatial patterns. Units that are more numerous, smaller, faster, and have lower transformity converge inputs to centers of energy transformation that have higher transformity, turn over more slowly, and affect larger territories. These centers feed back materials, controls, and special services outward. For example, the rural productivity of

TABLE 9.2. EMERGY Values in the Mississippi River System

Item*	Solar Empower ($\times 10^{22}$ sej/yr)	Em$/yr† ($\times 10^{9}$ 1993 $)
Original River:		
Chemical potential of fresh water	11.5	81.5
Geopotential of river	10.0	70.9
Water used by the original floodplain	7.9	56.0
Modified river in 1986		
Water used by developed and drained floodplain	3.0	21.2
Sediments carried	230.	1631.
Sediments discharged to Gulf of Mexico	127.	900.
Fisheries production	0.99	7.0
Loss of coastal land	1.4	9.9
River energy used by barges‡	0.43	3
Purchased inputs for economic use		
In agriculture, forestry, and crayfish culture	0.86	6.1
In river transportation^δ	0.96	6.8
(If railroad substituted for river transport)	(4.61)	(32.7)
In oil and gas production	5.2	36.8
In urban economy	8.2	58.2
Oil and gas production for use elsewhere	55.2	391.5

* Some items are included within others.
† Solar empower in column 2 divided by 1.41×10^{12} sej/1993 U.S. em$.
‡ If fraction of river required per vessel considering turbulent eddies is 100 times vessel displacement.
δ Includes fuel use, shipping goods and services, and operation of locks.
Source: Condensed from Diamond (1984) and H. T. Odum et al. (1987c); update and elaboration of earlier studies by Young et al. (1974); Bayley et al. (1977); and Zucchetto et al. (1980).

farms converges products to the centers of energy transformation, which feed back to the countryside recycled nutrients, law and order, and manufactured goods. The zones of empower distribution may be quite different from those of energy. Intense flows of high-transformity information may be high in centers of knowledge and politics, where the energy flow may not be large.

By representing the energy-transformation hierarchy, energy systems diagrams also represent the spatial concentration of EMERGY flows. Because of their spatial convergence (moving left to right), the EMERGY flows per area in Figure 9.5a are more concentrated in the hierarchical centers. We define the following measures:

Areal empower density = EMERGY flow per area per time
Areal EMERGY storage density = EMERGY storage per area

Maps of areal empower density show where important transformations are centered and where real value should be protected. Cities of this century

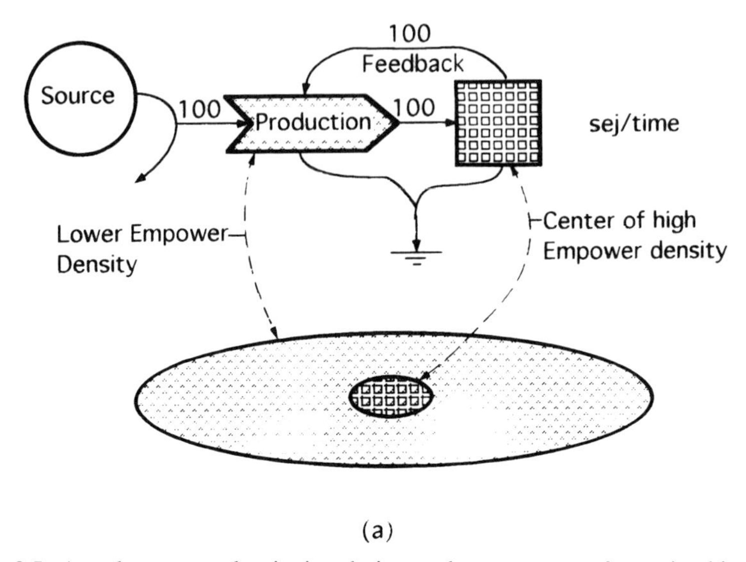

Figure 9.5. A real empower density in relation to the energy transformation hierarchy. **(a)** Concept diagram showing EMERGY concentration at higher centers of hierarchy. **(b)** Empower density map of Jacksonville, Fla. (Whitfield, 1994).

are points where purchased EMERGY of fuels, goods, and services enters the economy with a very high EMERGY flow. The areal empower density of these cities is higher than that of the agrarian-based cities. Table 9.3 lists areal empower densities for different areas. Lowest values are found in rural countries, highest values in developed cities. Higher-transformity systems turn over slowly, storing more EMERGY per area. For example, cathedrals and capitols accumulate EMERGY over long periods. Many of the structures in centers are higher-EMERGY symbols shared in knowledge by all the people of the larger landscape (see Chapter 12).

The transformity of flows also increases towards their hierarchical centers, since there are more transformations in the process. One hypothesis that follows from the principle of energy matching is that activities may be appropriately located according to their transformities. Functions operate best when there are interactions of flows in which transformities differ by an order of magnitude. Plans for a landscape must make hierarchical sense. For example, informational, financial, and control functions are compatible at city centers, whereas farms, forests, and environmentally surrounded housing developments are compatible away from the centers.

Evaluating Purchased Inflows to an Area

As already explained in Chapter 5, the inflows that are purchased outside and brought into an area have two components, each of which has to be evaluated. One is the EMERGY contained in the available energy brought in. The other is the EMERGY of the human services involved in bringing in the resource.

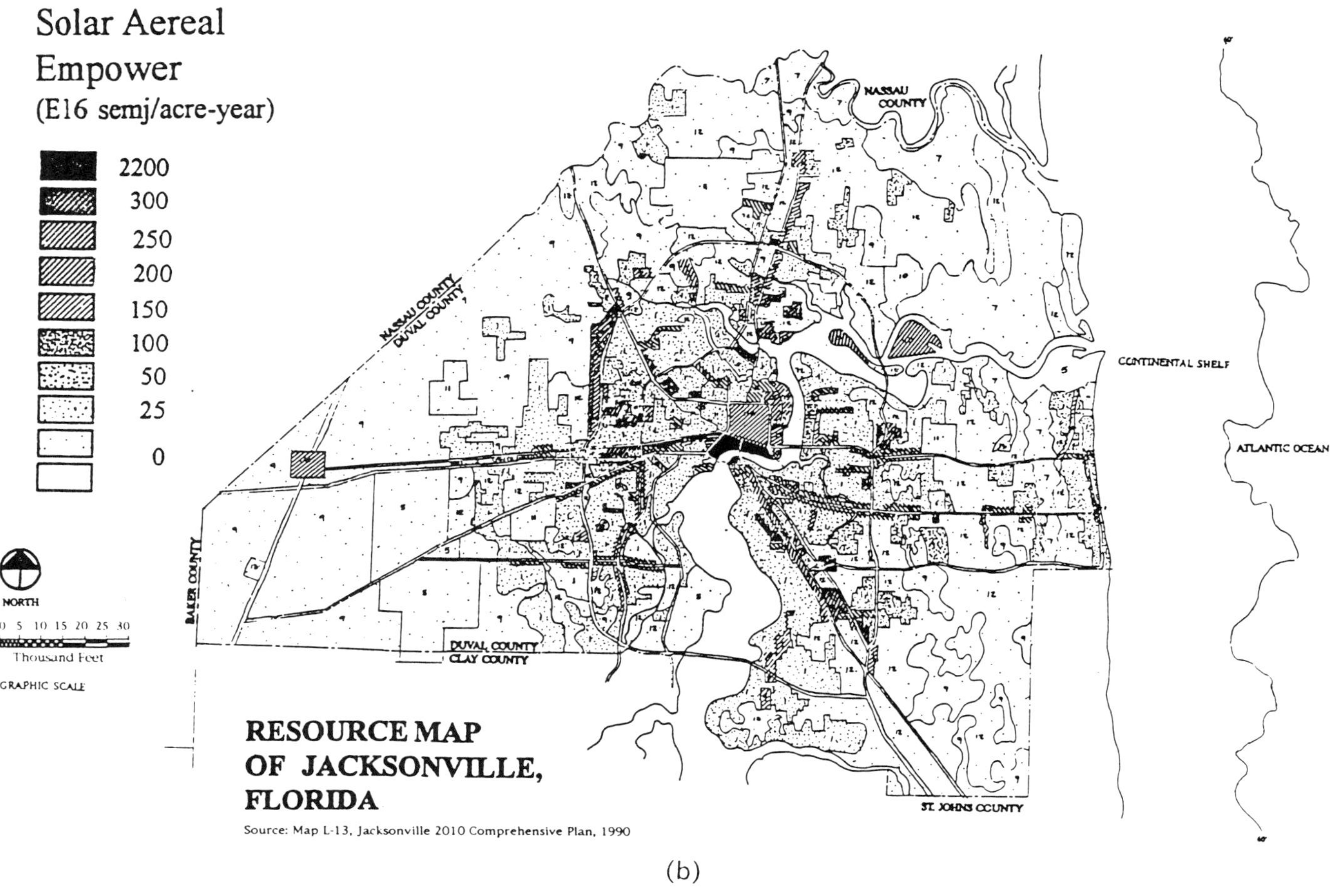

(b)

Figure 9.5. *(Continued)*

TABLE 9.3. Areal Empower Densities

Note	System	Solar empower/area/time* 10^{11} sej/m^2/yr
1	Wilderness	0.1–1
2	Rural nations	1.4–3
3	U.S.A.	8
4	Urban nations	10–100
5	Developed city	150–54,000
6	Power plant	157,000

* Solar empower is the solar EMERGY use per unit time in solar emjoules per year.
[1] Chapter 7. [2,4]Odum and Odum (1983). [3] Chapter 10.
[5] Jacksonville, Fla. (Whitfield, 1994).
[6] $(1000 \times 10^6\ \text{W})(1\ \text{J/sec/W})(3.15 \times 10^7\ \text{sec/yr})(2 \times 10^5\ \text{sej/J})/(4 \times 10^4\ \text{m}^2)$.

Since data for estimating what is purchased and brought in are rarely available for development areas, a nomogram (Figure 5.2) was provided by M. T. Brown (1980) for evaluating input of goods and services from an area's estimated income.

Transportation Infrastructure

The coupling of the environmental-landscape EMERGY with the attracted EMERGY of economic development requires infrastructure, especially the network of roads that facilitates the converging of flows to cities and the diverging flows from cities to the rural landscape. Development traditionally starts by building infrastructure (roads and utility lines). Development is possible if inputs to an area that the development attracts contribute more net EMERGY than is required for the infrastructure development. Sometimes infrastructure development is done without regard to the whole system, so that the EMERGY disrupted is as great as the EMERGY to be attracted by the construction. Any sector of the economy may be evaluated for its EMERGY requirements using the standard procedures of Chapter 5. For example, Figure 9.6 summarizes an evaluation of the highway system of Texas (Lyu, 1986). That study found a turnover time of 50 yr (with depreciation = 2%/yr).

Individual Quality of Life

Important to the thinking of planners and people voting on development alternatives in the political process is their perception of the quality of life. The human being is a complex hierarchy whose organization levels range from the fast, small biochemistry of living cells to the concepts of social behavior and culture of the mind, which have a slow turnover time and require

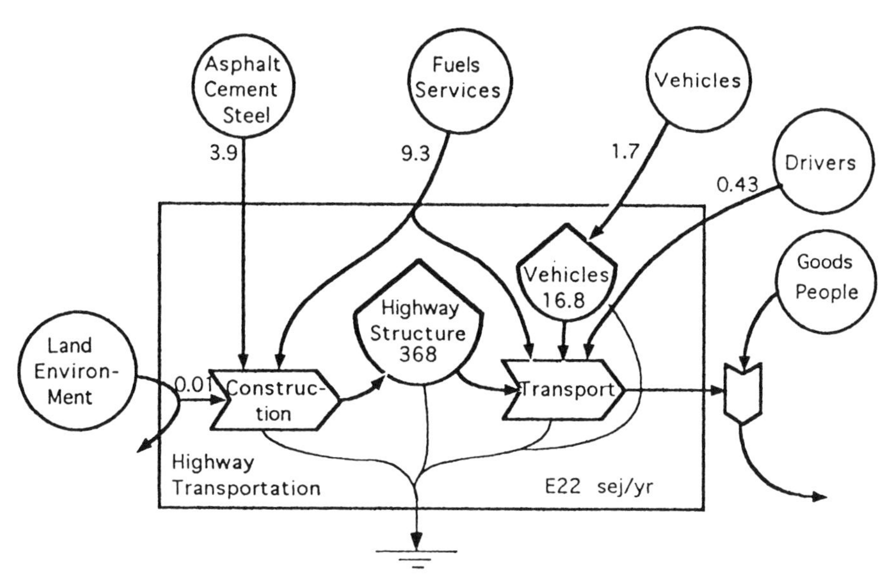

Figure 9.6. Summary of EMERGY evaluation of the highway system of Texas (Data from Lyu, 1986).

a large territory for support and operation. The principle of EMERGY matching may be applied to the human individual, with greatest EMERGY coming when there is a good match of the small-, medium-, and large-scale inputs to the human welfare. Three kinds of inputs of different transformity that may be required for human individual benefit are shown interacting in Figure 9.7: a, environmental influences; b, exergy-level support (Exergy is the available energy of medium transformity sources such as fuels, electricity, transport.); and c, informational exchange. Also shown is the EMERGY diversion (stress) of negative influences (d).

For existing and proposed developments, the EMERGY flow to each person in a development may be evaluated for the four categories in Figure 9.7. For the environmental component, the EMERGY flow in the area's parks, waters, greenbelts, and so forth is evaluated. For the EMERGY in the exergy support, fuel and electric use per person of the area are evaluated. For EMERGY in the informational exchange, the informational EMERGY flow per person in the zone is evaluated using procedures from Chapter 12. The negative effects may be evaluated either from known losses of productive function or indirectly as the EMERGY flows of the stress flow. *A priori,* we may predict that the good quality of an individual life is one with a high total EMERGY flow and a good matching (i.e., similar EMERGY flows) of the environmental, supporting, and informational contributions.

For example, affluent suburban housing developments and their commuting access to various centers have high values of EMERGY flow for the environment,

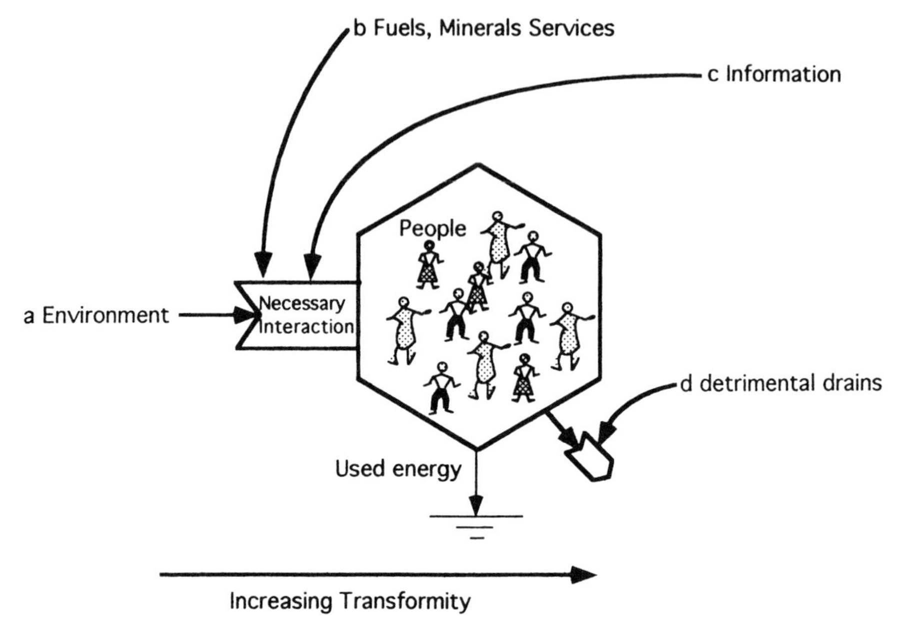

Figure 9.7. EMERGY requirements for the welfare of human individuals, including three categories of different transformity (a, b, c).

transportation-electricity, and information (in jobs and education). In calculating the individual EMERGY index, it might be desirable to subtract those disruptive EMERGY flows of individuals' lives due to the special conditions of the development. For example, the EMERGY per person that is drained due to increased accidents, crime, pollution, noise, and so on could be subtracted.

Ecotourism

The needs people from developed areas have for environmental experiences causes ecotourism development in many areas. Wilderness and special environmental values attract the hotels, restaurants, and transportation infrastructure for economic use. In order for the use to be compatible with the attraction area, a low EMERGY investment ratio has to be maintained. In other words, large undeveloped areas with scenic panoramas have to be sustained. Without control, development tends to expand, increasing investment ratios and displacing the environment attractions. Feedback of EMERGY for reinforcement and protection is required. Because of the money tourists bring from urban centers, local people and previous low-energy economies tend to be displaced. Because tourists from developed areas bring money with lower EMERGY/money ratios than that of the host area, they contribute much less real wealth than they receive (see the discussion of the EMERGY of foreign exchange in

Chapter 11). However, EMERGY accounting can be used to ensure that there is equity in the contributions of the ecotourism developments to the local area.

COST-BENEFIT VERSUS EMERGY EVALUATIONS

Many developments have been judged according to their money costs and benefits. As diagrammed in Figure 9.8, money spent on development is compared with the money that will circulate after the development is operating. Since the money spent on development might have been earning interest in alternate investments, this potential interest is added to the cost for comparison purposes. The potential interest is called the *discount* and varies with cycles in economic investment.

Unfortunately, the money cost-benefit procedure does not consider the resources to determine whether there will be a net benefit in real wealth (EMERGY) or not. A development based on maximizing profit through one activity pulls resources away from the general indirect support of the economic system that existed before development. Figure 9.8 shows the monetary benefits (with only the human services involved) and the EMERGY benefits (representing all benefits). An EMERGY evaluation of the pathways in Figure 9.3 is required to determine the overall benefits and cost to the whole system of environment and economy. A monetary cost-benefit method is not even a good predictor of economic success of the proposed development, because it doesn't evaluate whether the new resources to be processed are large net contributors or not. To judge success of future production, the monetary cost-benefit analyst has only the empirical results of similar developments in the past to work with. For proposed developments where there is little or no previous experience, the calculation will have little basis.

Because people think of value in terms of money (paid only to people), they might not realize that environmental resources, including the storage reserves, are already in use by the environmental systems on which human society depends. Any development that diverts resources from its existing use is taking it away from its support of the economy in other sectors. Thus, no development should be authorized until the proposed new alternative is proven to generate more EMERGY production and use. Any monetary cost-benefit analysis should be accompanied by an EMERGY evaluation of the pathways in Figure 9.8.

EMERGY IN DEVELOPMENT STAGES

The rapid growth of housing areas in Florida provides many examples of cases in which empower was maximized due to overdevelopment, often with loss of EMERGY value in the process. Developers in the 1970s were able to sell many condominium units by first building, for display, one high rise condominium

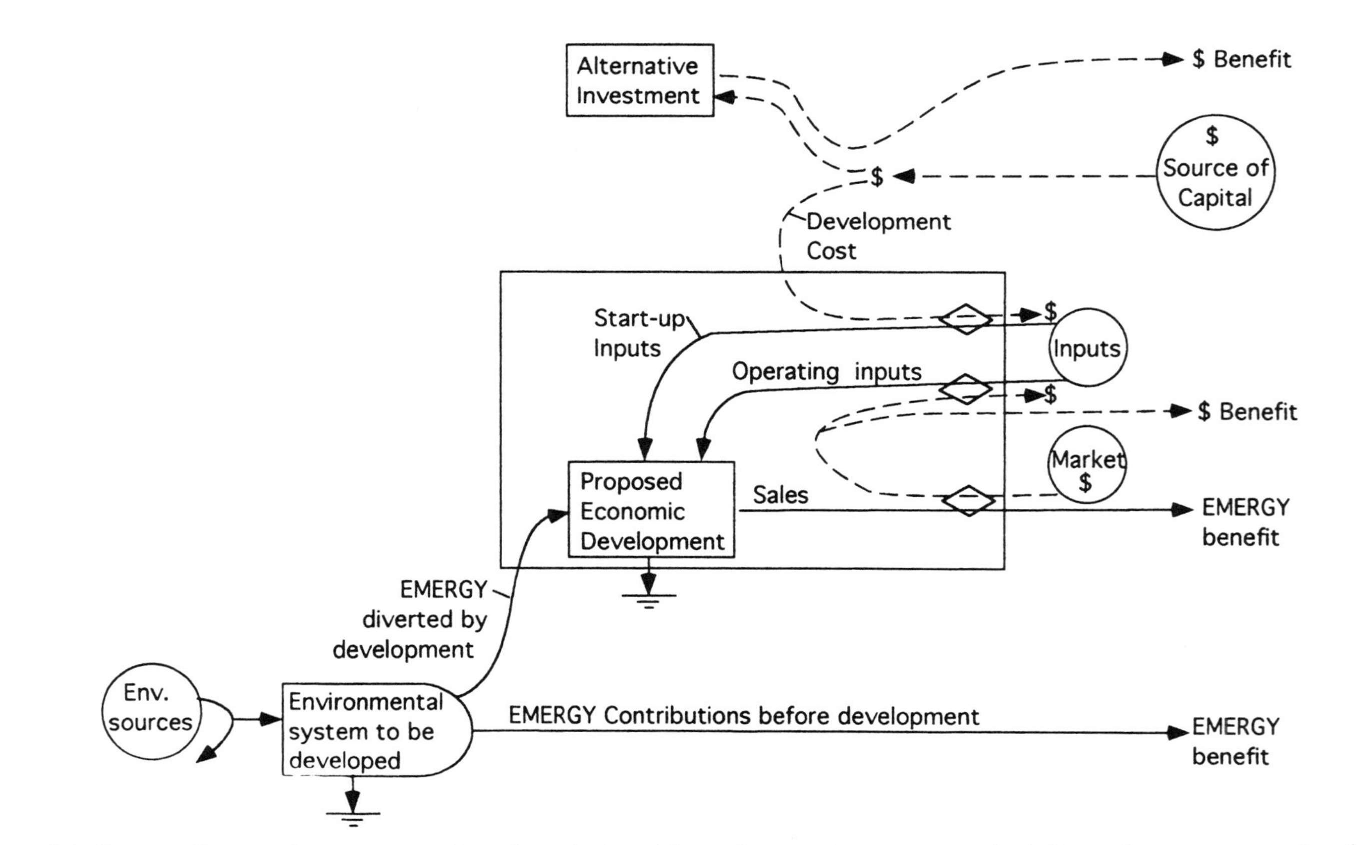

Figure 9.8. Systems diagram of monetary cost-benefit analysis and the pathways to be EMERGY-evaluated to make a more general evaluation of net benefit.

surrounded by large natural areas. Initially, therefore, the development had a low investment ratio, and the EMERGY of the natural areas was a significant attractant to bring in buyers. By the time the owners arrived to occupy their units, many more high rise condominia had been built, and the necessary parking lots for a large population had obliterated the natural areas, increasing the investment ratio, losing many of the beneficial qualities of environmental-economic matching.

With Marco Island as a case history, D. L. Stellar (1976) evaluated these changes with EMERGY evaluations (in coal emjoules), finding an optimum development density for maximum empower. M. G. Sell (1977), using simulation models, simulated the EMERGY budgets (in coal emjoules) of increasing housing developments, finding an optimum development density for maximum empower that took losses due to overdevelopment and hurricanes into account. M. Miller (1975) arranged the stages of successive development of housing in Collier County, Fla., as an energy hierarchy in which land was moved with value added successively in steps to more intense development, population, transformity, and empower density. In some ways the stages in economic development in a previously undeveloped areas were like the successional processes observed in the ecosystems at a smaller scale.

SUMMARY

Concepts and case-history examples provided EMERGY indices and guidelines for maximizing the environmental-economic prosperity of economic development through a symbiotic mix of free and purchased resources. Sustainable developments show a spatial pattern linking rural areas of lower economic activity with urban centers of high empower density. Maps showing EMERGY and transformity may be helpful in recognizing the spatial dimensions of an energy hierarchy and in planning the locations for enterprises of different kinds. Developments of various kinds may be appropriately located in the gradient from development centers to rural environs according to their transformity and EMERGY investment ratios.

CHAPTER 10

EMERGY OF STATES AND NATIONS

EMERGY evaluation of states, nations, and their resource basis gives large-scale perspective to appraisal of environmental areas, and helps select policies for public benefit. EMERGY evaluation of regions, states, and nations begins with systems diagramming (see Chapter 5 and Appendix A). Then a summary diagram is aggregated to show several categories of internal and external contribution (Figure 10.1). Finally, indices are calculated and suggestions made for maximum empower and policy. In this chapter, as an example, evaluation procedures are given for the United States.

Table 10.1 lists the data that are often needed to prepare a state or national EMERGY evaluation. Tables 10.2 and 10.3 have 1983 EMERGY flows for the United States that go with Figure 10.2. Table 10.4 has EMERGY storages of the United States. EMERGY values are summarized in a summary diagram (Figure 10.3) and two summary tables (Table 10.5, Table 10.6).

EMERGY Flows across the Boundary

First, the boundaries of the nation to be evaluated are defined; in the case of the United States, the lateral boundaries are usually at the edge of the submarine continental shelf. The upper boundary may be 100 m above the earth and water surface, and the lower boundary 2 m below the earth surface or floor of the lakes or seas. Identifying and evaluating the flows across the boundary come next. These flows include environmental flows, such as sun, wind, rain, rivers, tides, waves, and geologic processes; imported raw materials such as minerals; imported rural products such as food, fiber, and wood; imported equipment and machinery; imported workers and purchased services; imports

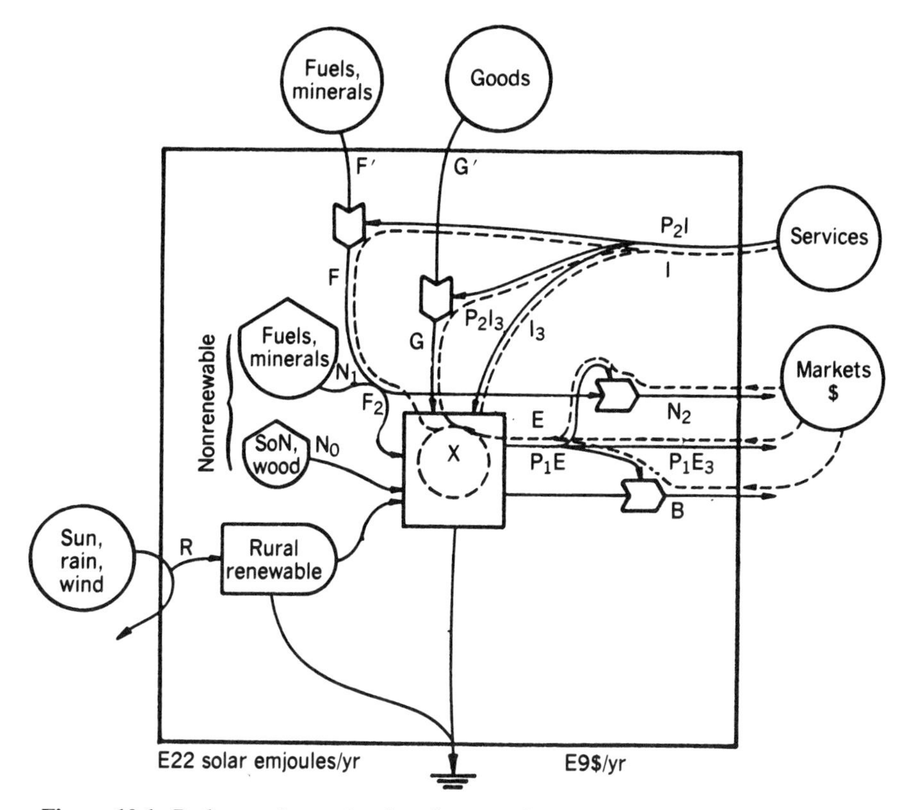

Figure 10.1. Pathways for evaluating the overall energy use of a state or nation.

of military operations; inflows of information; exchanges of money; and educational exchanges. Figure 10.2 for the United States is a typical diagram for a state or nation that has most of these boundary inflows.

Summary Diagram

Figure 10.1 is a summary diagram for representing inflows and outflows from states and nations. It includes a summary of money flows and shows the gross economic product. It includes each class of inflows that contributes EMERGY to the system's operation. On the left are the environmental resource inputs, renewable resource R and nonrenewable resources that originate within the boundaries (N_0, N_1, and N_2). N_0 designates the rural resources, such as soil and forest biomass, if their storages are being used faster than they are regrowing. N_1 represents the reserves of fuels and minerals that are renewed only over longer periods of geologic time.

Export pathway N_2 is shown for flows that pass resources through the system without appreciable transformation use. Examples are minerals that

TABLE 10.1. Data Needed for EMERGY Analysis of a State or Nation

Coarse land-use map
Table of areas (forest, pasture, wilderness, urban area, continental shelf)
Average elevation
Gross national product (GNP) or total income
Annual insolation
Fuel consumption
Population, immigration, emigration
Money paid for main imports
Tonnages of main imports
Money received from main exports and main purchasing countries
Tonnages exported
Total money exchanged in addition to trade (loans, investments, interest, transfer payments, foreign aids, military subsidies)
Rainfall, percent runoff, and elevations of main watersheds
Discharges of major rivers entering or leaving national boundaries
Estimated evapotranspiration
Mean winds, winter and summer
Rate of land erosion and isostatic replacement
Organic content in soils of agricultural areas and forest
Percentage of the economy in main sectors (e.g., health, education, government, defense, etc.)
Electric generation, use, import, export
Length of coastline facing incoming water waves
Mean tidal height and depth of tidal gauges
Mean wave height
Economic statistical abstract if available

are mined and exported without further processing. The EMERGY of pass-through commodities is not included in totals representing EMERGY contribution to an economy.

Imports are shown from the top and right including the EMERGY of fuels F, goods that have EMERGY in addition to services involved G, and total imported service EMERGY P_2I. The EMERGY of services is the product of the dollars of imports I and the EMERGY/money ratio P_2 of the country from which the imports come. The money paid for the imports has buying power abroad according to the EMERGY/money ratios of the countries sending the imports. The diagram shows money flows with dashed lines. Two money pathways for the services of bringing fuels and goods turn and converge to be included in the total services P_2I. I_3 is for the dollars of services other than those for delivering fuels and goods, and these services have EMERGY flow P_2I_3.

The exports on the lower right also have pathways for fuels, goods, and services like those discussed for imports. EMERGY of goods B and EMERGY of fuel-mineral export N_2 include EMERGY of services required to process and deliver. In Figure 10.1 the dollar flows coming in from markets on the right for all the exports are dashed lines that converge together as total dollars

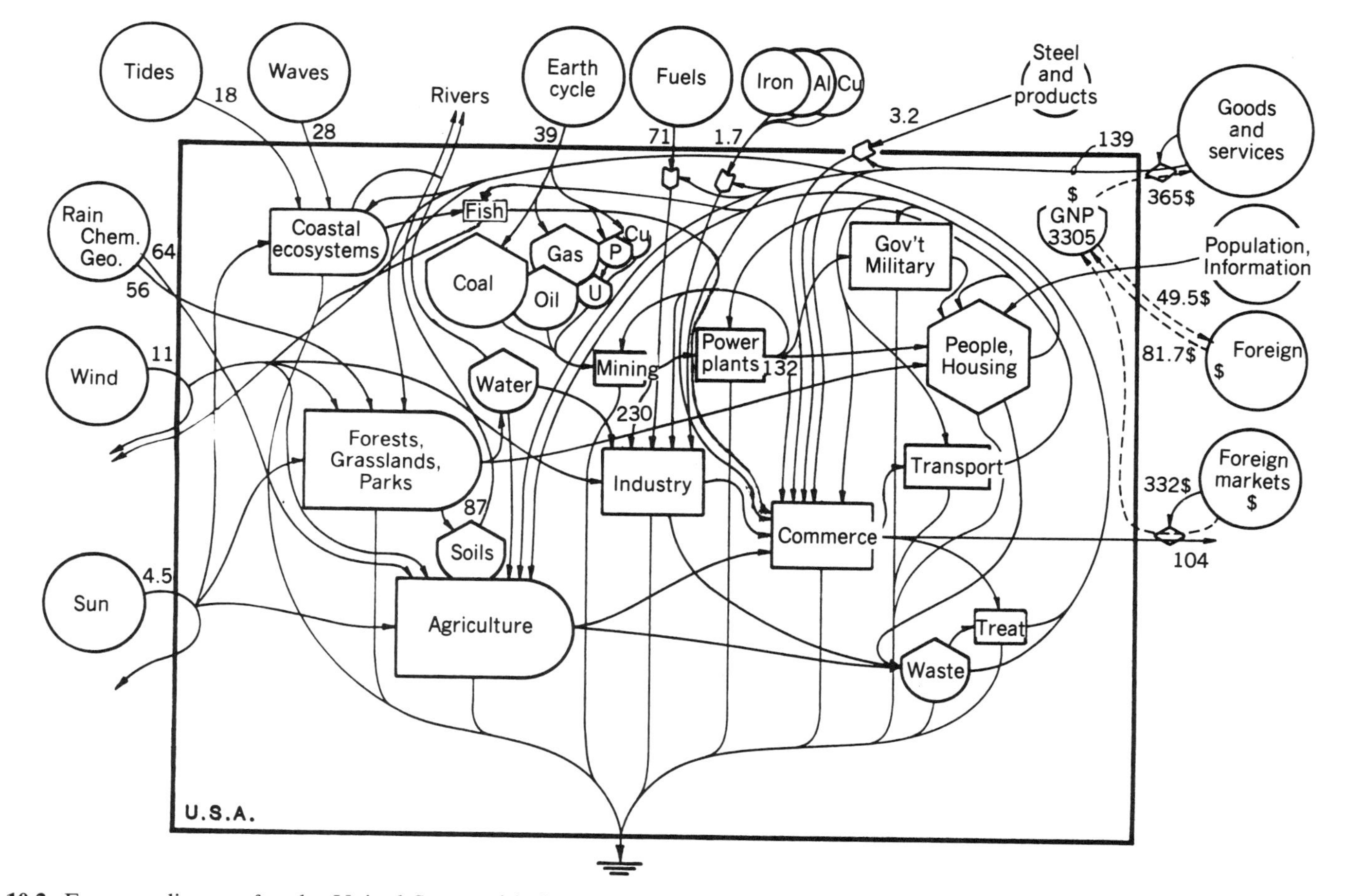

Figure 10.2. EMERGY diagram for the United States with flows evaluated in Table 10.2 (Odum et al., 1987a). Reprinted with permission of L.B.J. School of Public Affairs, The Univ. of Texas, Austin.

TABLE 10.2. EMERGY Flows of the United States in 1983*

Note	Item	Raw Units	Transformity (sej/unit)	Solar EMERGY ($\times 10^{22}$ sej)	Em$† ($\times 10^9$ 1983 U.S.$)
		Renewable Sources			
1	Sunlight	4.48×10^{22} J	1/J	4.48	18.66
2	Rain, chemical	4.17×10^{19} J	15,444/J	64.40	268.33
3	Rain, geopotential	6.33×10^{19} J	8888/J	56.21	234.21
4	Wind, kinetic	1.63×10^{20} J	663/J	10.80	45.02
5	Waves	1.09×10^{19} J	25,889/J	28.22	117.57
6	Tide	7.63×10^{18} J	23,564/J	18.0	74.91
7	River water				
8	Earth cycle	1.36×10^{19} J	29,000/J	39.44	164.33
9	Hydroelectricity	1.19×10^{18} J	159,000/J	18.92	78.83
10	Wood consumption	4.41×10^{18} J	34,900/J	15.39	64.12
		Imports and Outside Sources			
11	Natural gas	1.01×10^{18} J	48,000/J	4.84	20.2
12	Oil, crude	7.63×10^{18} J	53,000/J	40.43	168.49
13	Petroleum products	3.95×10^{18} J	66,000/J	26.07	108.62
14	Iron and steel products	1.78×10^{7} Tn	1.78×10^{15}/Tn	3.16	13.20
15	Iron ore	1.32×10^{13} g	8.55×10^{8}/g	1.12	4.70
16	Aluminum ore (bauxite)	7.5×10^{12} g	8.5×10^{14}/Tn	0.64	2.67
17	Goods and services	$3.65 $\times 10^{11}$	3.8×10^{12}/$	138.70	577.91
		Nonrenewable Sources from Within U.S.A.			
18	Iron-ore production	3.76×10^{13} g	3.55×10/g	3.21	13.4
19	Coal production	2.49×10^{19} J	39,800/J	99.10	412.9
20	Coal consumption	2.34×10^{19} J	39,800/J	93.13	388.1

21	Oil production	2.36×10^{19} J	53,000/J	125.08	521.2
22	Oil consumption	3.49×10^{19} J	53,000/J	184.97	770.7
23	Natural-gas production	1.76×10^{19} J	48,000/J	84.48	352.0
24	Natural-gas use	1.85×10^{19} J	48,000/J	88.8	370.0
25	Uranium production	6.52×10^{18} J	1,790/J	1.16	4.86
26	Electricity use	8.82×10^{18} J	159,000/J	132.28	551.2
27	Nuclear electric use	1.05×10^{18} J	159,000/J	16.69	69.6
28	Phosphate fertilizer production	2.84×10^{12} g	2×10^{10}/g	5.68	23.7
29	Phosphate rock production	1.15×10^{13} g	1.4×10^{10}/g	16.1	67.1
30	Bauxite production	8×10^{11} g	8.55×10^{8}/g	0.068	0.28
31	Bauxite consumption	8.8×10^{12} g	8.55×10^{8}/g	0.75	3.13
32	Earth loss	5.87×10^{14} g	1.71×10^{9}/g	100.37	418.3
33	Net topsoil production	2.1×10^{18} J	62,500/J	13.12	54.7

* Total area: 9.4×10^{12} m^2.

† EMERGY flow in column 5 divided by 2.4×10^{12} sej/$ for U.S. in 1983.

[1] Sunlight: total area, 9.4×10^{12} m^2; Alaska, 1.53×10^{12} m^2, Continental shelf: total, 1.7×10^{12} m^2; Alaska, 1.07×10^{12} m^2 estimated (National Geographic, 1981); 48 states average net absorbed solar radiation, 110 kcal/cm^2/yr:

$$(8.5 \times 10^{12}\ \text{m}^2 \text{ without Alaska})(110\ \text{kcal/cm}^2\text{/yr})(1 \times 10^4\ \text{cm}^2\text{/m}^2)(4186\ \text{J/kcal}) = 3.9 \times 10^{22}\ \text{J/yr}$$

Alaska, 35% albedo; solar radiation absorbed, 65% of 3.35×10^9 J/m^2/yr (Budyko, 1974):

$$(2.6 \times 10^{12}\ \text{m}^2)(2.18 \times 10^9\ \text{J/m}^2\text{/yr} = 0.57 \times 10^{22}\ \text{J/yr}$$

Total: 4.48×10^{22} J/yr

[2] Rain, chemical potential energy: average, 35.4 in/yr from mean of 50 values, each a median for one of the 50 states:

$$(9.4 \times 10^{12}\ \text{m}^2)(35.4\ \text{in})(2.54\ \text{cm/in})(1 \times 10^{-2}\ \text{m/cm})(1 \times 10^6\ \text{g/m}^3)(4.94\ \text{J/g}) = 4.17 \times 10^{19}\ \text{J/yr}$$

Table 10.2. (*Continued*)

[3] Rain, geopotential energy: mean elevation, 763 m; mean rainfall, 0.9 m:

$$(9.4 \times 10^{12}\ m^2)(0.9\ m/yr)(1 \times 10^3\ kg/m^3)(9.8\ m/sec^2)(763\ m) = 6.33 \times 10^{19}\ J/yr$$

[4] Wind, kinetic energy: mean of 25 stations in U.S. (Swaney, 1978). Eddy diffusion coefficients: January, 22.3 m^2/sec; July 3.6 m^2/sec. Velocity gradients: January, 6.08×10^{-3} m/sec/m, July 1.78×10^{-3} m/sec/m

Winter: $(1 \times 10^3\ m)(1.23\ kg/m^3)(22.3\ m^2/sec)(1.577 \times 10^7\ sec/0.5\ yr)(6.08 \times 10^{-3}\ m/sec/m)(2)(9.4 \times 10^{12}\ m^2) = 1.50 \times 10^{20}\ J/0.5\ yr$

Summer: $(1 \times 10^3\ m)(1.23\ kg/m^3)(23.6\ m^2/sec)(1.577 \times 10^7\ sec/0.5\ yr)(1.78 \times 10^{-3}\ m/5/m)(2)(9.4 \times 10^{12}\ m^2) = 0.135 \times 10^{20}\ J/0.5\ yr$

Total: $(1.50 + 0.135) \times 10^{20}\ J/yr = 1.63 \times 10^{20}\ J/yr$

[5] Waves: Continental straight coastline, 6.4×10^6 m estimated (National Geographic, 1981); Alaska N + W, 1.97×10^6 m; S. Alaska and Aleutians, 1.07×10^6 m. Wave power: av. U.S. 40.5 kw/m; N. Pacific for S. Alaska: 81 kW/m.
Continental USA:

$$(40.5\ kW/m)(1 \times 10^3\ W/kW)(1\ J/s/W)(3.15 \times 10^7\ sec/yr)(6.46 \times 10^6\ m\ \text{facing shore}) = 8.18 \times 10^{18}\ J/yr$$

South Alaska:

$$(81\ kW/m)(1 \times 10^3\ W/kW)(1\ J/s/W)(3.154 \times 10^7\ sec/yr)(1.07 \times 10^6\ m) = 2.7 \times 10^{18}\ J/yr$$

Total waves:

$$(8.18 \times 10^{18}\ J/yr) + (2.7 \times 10^{18}\ J/yr) = 1.09 \times 10^{19}\ J/yr$$

[6] Tide: U.S. continental shelf, $6.38 \times 10^{11}\ m^2$; Alaska continental shelf, N. + W., $3 \times 10^{17}\ m^2$; S. shelf, $1.29 \times 10^{17}\ m^2$ estimated (National Geographic, 1981); tide height, continental USA, 1.2 m averaged (U.S. Coastal and Geodetic Survey, 1956).
Continental:

$$(6.38 \times 10^{11}\ m^2)(0.5)(706/yr)(1.2\ m)^2(9.8\ m/sec^2)(1.025 \times 10^3\ kg/m^3)(0.1) = 3.26 \times 10^{17}\ J/y$$

S. Alaska:

$$(1.29 \times 10^{11}\ m^2)(0.5)(706/yr)(4\ m)^2(9.8\ m/sec^2)(1.025 \times 10^3\ kg/m^3)(0.1) = 7.3 \times 10^{18}\ J/yr$$

[7] River water, elevated rivers.

[8] Earth cycle. U.S. surface area assigned heat flows based on ages, following method and data of Sclater et al. (1980). 20%, $2 \times 10^6\ J/m^2$; 40%, $1.5 \times 10^6\ J/m^2/yr$; 25%, $1.2 \times 10^6\ J/m^2/yr$; 15%, $1 \times 10^6\ J/m^2/yr$:

$$(9.4 \times 10^{12}\ m^2)(1.45 \times 10^6\ J/m^2/yr) = 1.36 \times 10^{19}\ J/yr$$

[9] Electricity, hydro, 1983: 3.32×10^{11} kWh (EIA/DOE, 1984b, p. 22):

$$(3.32 \times 10^{11}\ kWh/yr)(3.606 \times 10^6\ J/kWh) = 1.20 \times 10^{18}\ J/yr$$

[10] Wood consumption, 1983: $14{,}660 \times 10^6\ ft^3/yr$ (US, 1984):

$$(14{,}660 \times 10^6\ ft^3/yr)(2.7 \times 10^{-2}\ m^3/ft^3)(0.76 \times 10^6\ g/m^3)(3.8\ kcal/g)(4186\ J/kcal) = 4.41 \times 10^{18}\ J/yr$$

[11] Natural-gas imports, 1983: $(9.28 \times 10^8$ tcf)(tcf = trillion cubic feet)$(1.1 \times 10^9$ J/tcf) $= 1.01 \times 10^{16}$ J, 4.64\$/tcf (EIA/DOE, 1985a, pp. 7–9):

$$(9.28 \times 10^8\ tcf)(4.64\ \$/tcf) = 4.3 \times 10^9\ \$/yr.$$

[12] Crude-oil imports, 1983: $(3.33 \times 10^6$ bbl/day)(365.24) $= 1.216 \times 10^9$ bbl/yr; 28.99 \$/bbl (EIA/DOE, 1985b, pp. 2–3):

$$(1.216 \times 10^9\ bbl/yr)(6.28 \times 10^9\ J/bbl) = 7.64 \times 10^{18}\ J/yr$$
$$(1.2 \times 10^9\ bbl)(28.99\ \$/bbl) = 34.8 \times 10^9\ \$/yr.$$

[13] Petroleum products imports, 1983: (6.289 bbl/yr)$(6.28 \times 10^9$ J/bbl) $= 3.94 \times 10^{18}$ J, 28.99 \$/bbl (EIA/DOE, 1984c, p. 17):

$$(6.29 \times 10^8\ bbl)(28.99\ \$/bbl) = 18.3 \times 10^9\ \$/yr$$

Table 10.2. (*Continued*)

[14] Iron and steel products imports, 1983: 17.8×10^6 Ton, $\$7.1 \times 10^9$ (U.S., 1984, p. 702).
[15] Iron-ore imports, 1983: 13.2×10^6 Ton, $\$0.446 \times 10^9$ (U.S., 1984, p. 713).
[16] Bauxite (aluminum ore) imports, 1983: 7.5×10^6 Ton (U.S., 1984, p. 716); 29.17 $/Ton (U.S., 1984, p. 473).
[17] Goods-and-services imports, 1983: 365×10^9 (U.S., 1984, p. 800).
[18] Iron-ore production, 1983: 37.6×10^6 Ton (U.S., 1984, p. 713).
[19] Coal production, 1983: 785×10^6 st (short tons) (EIA/DOE, August 10, 1985), pp. 4–9):

$$(78 \times 10^6 \text{ st/yr})(3.18 \times 10^{10} \text{ J/st}) = 2.49 \times 10^{19} \text{ J/yr}$$

[20] Coal consumption, 1983: 736.67×10^6 st (short tons)(EIA/DOE, August 10, 1985c, pp. 4–9).

$$(736.67 \times 10^6 \text{ st})(3.18 \times 10^{10} \text{ J/st}) = 2.34 \times 10^{19} \text{ J/yr}$$

[21] Crude-oil production, 1983: 3.76×10^6 bbl (EIA/DOE, 1984c, p. 16).

$$(3.76 \times 10^6 \text{ bbl})(6.28 \times 10^9 \text{ J/bbl}) = 2.36 \times 10^{19} \text{ J/yr}$$

[22] Crude-oil consumption, 1983: 15.23×10^6 bbl/day (EIA/DOE, 1984c, p. 16):

$$(15.23 \times 10^6 \text{ bbl/day})(365 \text{ days/yr})(6.28 \times 10^9 \text{ J/bbl}) = 3.49 \times 10^{19} \text{ J/yr}$$

[23] Natural-gas production, 1983: 16.3×10^{15} btu (U.S., 1984, p. 554):

$$(16.3 \times 10^{15} \text{ btu})(1054 \text{ J/btu}) = 1.71 \times 10^{19} \text{ J/yr}$$

[24] Natural-gas consumption, 1983: 17.4 btu (U.S., 1984, p. 554):

$$(17.4 \times 10^{15} \text{ btu})(1054 \text{ J/btu}) = 1.83 \times 10^{19} \text{ J/yr}$$

[25] Uranium production, 1983: 6.51×10^{18} J (EIA/DOE, 1984a, p. 195).
[26] Electricity, total, 1983: 2.31×10^{12} kWh (EIA/DOE, 1984b, p. 22):

$$(2.31 \times 10^{12} \text{ kWh/yr})(3.606 \times 10^{6} \text{ J/kWh}) = 8.32 \times 10^{18} \text{ J/yr}$$

[27] Electricity, nuclear, 1983: 2.94×10^{11} kWh (EIA/DOE, 1984b, p. 22):

$$(2.94 \times 10^{11} \text{ kWh})(3.606 \times 10^{6} \text{ J/kWh}) = 1.06 \times 10^{18} \text{ J/yr}$$

[28] Phosphate fertilizer production, 1983: 8.6×10^6 Tn (metric tons), 33% phosphorus (U.S., 1984):

$$(8.6 \times 10^{6} \text{ Tn})(0.33) = 2.84 \times 10^{6} \text{ Tn}$$

[29] Phosphate rock production, 1983: 46.9×10^6 Tn, of which 14.4×10^6 Tn is P_2O_5. (U.S., 1984, p. 711); 66/80, 0.8 is P in P_2O_5:

$$(14.4 \times 10^{6} \text{ Tn})(0.8) = 11.5 \times 10^{6} \text{ Tn/yr}$$

[30] Bauxite production, 1983: 0.8×10^6 Tn (U.S., 1984, p. 716).
[31] Bauxite consumption, 1983: 8.8×10^6 Tn (U.S., 1984, p. 716).

Source: Odum et al., 1987a. Reprinted with permission of L.B.J. School of Public Affairs, The Univ. of Texas, Austin.

TABLE 10.3. EMERGY of Exports and Other Flows of the United States in 1983*

Note	Item	Raw Units	Transformity (sej/unit)	Solar EMERGY ($\times 10^{22}$ sej/yr)	Em$† ($\times 10^9$ 1983 U.S.$)
			Exports		
1	Oil	3.76×10^{17} J	53,000/J	1.99	8.30
2	Petroleum products	1.32×10^{18} J	66,000/J	8.71	36.3
3	Coal	2.48×10^{18} J	39,800/J	9.87	41.12
4	Coal	1.9×10^{17} J	68,000/J	1.29	5.38
5	Iron and steel products	1.4×10^6 Ton	1.78×10^{15}/Ton	0.24	1.03
6	Phosphate rock	3.46×10^6 Ton	1.41×10^{16}/Ton	4.89	20.38
7	Phosphate fertilizer	2.18×10^6 Ton	2×10^{16}/Ton	4.36	18.16
8	Wood	1.4×10^{18} J	34,900	4.88	20.35
9	Nitrogen products	6.75×10^6 Ton	4.19×10^{15}/Ton	2.82	11.78
10	Goods and services	3.32×10^{11} $	2.4×10^{12}/$	79.68	332.
			Dollar Flows		
11	Gross national product	3305×10^9 $	2.4×10^{12}/$	793.2	3305.
12	U.S. assets abroad	4.95×10^{10} $	2.4×10^{12}/$	11.88	49.5
13	Foreign assets in U.S.	8.17×10^{10} $	2.4×10^{12}/$	19.61	81.7

* U.S. area: 9.4×10^{12} m^2.

†Em$ flow: EMERGY flow in column 5 divided by 2.4 sej/$ for U.S. in 1983.

[1] Crude-oil exports, 1983: (164×10^3 bbl/day)(365.24) = 5.99×10^7 bbl (EIA/DOE, June 1985b, pp. 2–3):

$$(5.99 \times 10^7 \text{ bbl})(6.289 \text{ J/bbl}) = 3.76 \times 10^{17} \text{ J/yr}$$

$$(5.99 \times 10^7 \text{ bbl})(28.99 \text{ \$/bbl}) = \$1.7 \times 10^9$$

[2] Petroleum products exports, 1983: 5.75 bbl/day)(365.24 days = 2.10×10^8 bbl (EIA/DOE, 1984c, p. 17):

$$(2.1 \times 10^8 \text{ bbl})(6.28 \times 10^9 \text{ J/bbl}) = 1.32 \times 10^{18} \text{ J/yr}$$

[3] Coal exports, 1983: 77.8×10^6 st (short tons)(EIA/DOE, 1985c, pp. 4–9):

$$(77.8 \times 10^6 \text{ st})(3.18 \times 10^{10} \text{ J/st}) = 2.48 \times 10^{18} \text{ J/yr}$$

$$(77.8 \times 10^6 \text{ Ton}(35.50 \text{ \$/Ton}) = 2.76 \times 10^9 \text{ \$/yr}$$

[4] Corn exports, 1982: 49×10^6 ton (U.S. 1984, p. 658); 1983, \$6.5 $\times 10^9$ (U.S., 1984, p. 820):

$$(49 \times 10^6 \text{ Ton})(1 \times 10^6 \text{ g/Ton})(0.92 \text{ kcal/g})(4186 \text{ J/kcal}) = 1.9 \times 10^{17} \text{ J/yr.}$$

[5] Iron-and-steel-products exports, 1983: 1.4×10^6 Ton, \$1.6 $\times 10^9$ (U.S., 1984, p. 702):
[6] Phosphate-rock exports, 1983: 4.2×10^6 Ton P_2O_5 (U.S., 1984, p. 711); 0.8 P in P_2O_5; \$0.327 $\times 10^9$:

$$(6.2 \times 10^6 \text{ Ton})(0.8) = 3.47 \times 10^6 \text{ Ton of phosphate}$$

[7] Phosphate fertilizer exports, 1983: 6.6×10^6 Ton, \$89 $\times 10^6$ (U.S., 1984, p. 702); 33% P:

$$(6.6 \times 10^6 \text{ Ton})(0.33) = 2.18 \times 10^6 \text{ Ton/yr}$$

[8] Wood exports, 1983: 4700×10^6 ft^3; wood, rough and shaped, \$2.88 $\times 10^9$ (U.S., 1984, p. 820):

$$(4700 \times 10^6 \text{ ft}^3)(2.7 \times 10^{-2} \text{ m}^3/\text{ft}^3)(0.7 \times 10^6 \text{ g/m}^3)(3.8 \text{ kcal/g})(4186 \text{ J/kcal}) = 1.4 \times 10^8 \text{ J}$$

[9] Nitrogen-products exports, 1983, 7.5×10^6 st (short tons); 1982: \$1.05 $\times 10^9$ (U.S., 1984, p. 702):

$$(0.9 \times 10^6 \text{ g/st})(7.5 \times 10^6 \text{st}) = 6.75 \times 10^{12} \text{ g/yr}$$

$$(2.1 \times 10^8 \text{ bbl})(28.99 \text{ \$/bbl}) = 6.1 \times 10^9 \text{ \$/yr}$$

[10] Goods-and-services exports, 1983: \$332 $\times 10^9$ (U.S., 1984, p. 800).
[11] Gross national product, 1983: 3305×10^9 (U.S., 1984, p. 428).
[12] U.S. assets abroad, net (money out of U.S.), 1983: \$49.5 $\times 10^9$ (U.S., 1984, p. 801).
[13] Foreign assets in the U.S., net (money into U.S.), 1983: 81.7×10^9 (U.S., 1984, 801).

Source: Odum et al., 1987a. Reprinted with permission of L.B.J. School of Public Affairs, The Univ. of Texas, Austin.

TABLE 10.4. EMERGY Storages of the United States in 1983*

Note	Item	Raw Units	Transformity (sej/unit)	Solar EMERGY ($\times 10^{22}$ sej)	Em$ ($\times 10^9$ 1983 U.S.$†)
1	Phosphate	1.8×10^{14} g	1.4×10^{10}/g	253.8	1,057
2	Coal	1.17×10^{22} J	39,800/J	46,566.	194,025
3	Natural gas	2.17×10^{20} J	48,000/J	1,041.6	4,340
4	Uranium	2.95×10^{20} J	1,790/J	52.80	220
5	Petroleum	1.71×10^{20} J	53,000/J	906.3	3,776
6	Topsoil	7.35×10^{20} J	63,000/J	4,630.5	19,293
7	Wood biomass	4.72×10^{19} J	34,900/J	164.72	686
8	Groundwater	1.88×10^{20} J	41,000	770.8	3,211
9	Economic assets	6.6×10^{13} $	2.4×10^{12} $	15,840.	66,000
10	Population (person-years)	7.25×10^9 per-yr	3.1×10^{16}/per-yr	22,475.	93,645

* U.S. area: 9.4×10^{12} m^2.

† Em$: EMERGY flow in column 5 divided by 2.4×10^{12} sej/$ for U.S. in 1983.

[1] Phosphate rock: 1.8×10^9 Ton (U.S., 1982); 10% P:

$$(1.8 \times 10^9 \text{ Ton})(0.1) = 0.18 \times 10^9 \text{ Ton}$$

[2] Coal: 3.7×10^{11} Ton; brown coal and lignite, 0.3×10^{11} Ton (U.S., 1981):

$$(7 \times 10^6 \text{ kcal/Ton}(4186 \text{ J/kcal})(4.0 \times 10^{11} \text{ Ton}) = 1.18 \times 10^{22} \text{ J}$$

[3] Natural gas: 5.7×10^{12} m^3 at 9077 kcal/m^3 (U.N., 1981):

$$(5.7 \times 10^{12} \text{ m}^3)(9077 \text{ kcal/m}^3)(4186 \text{ J/kcal}) = 2.17 \times 10^{20} \text{ J}$$

[4] Uranium: 5.3×10^5 Ton (U.N., 1981):

$$(5.3 \times 10^5 \text{ t})(0.007)(1 \times 10^6 \text{ g/Ton})(7.95 \times 10^{10} \text{ J/g } ^{235}\text{U}) = 2.95 \times 10^{20} \text{ J}$$

[5] Crude petroleum: 3.8×10^9 Ton (U.N., 1981); 45×10^9 J/Ton (Slesser, 1978):

$$(3.8 \times 10^9 \text{ ton})(45 \times 10^9 \text{ J/Ton}) = 1.71 \times 10^{20} \text{ J}$$

[6] Topsoil: farm area, 4.22×10^{12} m^2, 11.2 Ton org./Acre (Brady, 1974); forest and miscellaneous area, 4.79×10^{12} m^2, 17.5 Ton/Acre
Organic matter:

$$(4.22 \times 10^{12} \text{ m}^2)(11.4 \text{ Ton/A})(1 \times 10^6 \text{ g/t}) + (4.79 \times 10^{12} \text{ m}^2)(17.5 \text{ Ton/A})(1 \times 10^6 \text{ g/Ton}) = 3.25 \times 10^{16} \text{ g}$$

$$(3.25 \times 10^{16} \text{ g})(5.4 \text{ kcal/g})(4186 \text{ J/kcal}) = 7.35 \times 10^{20} \text{ J}$$

[7] Wood biomass (U.S., 1982).
[8] Groundwater: (U.S., 1982):

$$(9.4 \times 10^{12} \text{ m}^2)(0.05 \text{ porosity})(100 \text{ m})(1 \times 10^6 \text{ g/m}^3)(4 \text{ J/g}) = 1.88 \times 10^{20} \text{ J}$$

[9] Economic assets, 1983: gross national product × 20 (5%/yr depreciation: GNP $3305 × 10^9$ (U.S., 1985):

$$(\$330 \times 10^9)(20) = \$6.6 \times 10^{13}$$

[10] Population in 1983: 234×10^6 people; median age 31 (U.S., 1985):

$$(234 \times 10^6 \text{ people})(31 \text{ yr}) = 7.25 \times 10^9 \text{ people-yr.}$$

Source: Odum et al., 1987a. Reprinted with permission of L.B.J. School of Public Affairs, The Univ. of Texas, Austin.

TABLE 10.5. Summary Flows for the United States, 1983 (see Figure 10.3)

Letter in Fig. 10.3	Item	Solar EMERGY ($\times 10^{22}$ sej/yr)	Dollars ($\times 10^9$ \$/yr)
R	Renewable sources used (rain, tide, etc.)	82.4	
N	Nonrenewable sources flow within the U.S.:	534.6	
	N_0 Dispersed rural source	97.6	
	N_1 Concentrated use	420.2	
	N_2 Exported without use	25.5	
F	Imported fuels and minerals	714.2	
G	Imported goods	6.5	
I	Dollars paid for imports		365.
P_2I	EMERGY value of goods and services imports	138.7	
I_3	Dollars paid for imports minus goods		301.
P_2I_3	Imported services	114.	
E	Dollars paid for exports		332.
P_1E	EMERGY value of goods and services exports	73.	
B	Exported products transformed within US	19.	
E_3	Dollars paid for exports minus goods		309.
P_1E_3	Exported services	68.	
X	Gross national product		3305.
P_2	World EMERGY/\$ ratio, used for imports	3.8×10^{12} sej/\$	
P_1	US EMERGY/\$ ratio, used for U.S. and exports	2.4×10^{12} sej/\$	

R. Renewable sources used: rain + tide. 82.4×10^{22} sej/yr (Table 10.2)

N. Nonrenewable sources: $N_0 + N_1 + N_2 = 534.6 \times 10^{22}$ sej/yr.

N_0. Dispersed rural sources (Table 10.2): earth loss (100×10^{22} sej) + soil formation (13×10^{22} sej) − wood consumption (15.39×10^{22} sej) = 97.6×10^{22} sej/yr.

N_1. Concentrated use, 1983 (Table 10.2): $\times 10^{22}$ sej/yr: Oil (185) + coal (914.2) + gas (88.8) + nuclear electricity (16.69) + hydroelectric (18.92) + phosphate fertilizer and rock (12.53) + iron ore (4.33) + bauxite (0.8 + 420.17).

N_2. Exported without use, 1983 (Table 10.3): $\times 10^{22}$ sej/yr: oil (10.7) + coal (9.87) + phosphate rock (4.89) = 25.5×10^{22} sej/yr. \$ in N2: oil (7.8) + coal (2.8) + *P* rock (0.33) = 10.9×10^9 \$/yr.

F. Imported minerals and fuels, 1983 (Table 10.2): $\times 10^{22}$ sej/yr: oil (66.5) + gas (4.8) + iron ore (1.1) + bauxite (0.6) = 73.0×10^{22} sej/yr. \$ in F, $\times 10^9$ \$: oil and products (52.2) + gas (4.3) + iron ore (0.45) + bauxite (0.22) = 58.17×10^9 \$/yr.

G. Imported goods, 1983: iron and steel.

I. Dollars paid for imports, 1983: 365×10^9 (Table 10.2).

P_2I. EMERGY value of goods-and-services imports:

$$(365 \times 10^9 \text{ \$/yr})(3.8 \times 10^{12} \text{ sej/\$}) = 138.7 \times 10^{22} \text{ sej/yr}.$$

received for export E that connects with the gross economic product circle X. The total EMERGY of services exported is the product of the exports expressed in dollars E times the EMERGY/$ ratio P_1 of the exporting country (P_1E). E_3 is for dollars received for exports other than that for fuels and high-EMERGY goods sold. The EMERGY exported in exchange for sales E_3 is P_1E_3.

Published examples of evaluated summary diagrams for countries include 11 countries (H. T. Odum and Odum, 1983); Switzerland (Pillet and Odum 1984), Taiwan (Huang and Odum, 1991), and Italy (Ulgiati et al., 1994). Examples of state evaluations are Texas (H. T. Odum et al., 1987a), Florida (H. T. Odum et al., 1993), and Alaska (Woithe, 1992).

ANNUAL EMERGY FLOWS OF THE UNITED STATES

After an energy systems diagram was drawn for the United States (Figure 10.2), evaluation tables were developed for flows (Table 10.2 and 10.3) and

---

I_3. Dollars paid for imports minus dollars in goods (G) and minerals and fuels (F), 1983:

$$365 - 64 = \$301 \times 10^9.$$

P_2I_3. Imported services minus those in F and G, 1983:

$$(3.8 \times 10^{12} \text{ sej/\$})(\$301 \times 10^9) = 114 \times 10^{22} \text{ sej/yr}$$

E. Dollars paid for exports, 1983: 332×10^9 $/yr (Table 10.3).

P_1E. EMERGY value of goods-and-services exports:

$$(332 \times 10^9 \text{ \$/yr})(2.2 \times 10^{12} \text{ sej/\$}) = 73 \times 10^{22} \text{ sej/yr}$$

B, Exported products transformed within the country, 1983 (Table 10.3), $\times 10^{22}$ sej/yr: phosphate fertilizer (4.4) + corn (1.3) + iron and steel products (0.25) + nitrogen products (2.8) + wood (4.9) = 13.7×10^{22} sej/yr. $ in B: $\times 10^9$ $: fertilizer (0.9) + corn (6.5) + iron and steel (1.6) + nitrogen (1.05) + wood (2.88) = 12.93×10^9 $/yr.

E_3. Dollars paid for exports minus dollars in goods (B) and raw exports (N2); 1983:

$$332 \times 10^9 \text{ \$} - 23 \times 10^9 \text{ \$/yr.} = 309 \times 10^9 \text{ \$/yr.}$$

P_1E_3. Exported services, 1983:

$$(2.2 \times 10^{12} \text{ sej/\$})(309 \times 10^9 \text{ \$/yr}) = 68.0 \times 10^{22} \text{ sej/yr.}$$

X. Gross national product, 1983: 3305×10^9 $/yr.

P_2. EMERGY/$ of imports, using 3.8×10^{12} sej/$.

P_1. EMERGY/$ ratio of U.S. and its exports, 1983: 2.4×10^{12} sej/$.

Source: Odum et al., 1987a. Reprinted with permission of L.B.J. School of Public Affairs, The Univ. of Texas, Austin.

storages (Table 10.4). Summary Table 10.5 was used to prepare a summary EMERGY diagram (Figure 10.3). Then national indices were calculated (Table 10.6).

Any resource flow that has EMERGY not yet used requires two items of data: (1) the weight or energy content is required to evaluate the EMERGY of

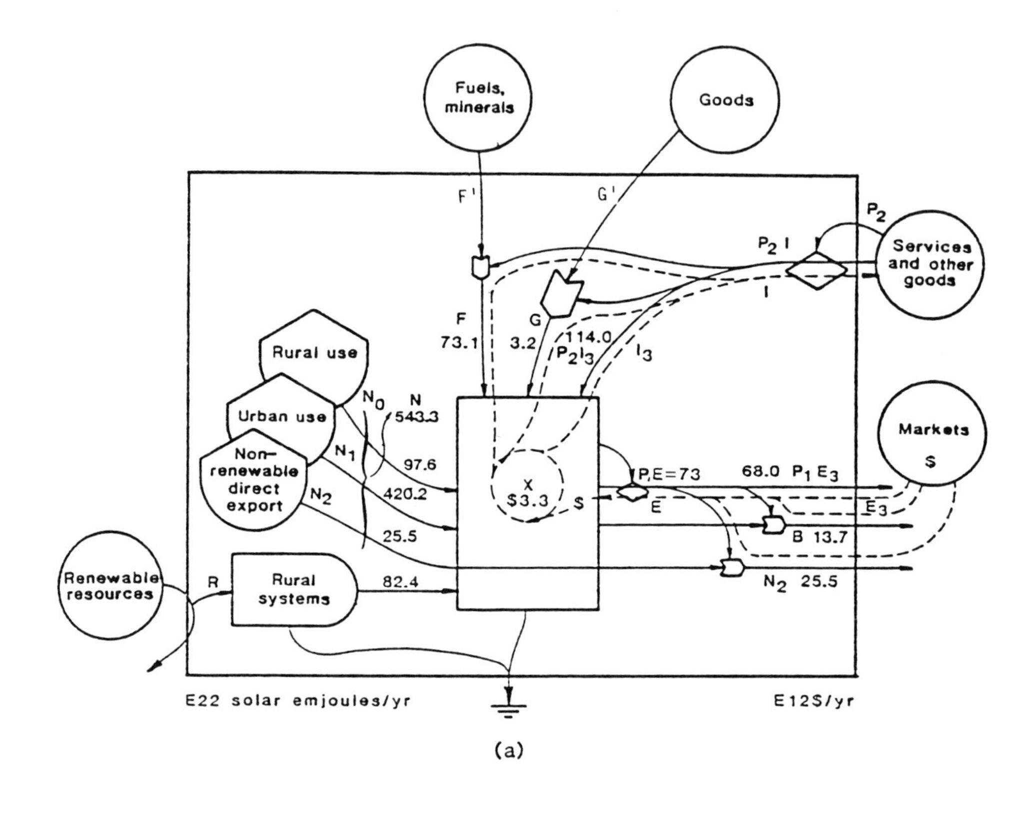

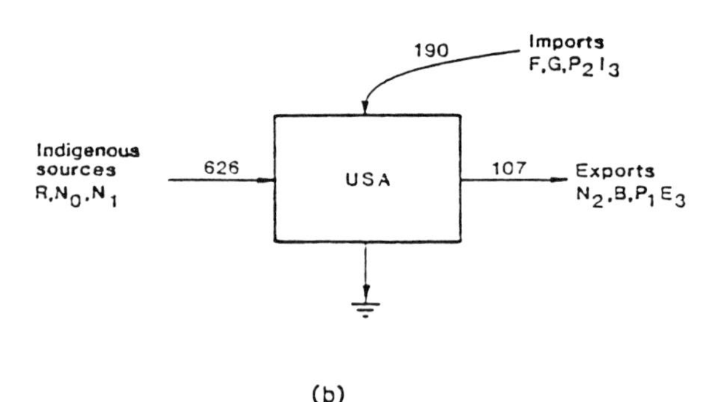

Figure 10.3. Summary diagram of EMERGY flows of the United States in 1983. Flows are in solar EMERGY units. Letters refer to flows in Table 10.5 (Odum et al., 1987a). Reprinted with permission of L.B.J. School of Public Affairs, The Univ. of Texas, Austin.

TABLE 10.6. Indices Using EMERGY for Overview of the United States

Item	Name of Index	Expression	Quantity
1	Renewable EMERGY flow	R	82.4×10^{22} sej/yr
2	Indigenous nonrenewable reserves	N	534.6×10^{22} sej/yr
3	Flow of imported EMERGY	$F + G + P_2I_3$	193.6×10^{22} sej/yr
4	Total EMERGY inflows	$R + N + F + G + P_2I_3$	810.6×10^{22} sej/yr
5	Total EMERGY used, U	$N_0 + N_1 + R + F + G + P_2I_3$	785.1×10^{22} sej/yr
6	Total exported EMERGY	$B + P_1E_3$	87.0×10^{22} sej/yr
7	Fraction of EMERGY used derived from home sources	$(N_0 + N_1 + R)/U$	0.76
8	Imports minus exports	$(F + G + P_2I_3) - (N_2 + B + P_1I_3)$	84.1×10^{22} sej/yr
9	Ratio of exports to imports	$(N_2 + B + P_1E_3)/(F + G + P_2I_3)$	0.57
10	Fraction used, locally renewable	R/U	0.10
11	Fraction of use purchased	$(F + G + P_2I_3)/U$	0.25
12	Fraction used, imported service	P_2I/U	0.18
13	Fraction of use that is free	$(R + N_0)/U$	0.22
14	Ratio of concentrated to rural	$(F + G + P_2I_3 + N_1)/(R + N_0)$	3.4
15	Use per unit area (9.4×10^{12} m^2)	U/(area)	8.4×10^{11} sej/m^2
16	Use per person (population = 234×10^6)	U/(population)	3.4×10^{16} sej/person
17	Renewable carrying capacity at present living standard	(R/U)(population)	23.4×10^6 people
18	Developed carrying capacity at same living standard	$8(R/U)$(population)	187.2×10^6 people
19	Ratio of use to GNP, EMERGY/$ ratio (GNP: 3.3×10^{12})	$P_1 = U$/(GNP)	2.4×10^{12} sej/$
20	Ratio of electricity to use (Elec: 132.3×10^{22} sej/yr)	(elec)/U	0.17
21	Fuel use per person (3.43×10^{24} sej)	(fuel)/(population)	1.5×10^{16} sej/person

Source: Odum et al., 1987a. Reprinted with permission of L.B.J. School of Public Affairs, The Univ. of Texas, Austin.

available energy and (2) the money spent for the EMERGY of associated services. For example, in the summary diagram (Figure 10.3), notice that imported fuels and goods have two EMERGY values: one that includes its services (F and G), and the other with the services subtracted (F' and G'). Therefore, evaluation of imports can be made by adding F', G', and total services (P_2I), or evaluation can be made by adding F, G, and other services (P_2I_3). Since the service components of fuels and goods is usually small, there is usually little difference between F and F' or G and G'. Similarly, total services exported are P_1E; whereas those services other than those with goods and mineral-fuels are P_1E_3. Exported goods B and fuels and minerals N_2 each have a component of their EMERGY derived from services.

Further Aggregation and the EMERGY/Money Ratio

As shown in Figure 10.3b, the system in Figure 10.3a is aggregated further into three flows: indigenous resources used, imports and exports. An EMERGY/$ ratio is calculated by summing the indigenous resources used plus imports used and then dividing by the gross economic product expressed in international U.S. dollars according to the money exchange market for the time. Item U (line item 5 in Table 10.6 was divided by the gross economic product X to obtain the solar EMERGY/money ratio P_1. Exports are not subtracted, since only EMERGY that is used within the country contributes to the national empower. Figure 4.2 is another style of diagram summarizing national flows. The U.S. value in Table 10.7 is for 1980. (See Appendix D.)

Money Flows across the Boundary

In addition to the trade exchanges in Figure 10.1, an accounting of all money flows across the boundaries is required, which may be supplied in economic statistical abstracts. These may be added to the overview diagram with dashed lines. Included should be dollars inflowing or outflowing with tourists and retirees, dollars exchanging due to immigrants and emigrants and their transactions, dollars in loans and investments from outside or from inside, plus their repayments and interest, foreign-aid programs, payments to support military operations, purchases of businesses, and money shifts by multinational companies such as profit earnings.

Use of Money Data to Evaluate Incoming Goods and Services

To evaluate the EMERGY inflow in the form of services coming in from across the boundary, we may use money data on the particular state or nation that is the source of the services. The data on imports expressed as international dollars spent during a particular year are multiplied by the EMERGY/money ratio (in sej/$) of the state or nation where the services

TABLE 10.7. National Activity and EMERGY/Money Ratio

Nation	EMERGY Used Per Year* ($\times 10^{20}$ sej/yr)	Gross National Product† ($\times 10^{9}$ \$/yr)	EMERGY/Money Ratio ($\times 10^{12}$ sej/\$)
Liberia	465	1.34	34.5
Dominica	7	0.075	14.9
Ecuador	964	11.1	8.7
China	71,900	376	8.7
Brazil	17,820	214	8.4
India	6,750	106	6.4
Australia	8,850	139	6.4
Poland	3,305	54.9	6.0
Soviet Union	43,150	1,300	3.4
USA	83,200	2,600	3.2
New Zealand	791	26	3.0
West Germany	17,500	715	2.5
Netherlands	3,702	16.5	2.2
World	23,200	11,600	2.0
Spain	2,090	139	1.6
Japan	15,300	715	1.5
Switzerland	733	102	0.7

* Calculated with procedure in this chapter.
† U.S. international dollars.
Source: Data for 1980–1987. See footnotes in Table 9.1.

were purchased (Table 10.7). For example, the EMERGY inflow represented by services purchased from Brazil in 1980, expressed in dollars, can be calculated by multiplying that dollar amount by 8.4×10^{12} sej/\$.

Since statistics are not often readily available on sales and purchases across boundaries of regions and states, some estimate may be obtained from Figure 5.2.

For analysis of states and regions within a country, there are also taxes, and various payments from the central government (transfer payments) to be included. These include pensions, social security, government aid programs, federal payrolls, and so on. There are also investments, payments of interest, repayments of loans, and profits to outside owners crossing state boundaries. There may also be an "underground" economy; especially where illegal drugs are sold, the funds crossing boundaries illegally are large and hard to evaluate.

National Parameters and Indices

Table 10.5 identifies each flow in the summary diagram. Each of these was calculated by summing data in the initial EMERGY analysis tables (Tables 10.2

and 10.3). Using the items in Table 10.5, various national parameters and indices are calculated, which may be useful for insights and comparisons. These are defined in Table 10.6, which includes the formula for each, and an example of each for the United States in 1983.

ANALYSIS OF IMPORTANT SUBSYSTEMS

The distinctive resource signature of each country causes the subsystems that utilize the most prominent resources to be special for that area. In the diagramming and evaluating process it becomes apparent which special economic subsystems are important. For example, for Liberia, iron mines, rubber plantations, and rainforest woods are important. It is appropriate to perform EMERGY evaluations on the dominant subsystems as explained in Chapter 5, especially if there are controversial issues about these operations and their transformities.

COMPARISONS WITH OTHER STATES AND NATIONS

After the indices in Table 10.6 are calculated, each may be compared with those of other countries to help understand the place that nation or state holds among the various kinds of economies of the world.

EMERGY/Money Ratio

The total annual EMERGY use by a nation measures its annual wealth. In different countries, money buys different amounts of real wealth even when the currencies are compared on a current international exchange basis. The amount of real wealth that circulating money buys is indicated by the EMERGY/money ratio. Table 10.7 gives the solar EMERGY budgets of several nations and the gross economic products expressed in U.S. dollars according to the money exchange rates current at the time. The last column is the solar EMERGY/money ratio for each country. An average EMERGY/money ratio for the world was estimated by dividing the total EMERGY flow of the earth (including renewable and non-renewable uses) by the world gross economic product in U.S. dollars.

The global annual EMERGY flow for the geobiosphere (9.44 E 24 sej/yr) estimated in Chapter 3 (Figure 3.2) was for the long-range average. For evaluating the EMERGY resources supporting the world economy in the short-run present (Figure 10.5 and Tables 10.7 and 10.8), the EMERGY of the intense global use of fuel reserves was included (about 13.8 E24 sej/yr). Figure 10.6 shows the estimation of global EMERGY/money ratio.

Rural countries have higher EMERGY/$ ratios because more of the wealth goes directly from the environment to the human consumer without money being paid. For example, a family in a remote rainforest gets most of its food, clothing, shelter, recreation, and so forth directly from nature without money being involved.

As economic development and urban concentration increase, more and more money is circulated for the same EMERGY use, and the EMERGY/money ratio decreases. For example, note the decreasing ratio with development in the United States, Taiwan, and the world (Figures 10.4 and 10.5). The more the economic urban centers develop, the lower the EMERGY/money ratio. A low ratio suggests a high position in the economic hierarchy (toward the bottom of Table 10.7).

EMERGY per Person

In Table 10.8 annual EMERGY use per person is calculated for several countries. A high EMERGY/person ratio suggests a high standard of living, given in more general terms than income, which does not include the unpaid, direct wealth to people from the environment or from public information. A person living a subsistence life in a rural setting may have a higher EMERGY than a person with more income who buys most of their needs in a city. Where the inputs were of a similar kind, Ballentine (1976) found a good correlation between the EMERGY/money ratio and the consumer price index.

Percent Electric Power

Electric power uses by different countries are compared in Table 10.9. Developed countries have a high proportion of their annual EMERGY going into electric power, especially those with cheap hydroelectric resources. Countries with a high proportion of electrical EMERGY tend to have high levels of technology. Electric power is one of the principal ways that basic resources are transformed into technology and information, concentrating EMERGY in the hierarchical centers of society, which includes the headquarters of government, finance, and industry, often in the center of cities. Countries with renewable sources of hydroelectric power may have an advantage in being information centers.

Economic/Environmental Ratio (Average Investment Ratio)

Table 9.1 compares the economic/environmental ratios for several countries. This is the quotient of the EMERGY brought into a local area by purchase divided by the EMERGY supplied free by the local environment. A high ratio suggests a fully developed economy, a high level of environmental stress, and

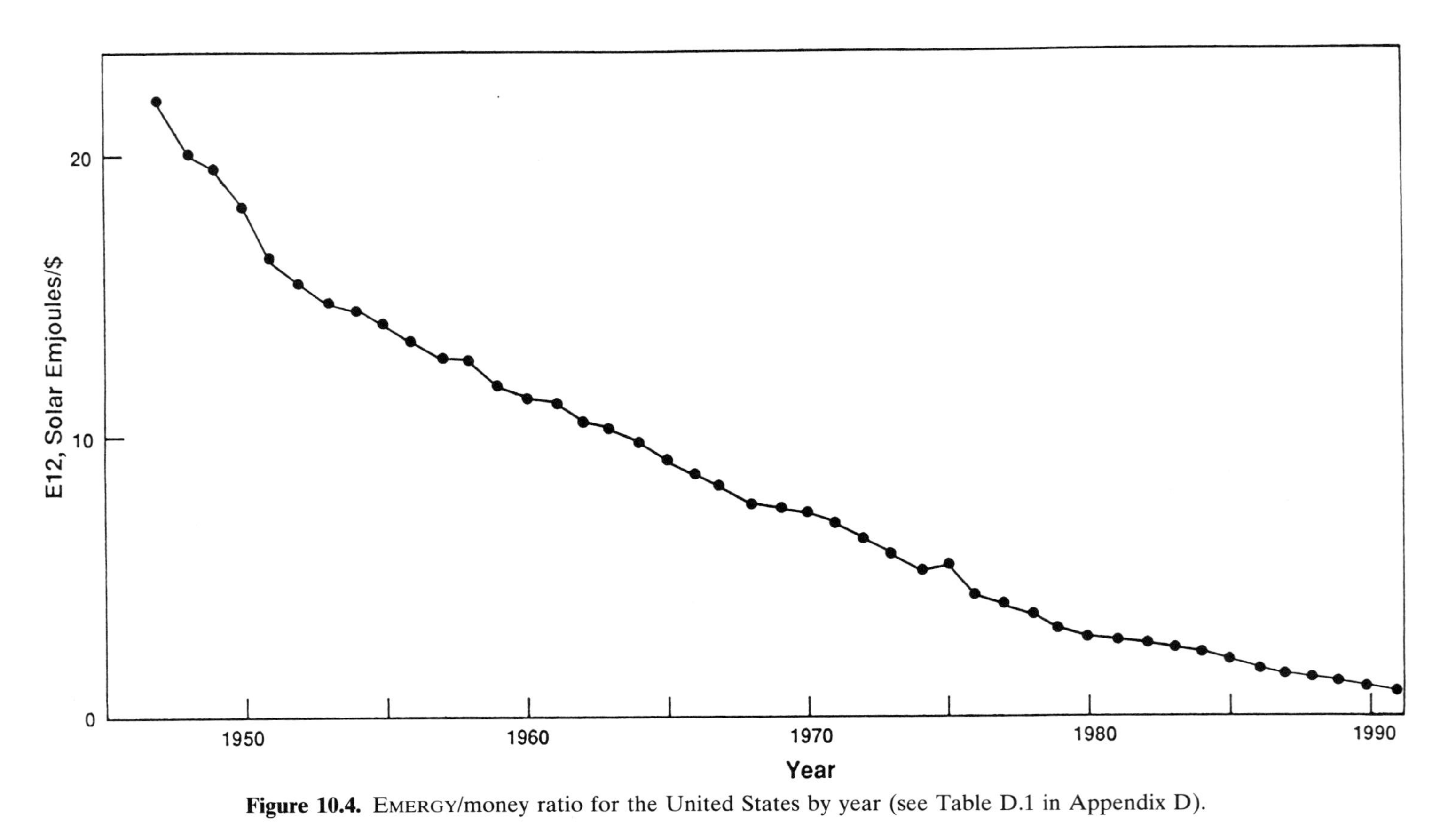

Figure 10.4. EMERGY/money ratio for the United States by year (see Table D.1 in Appendix D).

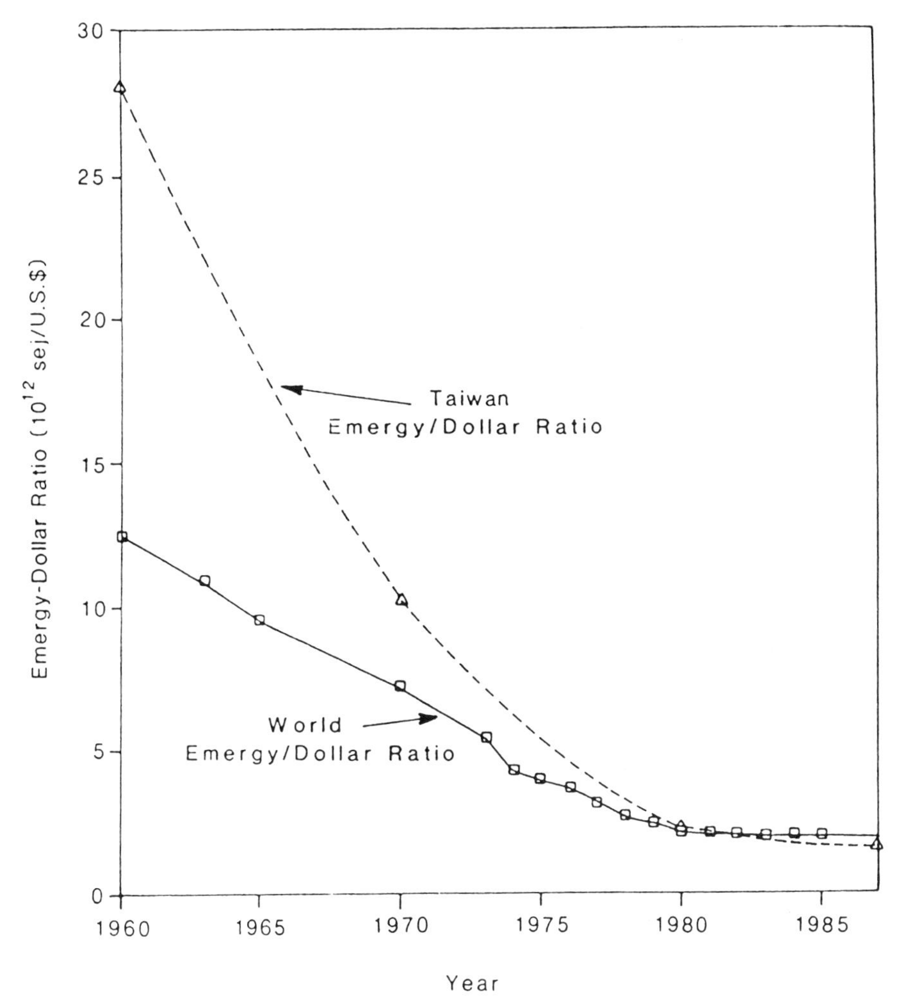

Figure 10.5. EMERGY/money ratios for Taiwan and the world (Huang and Odum, 1991). Reprinted with permission of Academic Press.

difficulty for environment-using industries (agriculture, forestry, and fisheries) competing with less developed areas.

EMERGY Signature

Each state of a country has a characteristically different set of EMERGY flows and storages that constitute its EMERGY signature (see Chapter 7, Figure 7.1). Both economic and environmental self-organization tend to reinforce those transformations that make use of the dominant EMERGY sources available. The occupations and culture that develop are molded in part by the resource use activity that predominates because of the EMERGY signature.

TABLE 10.8. Share per Person of Annual Solar EMERGY Use

State	EMERGY Used ($\times 10^{20}$ sej/yr)	Population ($\times 10^{6}$)	Per Capita EMERGY Use* ($\times 10^{15}$ sej/person/yr)
Australia	8,850	15	59
USA	66,400	227	29
West Germany	17,500	62	28
Netherlands	3,702	14	26
New Zealand	791	3.1	26
Liberia	465	1.3	26
Soviet Union	43,150	260	16
Brazil	17,820	121	15
Dominica	7	.08	13
Japan	15,300	121	12
Switzerland	733	6.4	12
Ecuador	1,029	9.6	11
Poland	3,305	34.5	10
Taiwan	1,340	17.8	8
China	71,900	1100	7
Spain	2,090	134	6
World	202,400	5000	4
India	67,500	630	1

Source: Data from 1980–1987. See footnotes in Table 9.1.

TABLE 10.9. Proportion of National Empower That Is Electric Power

Nation	% Electrical
Switzerland	32
Sweden	23
Spain*	22
Taiwan	20
USA	20
Soviet Union*	19
Poland*	18
New Zealand*	15
Italy	14
India*	10
Germany*	10
Netherlands*	10
Brazil	8
Australia*	7
China	4.3
Liberia*	1
Dominica*	0.01

* Older evaluations, before 1983.

Sources: See footnotes in Table 9.1.

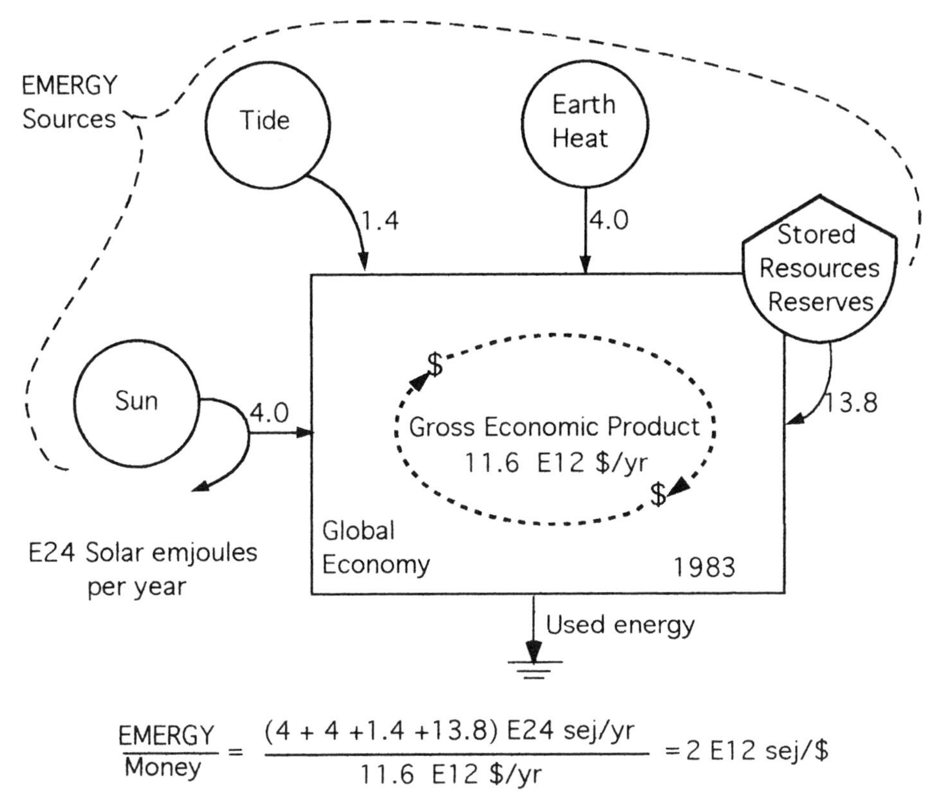

Figure 10.6. EMERGY inputs used to evaluate the EMERGY/money ratio of the world economy for 1983. See Figure 10.5.

SUMMARY

By evaluating EMERGY flows and storages, the stages of development, resource bases, needs, and past histories of states and nations can be understood and policies can be recommended for the future. Comparative studies show countries distributed in a global EMERGY hierarchy ranging from rural areas supplying resources to highly developed, fiscal, and information centers. EMERGY indices show great inequity in EMERGY use per person, uses of electric power, and buying power of their currency.

CHAPTER 11

EVALUATING INTERNATIONAL EXCHANGE

The global system of humanity and nature is increasingly unified by international trade, treaties, finance, military alliances, and shared information. In this chapter we place the window of systems attention on the relationships of nations, with EMERGY accounting of the benefits and losses of some international exchanges. Perhaps developing equity in shared information and international EMERGY exchange can increase global benefits and world peace.

Figure 11.1b is an energy systems diagram of the exchanges between two nations. These exchanges include the flows of trade goods and services, money, transactions by money traders, loans and payments, foreign aids, military services, information and technology transfer, investments, and so on. All of these are flows of EMERGY. The human services and information have higher transformity than most of the raw products and manufactured goods. Benefit to a nation is what is bought with money and well used, not the money itself. Furthermore, many important exchanges, such as information and population exchange, are not usually evaluated monetarily. Hence, EMERGY is the appropriate measure for evaluating benefits to nations from all types of international exchange.

The benefit from a foreign sale, purchase, or trade depends on the EMERGY exchange ratio of each trade (Figure 4.4). As shown in Figure 11.2, there are four pathways involved in exports and imports, two in each direction (two flows of the commodities and services, and two flows of money paid). An EMERGY table should be prepared evaluating the commodities purchased and sold and the monies paid and received.

The general energy hierarchy found in all of nature and in the economies of humanity is observed in the relationships of nations. Some are rural, and

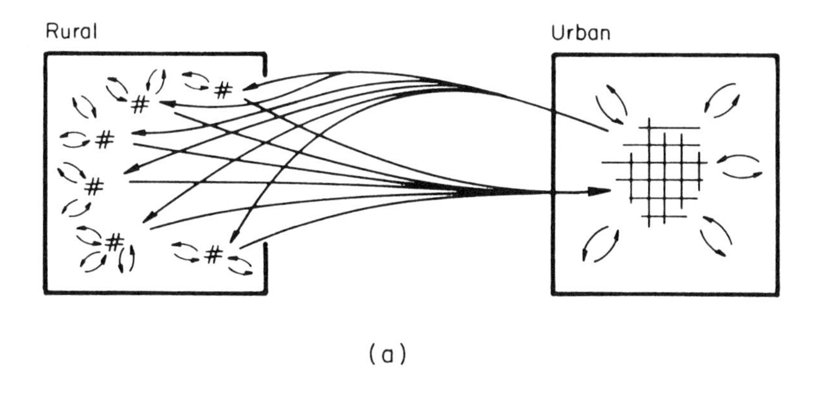

(a)

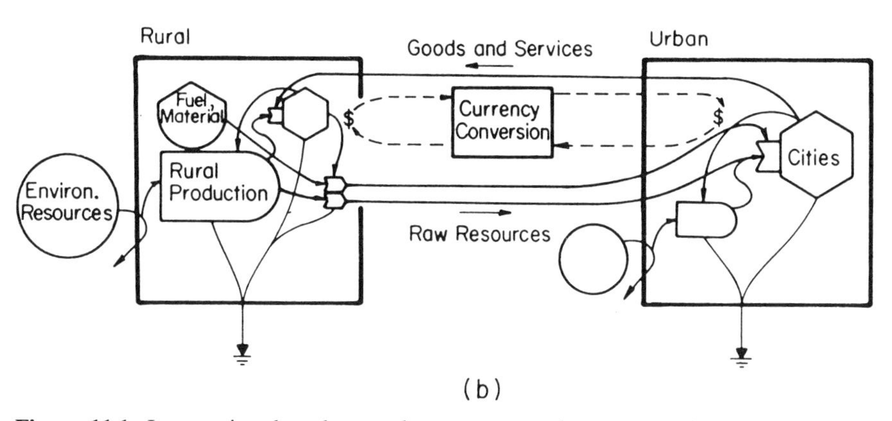

(b)

Figure 11.1. International exchanges between a rural country and an urban nation. **(a)** Spatial convergence and divergence. **(b)** Energy systems diagram.

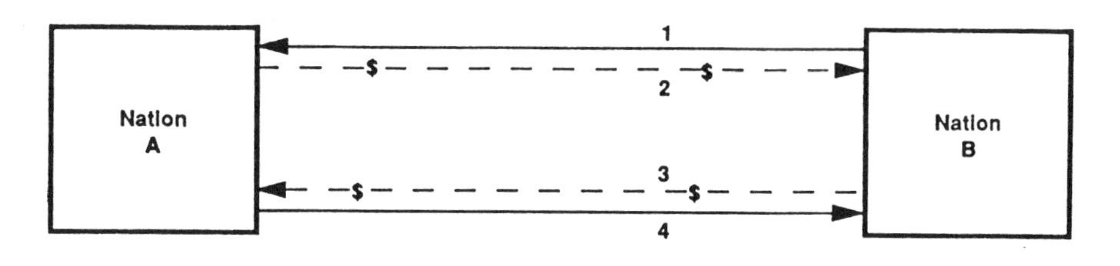

$$\text{Emergy Exchange Ratio} = \frac{\text{Emergy Received}}{\text{Emergy Sent}}$$

Figure 11.2. EMERGY exchange ratio for evaluating benefits of purchases and sales, foreign trade, or international loans. The ratio may be used for comparing 1 and 2, 3 and 4, 2 and 3, or all exchanges.

some are predominantly urban. Highly developed, predominantly urban countries are centers in the global hierarchy. Rural areas converge products to towns and receive services and materials in return. Small towns contribute to and receive from cities, and these to and from larger cities, and so on. Figure 11.1 shows exchange between a rural country on the left and an urban nation higher in energy hierarchy, on the right.

Earlier, with exchange considered in Figure 4.4 and Table 4.1, it was shown that much more EMERGY goes to the buyer of environmental commodities than is in the buying power of the money paid. Generally, a country loses wealth if it sells environmental raw products because the EMERGY of nature's work to make them is high, whereas the money received is only for some services to process them. Thus, developed nations tend to receive more EMERGY than they give in exchange.

On the larger scale of nations, the suppliers of raw commodities give to the buyer far more than they receive in exchange. Most of their exports are raw products that have a much higher yield to the buyer than is received by the seller. In contrast, the sales of finished, manufactured, high-technology, and military products have higher prices, so that the EMERGY of the money paid is more comparable with the EMERGY of the products sold.

EXCHANGE WITH UNEQUAL EMERGY/MONEY RATIOS

With international relationships the evaluation is more complicated because of the different EMERGY/money ratios of different countries. When converted to U.S. dollars, the currency of rural countries has higher EMERGY/$ ratios so that a dollar buys more real wealth than in urban countries. Table 10.7 has EMERGY/money ratios for several nations for comparison. Highly developed countries like West Germany, Holland, Japan, and Switzerland, have lower ratios than the less developed countries with more rural areas and unmonied, direct input of environmental resources to their people.

The inequity in foreign exchange is shown dramatically in Figure 11.3, where 1 international U.S. dollar is circulated between nations. In 1985 the buying power of the U.S. dollar (2.0×10^{12} sej/$) was about four times what an international dollar would buy in resource countries with 8.0×10^{12} sej/$. As shown on the left side of Figure 11.3, every dollar that circulates transfers four times more real wealth (EMERGY) to the United States than is received by the resource countries. Similarly, on the right, a dollar circulating between the United States and Japan (0.5×10^{12} sej/$) transfers 2 times more wealth to Japan than is received by the United States. Trade between the resource countries and Japan favored Japan with 8 times more wealth. As shown by the EMERGY/money ratios in Table 10.7, trade or financial transactions between Brazil, China, and Ecuador and the U.S. was very unequal.

When an environmental product is sold from a rural state to a more developed economy, there is a large net EMERGY benefit to the developed buyer

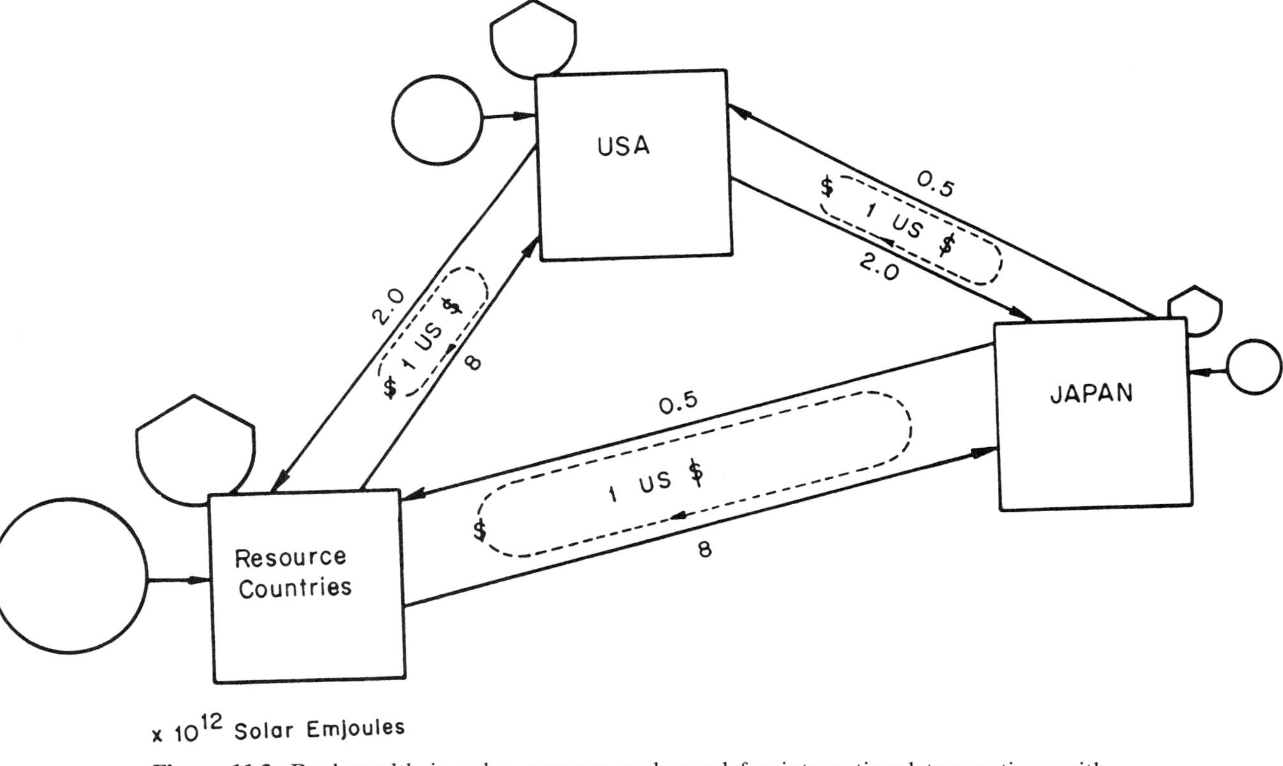

Figure 11.3. Real wealth in solar EMERGY exchanged for international transactions with international U.S. dollars.

for two reasons: (1) the EMERGY of environmental products is higher than that in the money paid for the processing services; and (2) the EMERGY/money ratio is much greater in the rural state supplying the product than in the purchasing economy. For example, the net benefit in selling Alaskan salmon to Japan sends 10 times more EMERGY to Japan than is received by Alaska (Brown et al., 1993). Alaska, in this case, has a high EMERGY/money ratio (23×10^{12} sej/$), 46 times that of Japan. Alaska in this relationship behaves like an undeveloped country.

Net EMERGY Yield of Imported Fuel

Policy on whether to use fuels from home sources or import from abroad can be determined by the net EMERGY yield ratios. If importing fuel contributes more net EMERGY, the foreign sources should be used until their ratios decline. The policy of using imported fuels first has the additional advantage of preserving home sources for emergency in case there are interruptions in foreign sources, as with war. Maintaining unused reserves at home while importing foreign fuels helps keep prices down.

The net EMERGY yield of importing a fuel is obtained by evaluating the EMERGY exchange ratio (Figure 4.4): the EMERGY of fuel received is divided by the EMERGY of buying power of the currency used to pay for the fuel (EMERGY/money ratios in Table 10.7). Different countries purchasing oil with international dollars get very different net EMERGY yields, because the EMERGY/money ratio of their currencies varies. Brazil, with nearly 4 times more EMERGY purchased with an international U.S. dollar, has a much lower net EMERGY yield ratio than the United States buying oil at the same price. The stimulus to the economy of foreign oil in 1980s was much greater in the United States than in Brazil, which may be one reason that alcohol for automobiles developed there.

Economic Development of a Cash-earning Export

Loans are often provided to underdeveloped countries to develop environmental resources for export so as to earn foreign exchange. These projects tend to drain the underdeveloped country (because environmental commodities benefit the purchaser more and because the EMERGY/money ratio of the underdeveloped country is higher than that in the country receiving the export).

For example, much of the mangrove estuaries of the coastal zone of Ecuador was converted to shrimp pond aquaculture, with most of the shrimp exported to the United States. The EMERGY characteristics of this aquaculture system (Odum and Arding, 1991) were given in Chapter 4 (Figure 4.8). A large dollar-earning export was substituted for the more subsistence-style economy that previously supported the people of the region. EMERGY evaluation of the shrimp export trade (Figure 11.4) showed 4.1 times more EMERGY going to the United States than received in goods and services by Ecuador. In other words, the coastal zone people were deprived of their EMERGY, the regional wealth was reduced to one fourth, and only those operating the aquaculture gained.

Aluminum and Transformed Electric Power

In elevated areas that have large hydropower potential and are far from industrial centers, hydroelectric power from dams has been used to make aluminum ingots, a process that uses much electric power (Figure 11.5a). The aluminum ingots are then sold to industrial centers as a way of transporting the original EMERGY of the hydropower. EMERGY evaluation shows uneven trade, with most of the benefit for the receiving country. For example, aluminum ingots made with New Zealand hydroelectric power were sold in Japan for $58 million, even though the contribution to the Japanese economy evaluated with EMERGY methods was close to a billion emdollars. For the benefit of New Zealand, the ingots should be kept at home and used to make final products for export, thus creating jobs in New Zealand.

In Figure 11.5b the hydroelectric power is used at home. This stimulates the whole economy by making everything that uses electricity or aluminum cheaper, making finished goods compete, and eventually getting much more from trade than from selling the raw resource below its real emdollar value. With prices of aluminum and electricity low, industrial plants with new jobs develop in New Zealand. Since increasing capital without inflation depends

Goods & Services to Ecuador:
(3.15 E8 $/yr)(2.4 E12 sej/US 1986 $) = 7.6 E20 sej/yr

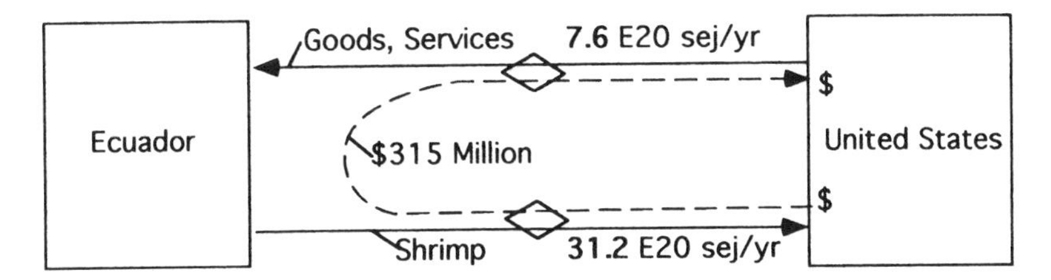

Shrimp from Ecuador:
(3 E10 g/yr)(0.2 dry)(6.7 Kcal/g)(4186 J/kcal)(18.2 E6 sej/J) = 31.2 E20 sej/yr

$$\text{Benefit Ratio: } \frac{\text{EMERGY to USA}}{\text{EMERGY to Ecuador}} = \frac{31.2}{7.6} = 4.1$$

Figure 11.4. Exchange of EMERGY and money with foreign sales of 30,000 t (metric tons) of shrimp from Ecuador.

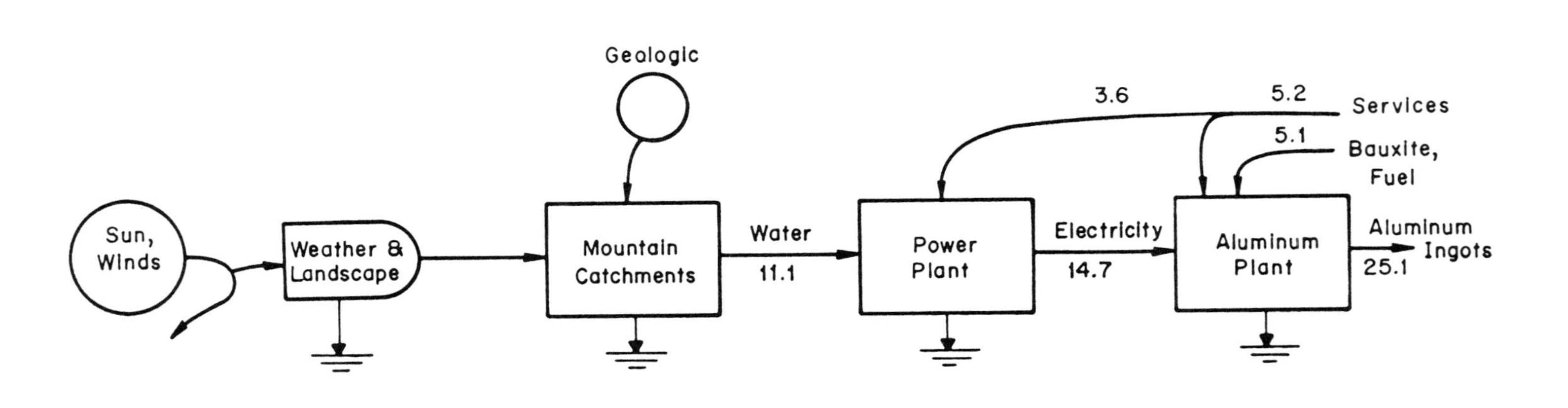

Figure 11.5. Environmental basis of hydroelectric power in New Zealand, and its use to manufacture aluminum for export or alternative uses (Data from H. T. Odum and E. L. Odum, 1983; H. T. Odum, 1984b). **(a)** System aggregated as energy transformation chain to show net EMERGY and exchange.

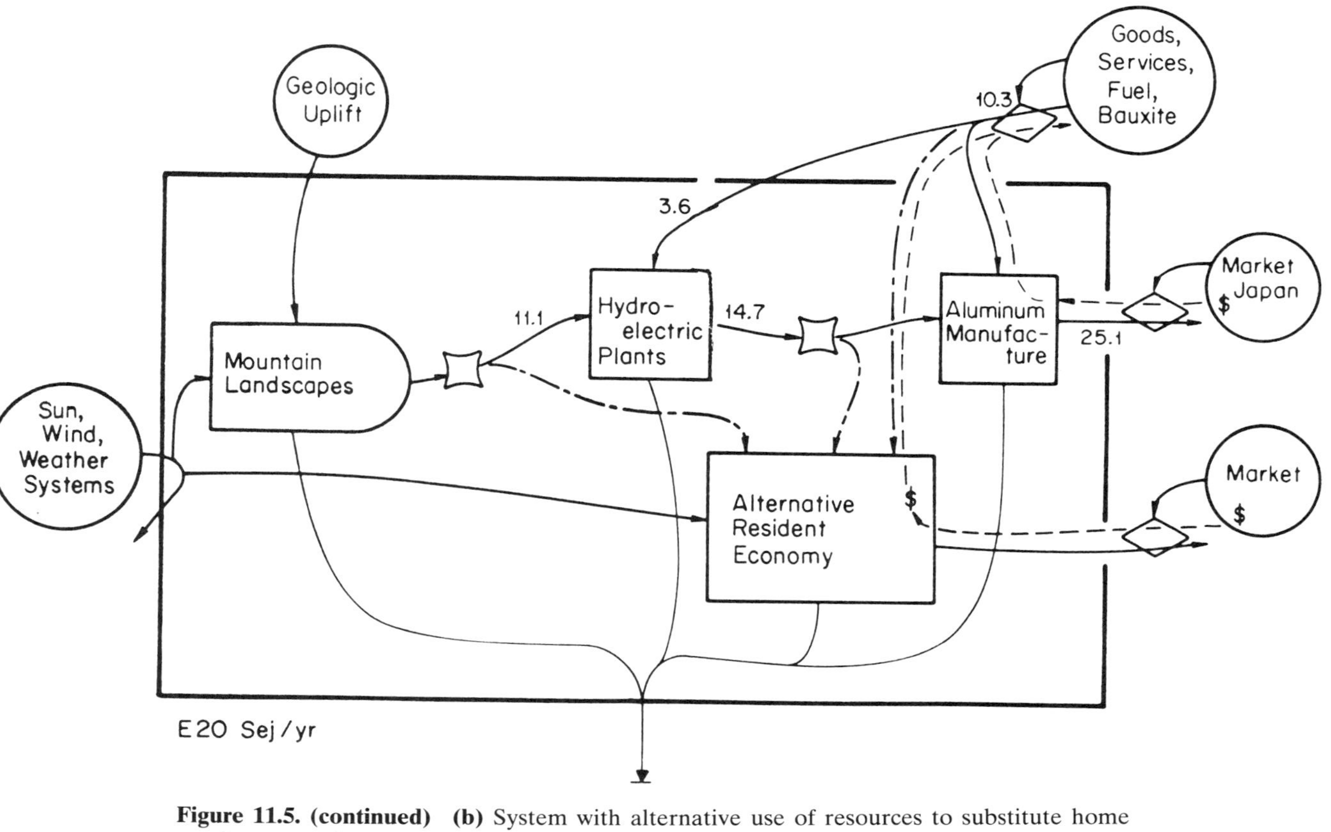

Figure 11.5. (continued) **(b)** System with alternative use of resources to substitute home use for unequal export.

on the EMERGY use, increasing the home use of EMERGY develops home capital faster than selling products for foreign exchange.

INTERNATIONAL FINANCE

Because of the differences in real buying power of a nation's money when converted to international dollars, loans and repayments in dollars are grossly in error as measures of equity. Those nations with high EMERGY/money values who borrow from those with low EMERGY/money values, pay several times more than they realize when they pay interest and repay the principal.

The EMERGY/money ratio is larger for rural and undeveloped countries where people are supported from the environment directly. They are using money less than those in urban areas. If one borrows money from a country, the buying power of that loan is given by the EMERGY/money ratio of that country. When the money is paid back, the buying power in the payback is that of the borrowing country.

If an undeveloped country with a 5-times-higher EMERGY/money ratio borrows from a developed country, the borrowed money purchases products, goods, and services from the developed country at the EMERGY/money value of the developed country. When it is time to pay back, the undeveloped country pays back from its own dollars with a much higher EMERGY/money ratio. In other words, the borrowing country ends up paying back 5 times more buying power than was borrowed and 5 times more interest. Much of the economic plight of undeveloped countries is due to these overpayments that come from using money instead of EMERGY as a measure of public wealth.

For example, in the 1980s Brazil paid back 2.6 times more real wealth than it received with a foreign loan. Where a 10% interest rate was intended, the interest was really 260%. Including interest, the total repayment was 286%. Neither lending nor borrowing country long prospers in this situation. The borrower finds its economy under a crushing load and can't repay. When repayments fail, the lending banks are forced to write off the debt or to accept a small fraction of the loan in final settlements. In one case, the repayment on the debt market was not far from what an emdollar-based repayment would suggest was equitable.

Scienceman (1987) suggests that a pool of international money could be established for trading. If each country transferred its currency into emdollars according to its EMERGY/currency ratio, then trading would be in proportion to EMERGY, a procedure that would help establish EMERGY equity and mutual prosperity among nations.

NATIONAL NET EMERGY BENEFIT FROM EXCHANGE

When the net benefits of international trade to countries are estimated in EMERGY units, benefit ratios may be compared (EMERGY imported to that exported in Table 11.1). There is a large range from 0.15 to 4.3. Countries such as Holland and West Germany are getting 4 times more EMERGY than they are returning in exchange. Little wonder that these countries have a high standard of living and that the countries supplying the commodities have a low standard.

EMERGY Self-Sufficiency

EMERGY self-sufficiency can be calculated as the percent EMERGY received from renewable and nonrenewable resources within the national border. Table 11.1 includes self-sufficiency calculations in solar EMERGY units. In general, the larger the country, the more self-sufficient it is. The more a country is an urban center at the top of the hierarchy of spatial organization of society elsewhere, the less it is self-sufficient. Being at the hierarchical center also increases benefit from international sales and fiscal transactions (Table 11.1).

EMERGY Exchange Equity

Allowing individual businesses to maximize their own profits in monetary terms often imbalances EMERGY trade equity. The dollar value of profits may

TABLE 11.1. National EMERGY Self-Sufficiency and Trade

Nation	% EMERGY from Within	EMERGY Benefit Ratio*
Netherlands	23	4.3
West Germany	10	4.2
Japan	31	4.2
Switzerland	19	3.2
Spain	24	2.3
U.S.A.	77	2.2
Taiwan	24	1.89
India	88	1.45
Brazil	91	0.98
Dominica	69	0.84
New Zealand	60	0.76
Poland	66	0.65
Australia	92	0.39
China	98	0.28
Soviet Union	97	0.23
Liberia	92	0.151
Ecuador	94	0.119

* Ratio of annual EMERGY inflow to annual EMERGY outflow. Evaluations are for different years 1980–1993.

Sources: See footnotes in Table 9.1.

be small compared with the emdollar value and public gross economic product given away to other countries. Free trade tends to result in unequal EMERGY exchange in favor of developed countries. Equity in trade can be achieved by treaty, adjusting imports and exports to balance EMERGY. If the EMERGY trade balance is uneven, the difference can be made up in education, military, or technology transfers duly evaluated for their EMERGY contributions. In this way, balances between nations can be equated while still allowing countries to be at different levels in the urban-rural hierarchy and national specialization.

EMERGY and Military Power

If the real basis for an economy is its rate of EMERGY use, and if its capability for meeting war and other emergencies is proportional to its EMERGY uses and reserves, then the military capabilities of nations may be estimated with EMERGY evaluations. Dalton (1986) evaluated the EMERGY use of the Defense Department of the United States in overseas military bases and operations and compared it to the net EMERGY benefits to the United States from trading partners that were within the worldwide defense perimeter maintained by the United States. In that time of peace, the ratio of EMERGY benefit to defense EMERGY was 4.5. The percentage of the U.S. budget spent on national defense, when estimated on an EMERGY basis, was about twice that figured on a dollar-expenditure basis. This is because of the heavy use of fuels, high-transformity technology, and strategic, scarce metals.

Mutliscale Evaluation of Development

EMERGY accounting provides guidelines for development projects in underdeveloped countries. Measures that maximize the EMERGY budget are believed to maximize economic wealth and vitality. However, there are several scales for the window of evaluation. Five levels of consideration are given in Figure 11.6: (1) the local development project and its people, (2) the regional economy

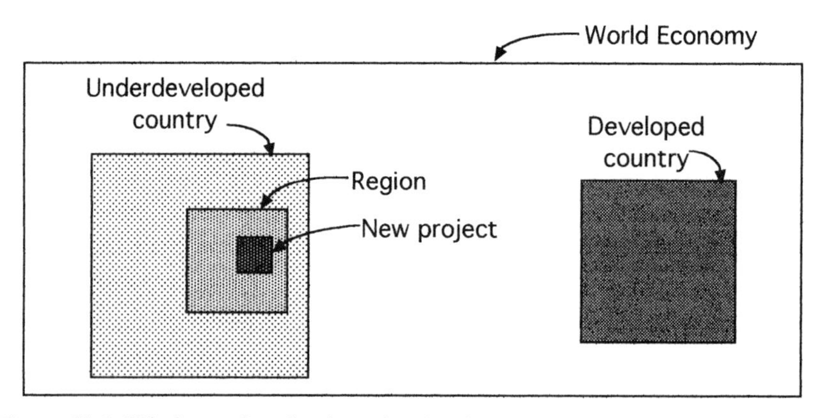

Figure 11.6. Windows of evaluation of a development project on different scales.

TABLE 11.2. Net EMERGY Benefit of Shrimp Aquaculture for Five Viewpoints (see Figure 11.6)

Note	Item	Million Em$/yr
1	Project output	+1092
2	Local region	− 602
3	Ecuador	−1100
4	Foreign economy (USA)	+ 710
5	Global system	+ 80

[1] EMERGY of shrimp produced including environmental and purchased sources.
[2] Net EMERGY including buying power of sales money plus purchased inputs brought into the area minus environmental losses minus exported shrimp.
[3] Net EMERGY including buying power of foreign sales plus new inputs developed minus exported shrimp minus environmental losses.
[4] Net EMERGY in imported shrimp minus payments and investments for shrimp.
[5] Net EMERGY in new inputs developed minus environmental losses.
Source: Data from Odum and Arding (1991).

of the development project, (3) the economy of the underdeveloped country, (4) the economy of the developed country, and (5) the world economy. Ideally, a project should increase the EMERGY of all scales. EMERGY evaluation can be made for each scale to determine which economies are having their EMERGY increased or decreased.

For example, for the five scales of evaluation, an EMERGY evaluation of shrimp aquaculture in Ecuador gave the results shown in Table 11.2 (Odum and Arding, 1991). Benefits and losses are in 1989 U.S. Em$. A plus means a net benefit in annual EMERGY use; a minus means a net loss. The evaluation showed large benefits for investors from capital-rich urban areas, and economy-damaging losses for all levels in Ecuador except for project operators.

SUMMARY

Uneven distribution of global EMERGY flow in free trade and finance between countries and in global military postures creates strains and distortions in the global economy. EMERGY accounting may help in developing international policies to develop empower equity, thus contributing to world peace.

CHAPTER 12

EVALUATING INFORMATION AND HUMAN SERVICE

Information may be the most important feature of many systems, including genetic biodiversity and the knowledgeable contributions of human beings. Much of the control of systems is with information. *Shared information* occupies large territory, has a slow depreciation and replacement time, and has high transformities. Much of the controversy over environment, economy, and policy concerns the information in ecosystems and the technological human society. Here EMERGY accounting is used to measure information and human service.

For our purposes, information is defined as the units, connections, and configurations of a system. Information is useful if it can make its system operate. The information may be in an operating system, or extracted and stored in compact form in a plan or code. Examples of useful information are the genetic codes of living organisms, the organization of an ecological system, and the cultural information of human societies. Information may be saved, copied, improved, transmitted, and used to make and operate new systems. Examples are the genetic information in seeds, the information in books, the messages by telephone, and the archeological artifacts in ancient deposits.

Even when isolated in compact form, information requires some form of energy as a carrier, such as that in the DNA of seeds, the paper of books, the electromagnetic waves of radio transmission, or the neuroelectrical processes of the brain. The EMERGY requirements for developing useful information for a functional, competitively surviving system are large, but the energy of its carrier is small. Consequently, the transformity of information (EMERGY per unit of energy) is large. EMERGY and transformity provide a scale for quantitatively measuring information and its utility. More EMERGY is required to regenerate useful information from scratch than to maintain it. Moreover, information may be characterized as something that requires fewer resources to save and copy than to make anew. This may be why systems store useful information once it is developed. EMERGY and transformity evaluations put information in perspective with other aspects of ecological and economic systems. EMERGY accounting that does not include this information is incomplete.

Knowledge and information are found in ecological and economic networks, the result of many complex transformations of energy. The transformities of knowledge, information, culture, art, and so forth are very high, but have a wide range of values. Information that has been found effective and retained for use has very large amplifier control effects commensurate with the high EMERGY required for its development and maintenance. Clearly, information transfer is not restricted to market transfer and cannot be evaluated with money.

Like other structures, information is thermodynamically away from equilibrium, and thus is continuously lost by dispersal and depreciation, usually described as the second law (Figure 2.3). Information is maintained by work processes, continually copying, correcting, replacing, and revising. Information is lost when its carrier disperses. Living organisms reproduce, copy, repair, revise, and reapply their information. Ecosystems and biogeochemical processes of the earth are now configured and controlled with living information. Large storages of information from eariler evolution are the basis for a stable and efficient biosphere.

In aggregated systems diagrams there is a small amount of information in the components, configurations, and connections, but this in no way represents the information of the system summarized there. However, the information of the real system may be represented in an aggregated way as an information storage (the "tank" symbol) with the usual depreciation pathway to a heat sink (Figure 12.1). Also shown there are pathways for the generation of new information and information duplication (making copies).

CIRCLE FOR MAINTAINING INFORMATION

The configuration for maintaining a stock of information copies, shown in Figure 12.1, is not sufficient to maintain information that is useful for long because errors develop and conditions for functions change. To keep information functional, a system is required to make more copies than needed, test each functionally, discard the erroneous and ill-adapted copies, and use the ones that make the system perform best for future copying.

Information develops errors through second-law processes. Error is reduced by making extra copies and discarding those that develop errors. Repair is accomplished largely by putting many copies into operation so that those that are still useful continue to be copied. Self-organization reinforces those that

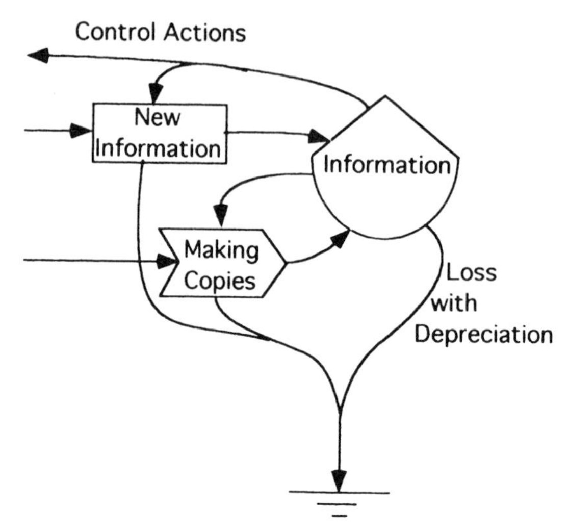

Figure 12.1. Use of "storage" symbol to represent information in systems diagrams.

still work by making copies of these. Those copies that don't work are not continued. If copies are not made and tested fast enough to eliminate error, information is lost. When organisms and other information-containing units reproduce, they copy their information storage, disperse the copies, and develop new systems with the copies so that the ones working best predominate and errors drop out.

In other words, a closed circle of information processing is necessary to maintain one unit of information (Figure 12.2). However, this *Information maintenance circle* or *cycle* is not entirely analogous to a material cycle because information is copied and lost (not conserved). One unit of information is maintained throughout the cycle, whether a part of the circle has one or many copies. All of the circle is required to maintain the one unit. For example, a whole population-reproduction cycle is required to maintain its genetic information. A whole industry is required to maintain its information technology.

EMERGY EVALUATION OF INFORMATION

The EMERGY required for an information flow or storage is the sum of the EMERGY contributions required and is determined from an appropriate energy systems diagram in the usual way. In the information circle there are several kinds of information that may be evaluated. These are given with examples from a rainforest in Tables 12.1 and 12.2.

(1) *EMERGY for an Information Copy.* Very little EMERGY is required to make one copy. For example, little EMERGY is required to make one copy of

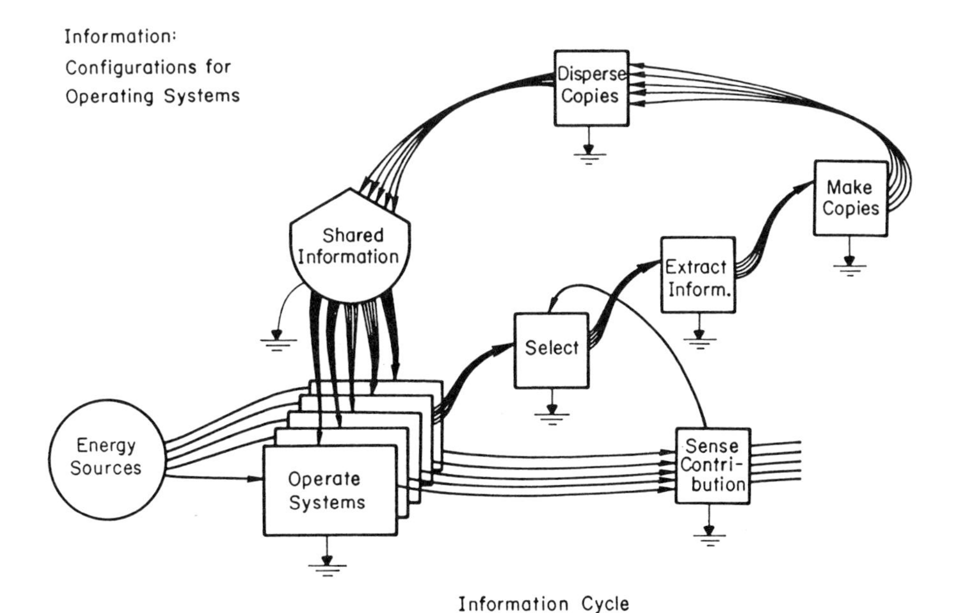

Figure 12.2. Main features of an information maintenance circle, including depreciation, extracting, copying, operating, testing, and selection.

a page of text on a photocopying machine; little EMERGY is required for simple biological reproduction. In Table 12.1 the first item is the EMERGY per hectare required to copy the information in rainforest tree leaves with a replacement time of 1.5 yr.

A copy can be made of an operating unit and, in the process, its contained information. For example, reproducing a tree duplicates the information as well as the operational part of the tree. Or, information that is isolated from its operational unit may be copied. For example, plans for a house may be photocopied.

(2) EMERGY for Extracting Information. More EMERGY is required to extract the information from the system it operates into a compact form suitable for transmission or transport. Many hours would be required to redraw the lost plans of a house from study of the house. When a tree reproduces, the genetic information that operates the whole tree is extracted in the form of seed and transmitted for dispersal and reproduction. Living systems make extracting easier by keeping copies of the plans in all its units. The second item in Table 12.1 evaluates average solar EMERGY of seed production in 1 ha of rainforest.

(3) *EMERGY for Sustaining Shared Information.* Even more EMERGY is required to keep a unit of information functional. The EMERGY comprises

TABLE 12.1. Kinds of EMERGY Evaluation of Information with Rainforest Examples

Note	Name	Forest Example	Solar EMERGY (sej)	Solar Transformity (sej/j)
1	EMERGY to copy units containing information	Duplicating tree leaves	9×10^{14}	7.0×10^{6}
2	EMERGY to isolate information	Growing seeds	6×10^{14}/ha/yr	1.13×10^{9}
3	EMERGY to sustain an information circle	Maintain a tree species population	7.7×10^{16}/yr	7.26×10^{11}
4	EMERGY to develop new information	Develop new tree species	7.7×10^{20}	4.8×10^{15}

[1] Annual EMERGY support to the forest (6.0×10^{14} sej/ha/yr) times the replacement time (1.5 yrs) = 9×10^{14} sej. DNA in El Verde forest from Canoy (1970):

$$(600 \text{ mg DNA/m}^2)(0.001 \text{ g/mg})(1 \times 10^4 \text{ m}^2/\text{ha})(5 \text{ kcal/g})(4186 \text{ J/kcal}) = 1.25 \times 10^8 \text{ J DNA/ha}$$

Transformity:

$$\frac{9 \times 10^{14} \text{ sej/ha in leaves}}{1.28 \times 10^8 \text{ J DNA in leaves}} = 7.03 \times 10^6$$

[2] Annual Solar EMERGY support. 6.0×10^{14} sej/ha/yr; DNA in seedfalls:

$$(292 \text{ J seed/m}^2/\text{day})(5 \times 10^{-4} \text{ J DNA/J seed})(1 \times 10^4 \text{ m}^2/\text{ha})(365 \text{ days/yr}) = 5.33 \times 10^5 \text{ J DNA/ha/yr}$$

Transformity: solar EMERGY (6.0×10^{14} sej/ha/yr) divided by DNA flux:

$$\frac{(6.0 \times 10^{14} \text{ sej/ha/yr})}{(5.33 \times 10^5 \text{ J/ha/yr})} = 1.126 \times 10^9 \text{ sej/J}$$

[3] One tree species out of 153 species occupying 19,648 ha of LuqLuillo forest:

$$(6 \times 10^{14} \text{ sej/ha/yr})(128 \text{ ha/species}) = 7.7 \times 10^{16} \text{ sej/yr/species}$$

One tree out of 3140 trees per 4 ha, larger than 10 cm diameter, occupies:

$$\frac{(4 \times 10^4 \text{ m}^2/\text{ha})}{3140 \text{ trees}} = 12.7 \text{ m}^2/\text{tree}.$$

DNA flow per tree:

$$(12.7 \text{ m}^2/\text{tree})(0.6 \text{ g DNA/m}^2)(5 \text{ kcal/g DNA})(4186 \text{ J/kcal})/(1.5 \text{ yr turnover}) = 1.06 \times 10^5 \text{ J/tree/yr}.$$

EMERGY supporting population divided by DNA in one tree:

$$\frac{7.7 \times 10^{16} \text{ sej/yr/species}}{1.06 \times 10^5 \text{ J/tree/yr}} = 7.26 \times 10^{11} \text{ sej/J DNA}$$

that necessary to maintain a population, including its processes of duplication, use, selection, and reproduction. To evaluate the EMERGY required to maintain a species of tree, the total EMERGY flow supporting the Luquillo rainforest area in El Verde, Puerto Rico, was divided by the number of tree species in the area (Tables 12.1 and 12.2). Each member of the population carries similar genetic information. The EMERGY to maintain information by maintaining the population can also be called the "EMERGY of shared information." The larger the population using common information the greater the EMERGY.

(4) *EMERGY to Develop Useful Information.* EMERGY to develop new, useful information is that required to make at least one copy of new information from its precursor. For example, the development EMERGY of a new species of tree is that EMERGY required to operate the population over the time period required for its evolutionary changes. For an example, Tables 12.1 and 12.2 used a new-species development of 10,000 yr. The EMERGY for new useful information development is the largest of the several EMERGY measures of information because more resources are required to make a unit anew than to copy, share, and select an old unit. The development EMERGY may be suitable for evaluating an endangered species or the last copy of a book.

Solar Transformity of Information

The transformity of information is the EMERGY per unit of energy carrying the information. The information categories requiring more EMERGY have higher transformities, as illustrated in Table 12.1. Thus, the EMERGY of the first or last copy is highest in the EMERGY hierarchy and belongs at the far right in systems diagrams. The high transformity is commensurate with the very great effects seeding of usable information can have.

After information has been copied, it may be shared. The same information then has a larger territory, a greater area of influence, and a slower depreciation rate. Considerable additional work must be done to develop the shared status. Thus, the shared information has a higher transformity for maintaining the same information in the shared status. An item of information that is shared—held by many units—has the greater EMERGY that copied and established it

---

[4] Species formation taken as 10,000 yr with sustenace EMERGY from note 3:

$$(7.7 \times 10^{16} \text{ sej/yr/population})(10{,}000 \text{ yr}) = 7.7 \times 10^{20} \text{ sej/population}$$

Transformity: Using DNA content of one individual,

$$\frac{(7.7 \times 20^{10} \text{ sej/pop})}{(1.59 \times 10^{5} \text{ J DNA/ind})} = 4.8 \times 10^{15} \text{ sej/J}$$

Sources: Data from El Verde, Puerto Rico (H. T. Odum, 1970); DNA data from Canoy (1970, 1972).

TABLE 12.2. EMERGY and Emdolar Value of Rainforest Trees at El Verde, P.R.: Annual Solar EMERGY per Hectare, 7.0×10^{14}sej/ha/yr*

Note	Item	EMERGY (sej)	Em$ (1990 U.S. $)†
1	Plantation monoculture, 10 yr, 1 ha	1.75×10^{15}	875
2	Mature forest, 300 yr old, 1 ha	2.1×10^{17}	105,000
3	Annual support of all 153 tree species, 19,648 ha	1.38×10^{19}	6.9×10^{6}
4	Mature forest, 300 yr, 19,648 ha	4.1×10^{21}	2.05×10^{9}
5	Sustain 1 species, 128 ha, 1 yr	9.0×10^{16}	45,000
6	Evolve 1 species, 128 ha, 10,000 yr	9.0×10^{20}	4.5×10^{8}
7	Evolve 153 tree species, 10,000 yr, 19,648 ha	1.37×10^{23}	6.9×10^{10}
8	Average tree, 12.7 m^2 area, age 50 yr	4.4×10^{13}	22
9	Dominant climax tree, 300 yr, 500 m^2 crown	5.25×10^{15}	2,625
10	EMERGY of species diversity per bit	5.78×10^{13}	28.9

* Preponderance of solar EMERGY that of rain used:

$$(2{,}140 \text{ g/m}^2 \text{ water transpired/day})(5 \text{ J free energy/g})(365 \text{ days/yr})(1 \times 10^4 \text{ m}^2\text{/ha} = 3.9 \times 10^{10} \text{ J/ha/yr}$$

Solar transformity of rain, 1.8×10^4 sej/J:

$$(3.9 \times 10^{10} \text{ J/ha/yr})(1.8 \times 10^4 \text{ sej/J}) = 7.0 \times 10^{14} \text{ sej/ha/yr}$$

† Solar EMERGY divided by 2×10^{12} sej/1990 U.S. $.

[1] Formation in 10 yr average solar EMERGY half of that at the end of growth. Plantation has half the metabolism of the mature complex forest:

$$(0.5)(7.0 \times 10^{14} \text{ sej/ha/yr})(0.5)(10 \text{ yr}) = 1.75 \times 10^{15} \text{ sej/ha}$$

[2] Average EMERGY used in formation taken as that after 100 yr:

$$(7.0 \times 10^{14} \text{ sej/ha/yr})(300) = 2.1 \times 10^{17} \text{ sej/ha}$$

[3] $(7 \times 10^{14}$ sej/ha/yr)(19,648 ha) = 1.38×10^{19} sej/yr/forest

[4] $(7 \times 10^{14}$ sej/ha/yr)(19,648 ha)(300 yr) = 4.1×10^{21} sej/forest

[5] $(7 \times 10^{14}$ sej/ha/yr)(128 ha/species) = 9.0×10^{16} sej/species

[6] $(7 \times 10^{14}$ sej/ha/yr)(128 ha)(10,000 yr) = 9.0×10^{20} sej/species

[7] $(7 \times 10^{14}$ sej/ha/yr)(19,648 ha)(10,000 yr) = 1.37×10^{23} sej/forest species

[8] Assumed that average tree over 10 cm s 50 yr:

$$(7 \times 10^{14} \text{ sej/ha/yr})(12.7 \times 10^{-4} \text{ ha})(50 \text{ yr}) = 4.4 \times 10^{13} \text{ sej/tree}$$

[9] $(7 \times 10^{14}$ sej/ha/yr)(0.05 ha)(300 yr)(0.5) = 5.25×10^{15} sej/tree

in many units, but its territory is now much bigger than any one carrier and its effect is larger, its time constant longer, and its depreciation less than those of any one carrier. The shared genetic information in the populations of birds and plants has broad territories due to bird movements and seed dispersal. Information shared by a species has a larger territory than the individual organisms.

EMERGY of Life Cycles

When organisms and other information-containing units reproduce, they copy their information storage; disperse the copies in spores, eggs, seeds, and so forth; use these copies to develop new systems; and expose these copied units to system-selective processes so that the ones working best predominate and errors drop out. A great opportunity exists for evaluating the information processing in different life cycles well known in biology.

For example, Figure 12.3 contains the life cycle of penaeid shrimp included in an EMERGY evaluation study (H. T. Odum and Arding, 1991). The solar EMERGY maintains the information cycle of the shrimp life history so that about 300,000 eggs (1.33×10^6 sej/egg) are released at sea for each two adults migrating out of the estuary. Thus, two adults with 1.0×10^{11} sej/adult are equivalent to 300,000 individual eggs and larvae. The result is that the solar transformities of individuals increase in the cycle from left to right as they grow. These transformities represent the EMERGY required for copies. The genetic information is conserved throughout the cycle, although different numbers of copies exist at different stages. The information carrier is the DNA, with about 2 mg DNA/g dry wt of adult.

The energy in a gene set is as follows:

$$(10 \text{ g/individual})(0.2 \text{ dry wt.})(2 \times 10^{-3} \text{ g DNA/g dry wt.})(5 \text{ kcal/g DNA})(4186 \text{ J/kcal}) = 83.7 \text{ J DNA/adult}$$

Solar transformity of maintaining shared population information in one individual gene set is as follows:

$$\frac{(1.0 \times 10^{11} \text{ sej/individual})(3.2 \times 10^{9} \text{ individuals/population})}{83.7 \text{ J DNA/individual}} = 3.8 \times 10^{18} \text{ sej/j}$$

---

[10] Shannon diversity 4.6 bits/individual (N. Sollins, 1970); 785 individual trees per hectare. The total tree bits in the whole sustaining forest is:

$$(785 \text{ ind/ha})(19{,}648 \text{ ha/forest})(4.6 \text{ bits/ind}) = 7.09 \times 10^{7} \text{ bits/forest}$$

Solar EMERGY for the whole forest from item 4 divided by forest bits:

$$\frac{(4.1 \times 10^{21} \text{ sej/forest})}{(7.09 \times 10^{7} \text{ bits/forest})} = 5.78 \times 10^{13} \text{ sej/bit of trees}$$

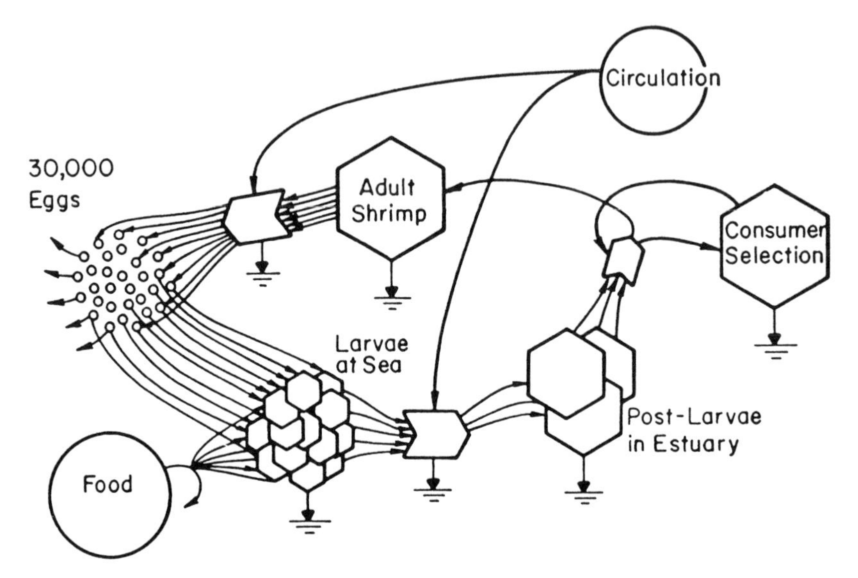

Figure 12.3. Life cycle of a population penaeid shrimp, an information maintenance circle that includes extracting, copying, reusing, and testing.

EMERGY Accounting of Evolutionary Products

The evolution of life and its forms over several billion years has accumulated very large EMERGY inputs into the genetic information storages of different taxa. Biological theory has long held that larger categories of taxonomy represent larger areas and times of evolution. For example, a species might require 10,000 yr to form from a precursor species and usually occupy a smaller area, whereas a phylum might require 1 billion yr and be spread over much of the world. Ager (1965) estimated that 1.5×10^9 species have formed in the history of the earth over a period of 2×10^9 yr of evolution. The solar EMERGY per species was estimated using an annual earth EMERGY budget of 9.44×10^{24} sej/yr.

$$\frac{(9.44 \times 10^{24}\ \text{sej/yr})(2 \times 10^9\ \text{yr})}{(1.5 \times 10^9\ \text{species})} = 1.26 \times 10^{25}\ \text{sej/species}$$

A computer simulation of a model of phylogenetic evolution (Figure 12.4a) and its EMERGY inputs was arranged (H. T. Odum, 1989). Individuals are produced ("interaction" symbols) and consumed (K) within the population (contained in the "production" symbol on the left). Small changes in species accumulate with time (microevolution). In proportion to the number of individuals that have lived, a new species is generated (calibrated for an emergent new species for each 10,000 yr). In proportion to the species number, larger changes are occasionally emergent at the next higher systematic level (genera).

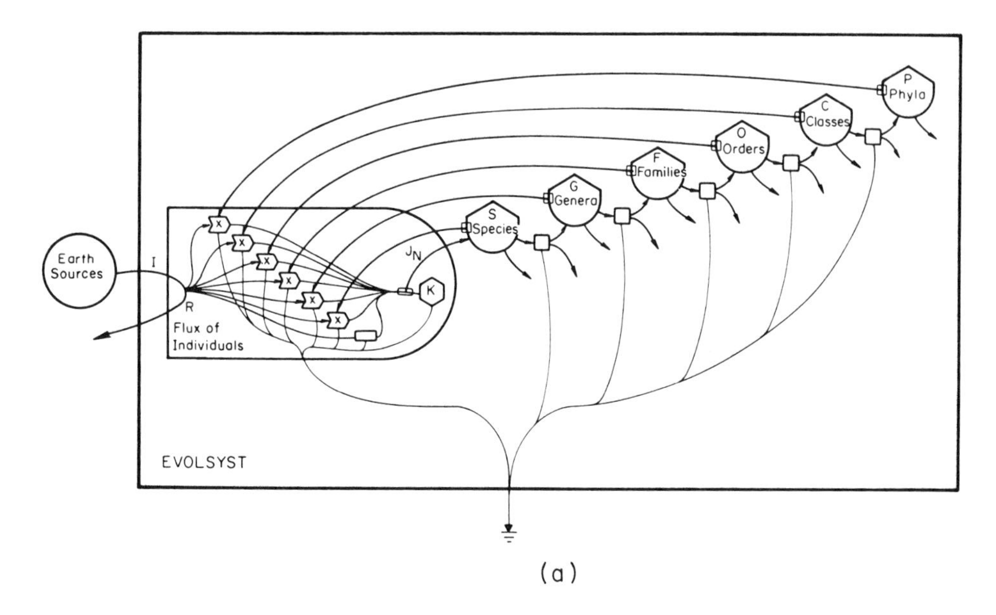

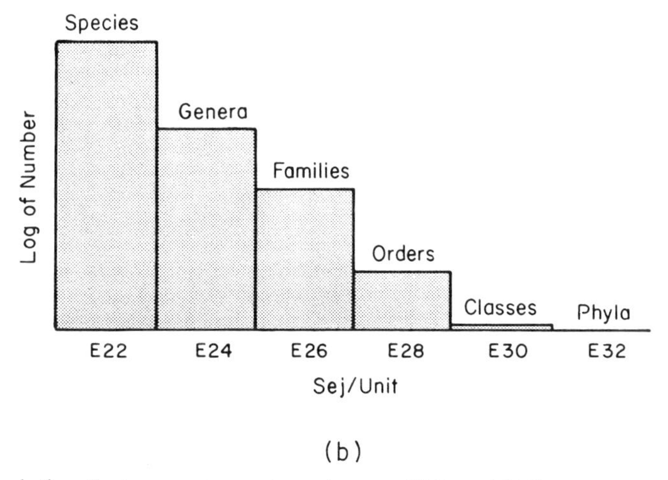

Figure 12.4. Evolution that generates systematic taxa (Odum, 1989). **(a)** Energy systems diagram of the flux of EMERGY into the population microevolution on the left (necessary to generate higher levels of innovation, i.e., macroevolution) and then into the larger territory on the right. **(b)** Distribution of taxonomic categories as a function of the logarithm of the solar transformity of their evolutionary formation calculated in a microcomputer simulation of the model. Reprinted with permission of the International Society for the Systems Sciences.

In proportion to the longer accumulation of genera, larger changes are emergent at the next higher level still (families), and so forth. In this way, macroevolutionary changes in jumps are the sum of the EMERGY values of the microevolutionary changes of the levels below. As each taxonomic category appears, some increase in the flux of individuals results from biodiversity interactions.

Figure 12.4b plots the solar transformities of the taxonomic categories that result from operating the model until 50 phyla are generated (in 5 billion yr). These transformities are even higher than those of geologic units in Chapter 4. Genetic information developed in interaction with environmental processes over geologic time represents the memory of the earth as a whole. Notice the way genetic information fits into the geologic space-time graph in Figure 3.5. It has the highest transformities and highest role as controlling the earth system.

EVALUATING HUMAN SERVICE

Late in evolution the human has emerged as the earth's information-processing specialist. With highly developed information organs, and with social mechanisms and institutions for group information processing, humans can now control living and nonliving operations. The planet is becoming a "noosphere," to use Vernadsky's term for the increasingly information-controlled shell of the earth.

In previous chapters EMERGY/money ratios were used to evaluate human services where data on wages and income were in currency units, but this procedure is only useful where the EMERGY flows of an average person are appropriate. In this chapter human service is evaluated by multiplying the energy expended by a human being by the transformity of that person's education and experience.

Another way to evaluate average human EMERGY is to divide the total national EMERGY flow by the number of people and the metabolism. For example, in an evaluation of the average human contributions within Ireland in the 19th century, the total annual solar EMERGY of the country was estimated from environmental data and divided by the number of people and their metabolism to obtain an average solar EMERGY per hour per person (H. T. Odum, 1986). Annual EMERGY flows per person were given in Table 10.8.

EMERGY Evaluation of Learning and Teaching

As with other products, the EMERGY of learned information is the sum of the interacting inputs. A person learns by an interaction of the inflow of information with his or her learning equipment (sensors and mental learning mechanisms). For example, the EMERGY that is stored in a child learning from his or her mother is a sum of what is transmitted from the mother and the EMERGY flows supporting the child (food, clothing, shelter, and so on). A human teaches by copying knowledge out to a receiver, without draining the person's knowledge. The copying process reinforces the person's own knowledge.

The solar EMERGY contained in human delivery of information is calculated as the product of the time of information delivery, times the metabolism per unit time, times the solar transformity estimated for the educational level of the teachings. For example, the mother teaching the child delivers more EMERGY per hour if the transformity of her education is higher. This procedure can be used to estimate the information in immigrants, the teaching in schools, and the input to high technology from the general public education of its employees.

Transformity of Educational Levels

The EMERGY accounting of human information is fairly new, and few known transformities are available. Figure 12.5 shows a wide range of solar transformity for different levels of education and experience among the hierarchical levels of human life in the United States. Included on the right are public leaders and writers who are symbols shared in the minds of many. They can contribute with a high transformity, representing the large solar EMERGY of a large territory of support, development, and identification. In a rough way the graph begins to evaluate what humans contribute.

EMERGY in High-Technology Industry

An example of EMERGY accounting of human service is the evaluation of high-technology industry. Part of high technology is new research and development, which runs as high as 50% of the expenditures in some defense industries. High-technology information is also involved in manufacturing, which is essentially copying information and coupling it to hardware, and has a somewhat lower transformity than developing new information (research) does. Both kinds of information were evaluated for a Texas high-technology industry (modified from an analysis by Allert Brown, 1986; H. T. Odum, 1987a). About one-fourth of the EMERGY basis of the company, especially in its research-and-development division, was based on prior public and private education and a high level of information in the society. The input of EMERGY from professionally trained people was generated from family education and public education sources, not directly paid for by the company.

EMERGY Accounting in Universities

Annual solar EMERGY contributions to a university are given in Figure 12.6 and Table 12.3 with faculty, students, and libraries important. The contributions from inputs rich in information are much greater than the EMERGY in the money paid for services. This example illustrates the hypothesis that more information circulates than is paid for with money. Should the library books

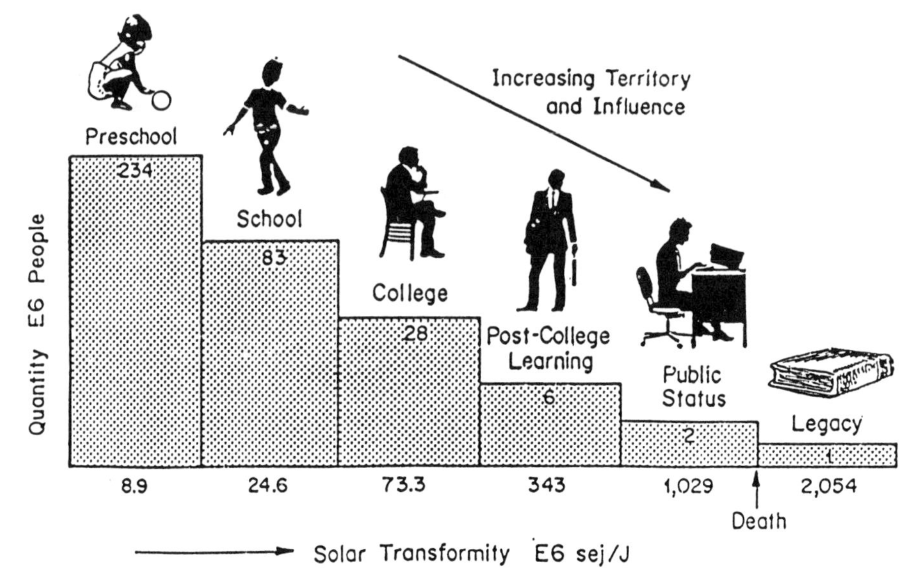

Figure 12.5. Hierarchical levels of education attained in human growth and development. Population of the United States in each category as a function of solar transformity. Transformities were calculated by dividing the total annual solar EMERGY use of the United States by the number of people in the category and the annual metabolic energy per person. Using EMERGY territories and the United States, the average per person is the total solar EMERGY budget of the USA in 1980 (785×10^{22} sej/yr) (Odum and Odum, 1993) divided by the population (3.4×10^{16} sej/person/yr). (H. T. Odum, 1987a)
Energy per individual:

$$(2500 \text{ kcal/day})(365 \text{ days/yr})(4186 \text{ J/cal}) = 3.82 \times 10^{9} \text{ J/ind/yr}.$$

The following table shows the effect of convergence on the EMERGY distribution:

Attainment	Numbers ($\times 10^6$)	EMERGY/ individual ($\times 10^{16}$ sej/ind/yr)	Transformity ($\times 10^6$ sej/J)
Total (preschool)	234	3.4	8.9
School	83	9.4	24.6
College grad	28	28.0	73.3
Post college ed	6	131.0	343.0
Public status	2?	393.0	1029.0
Legacies*	1?	785.0	2054.0

* Written or other memories in general use. The EMERGY divided by their work hours is their EMERGY delivery rate.

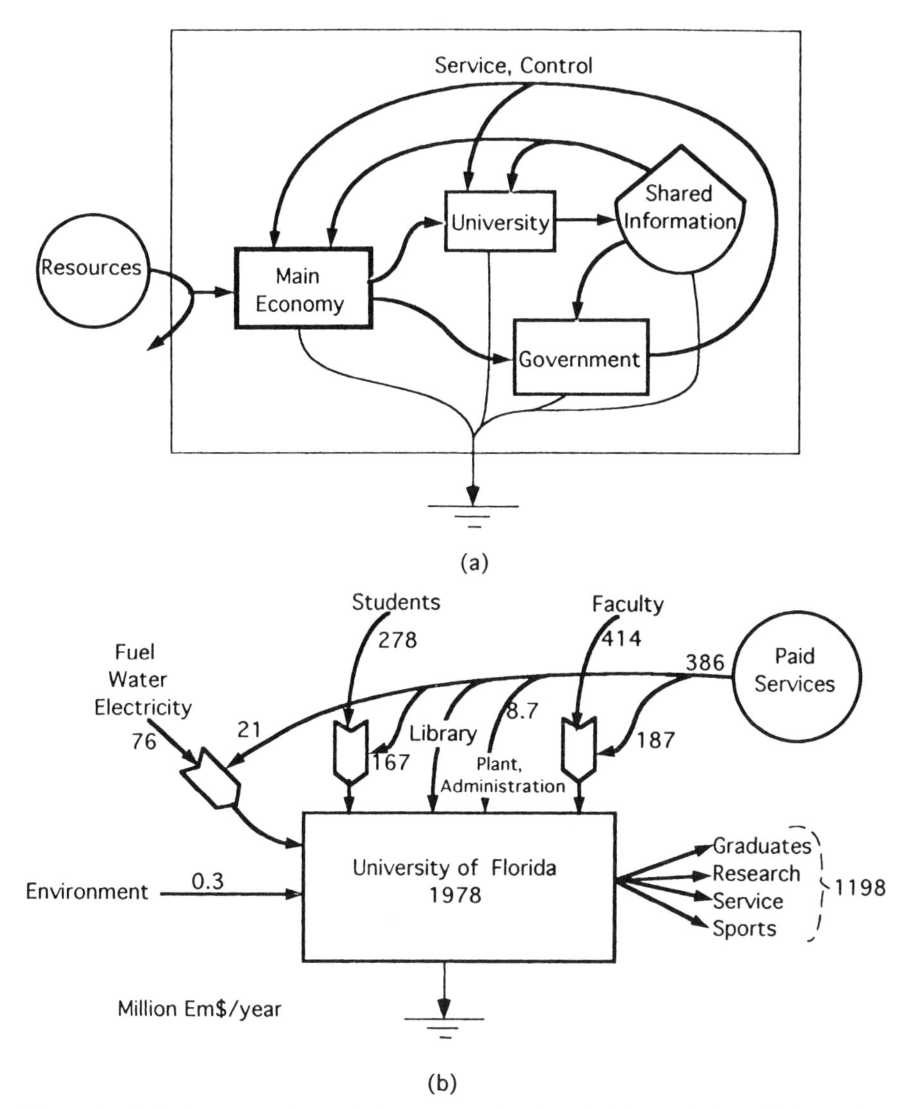

Figure 12.6. Systems overview of education and society. **(a)** Role of shared information generated by universities in the operation of society. **(b)** EMERGY flows of the Univ. of Florida.

be assigned the relatively small EMERGY of purchasing copies, or should the EMERGY of each book be given its share (split according to the number copies in existence) of the large EMERGY required to make the first copy?

A university president posed the following question: Should some unallocated money be used to build a cogeneration plant to save utility costs, or should these funds be used for academic purpose? To answer this question

TABLE 12.3. EMERGY Evaluation of University Inputs University of Florida, Gainesville, 1977–1978

Note	Item	EMERGY flow ($\times 10^{19}$ sej/yr)	Em$ ($\times 10^{6}$ $/yr)*
1	Environmental inputs, sun, wind, rain	0.076	0.26
	Fuels, gasoline, water, electricity		
2	EMERGY content	22.8	75.9
3	Paid services	6.4	21.2
4	Plant, administration (services)	8.7	29.0
	Students		
5	EMERGY use from prior education	83.3	278.
6	Paid support of students	50.2	167.
	Library Books		
7	Costs of replacing copies	0.53	1.8
8	Operation	1.45	4.8
	Faculty		
9	EMERGY of knowledge used	124.	414.
10	Paid services	56.	187.
	Maintainenance and Replacement Structure		
11	EMERGY in new building	0.78	2.6
12	Paid services	4.66	15.5
	Sum of inputs to university production.	358.896	1197.06

* Annual solar EMERGY used divided by 3×10^{12} sej/1981$.

[1] Energies used were multiplied by their respective solar transformities to obtain the rates of solar EMERGY use. The largest of the three is used, since they are all by-products of the same global solar EMERGY.

[2] Each type of energy used (in joules per year) was multiplied by its solar transformity (solar emjoules per joule) to obtain solar the EMERGY use per year.

[3,4,6–8,10] Where an input was paid for with money, dollars paid were multiplied by the 1977 solar EMERGY per dollar to obtain the solar EMERGY contributions of people (labor, services, prior services in goods, etc).

[5] This line contains the EMERGY of information in the students that they contribute to the university process from their prior education. The number of students was multiplied by the hours of intellectual activity, by the metabolic energy per hour, and by the solar transformity of the knowledge level in students entering the university.

[7] Library books are an information storage that is maintained by replacing old copies with new ones, a process that uses EMERGY of paper and human services. The much larger EMERGY of the first (or last) copy is what was required to generate the original book manuscript, but this is a property of the larger world information storage and is not included in this table.

[9] The contributions of knowledge by faculty from their information storages were evaluated by multiplying hours of intellectual efforts times metabolic energy per hour times the solar transformity (solar emjoules per joule) of the appropriate level of knowledge.

[11] New square feet were multiplied by weight per square feet and by solar transformity of cement.

Source: Adapted from Odum et al. (1978).

requires comparing the EMERGY savings of the plant with the EMERGY increase in the state economy due to the graduation of better trained potential leaders. To answer the question requires the difficult, still unfinished EMERGY evaluation of the academic feedback contribution of the university in providing high-transformity information to operate the whole state system (Figure 12.6a)

EMERGY in Culture

Considering culture as the shared information of a human society by which it operates, an EMERGY evaluation can be calculated based on the inputs and time required for cultural development. Table 12.4 contains a very preliminary calculation of the solar EMERGY in development of a native culture from one that went before, if 300 years was required. Although Papua New Guinea was in mind, the numbers are assumptions made to show how such evaluations might be done. The EMERGY of learned information of culture is compared with the larger EMERGY required to make racial genetic changes from a precursor population if 300,000 yr were required. An important part of cultural learned information is the interchange of information with surrounding areas. For example, countries prosper when educated humans immigrate. Ancient Rome was developed in part with information from Greece.

EMERGY in Television

When information is globally shared, it has a very large territory, that of the people sharing—nationally or worldwide. The processing of that information has centers, usually in the most important cities. Centers of business information, religious symbols, and high-technology ideas increasingly have become the highest level of the hierarchy of urban civilization. Combining the high transformity of information with large energies of electricity, the communication industry, through television, organizes the entire system through the many hours of people watching. Figure 12.7 and Table 12.5 show the EMERGY flows in the U.S. television system. The largest inputs to television reception is that of the viewers, who contribute the EMERGY of their own watching and listening attention. This is a good part of the total EMERGY flow of the nation. The possible waste of resources through television overemphasis was considered for the *Valdez* oil spill in Chapter 7 (Figure 7.8).

By means of electronic communication, information sharing has become worldwide. Shared information with very high transformities is changing society and biosphere. Unified world thinking is overriding the power of single nations, superseding power politics, preventing wars, forming a new international public opinion, and perhaps evolving a shared dedication to the environmental care of the planet.

Money and economic transactions are a special form of information. However, as suggested by drawing shared information to the right in Figure

TABLE 12.4. EMERGY Evaluation of a Native Culture

Note	Item	EMERGY Flow (sej/yr)	Energy flow (J/yr)	Solar transformity (sej/J)	Em$ ($\times 10^9$ $/y
		Annual Flow			
1	Renewable sources	1050×10^{20}			52.
2	Human metabolism	1050×10^{20}	1.33×10^{16}	7.7×10^{6}	
3	Information flow	1050×10^{20}	1.33×10^{15}	7.7×10^{7}	
4	Genetic flow	1050×10^{20}	6.3×10^{10}	1.7×10^{12}	
		Steady-state Storage			
		(sej)	(J)	(sej/J)	($\times 10^{12}$
5	Population, 3.5×10^{6}	3.36×10^{24}	9.9×10^{14}	3.4×10^{9}	1.
6	Culture information	3.47×10^{25}	9.9×10^{13}	3.5×10^{11}	17.
7	Human DNA, gene information	3.06×10^{28}	2.1×10^{12}	1.47×10^{16}	1530.

* EMERGY divided by 2×10^{12} sej/1990 U.S. $

[1] Renewable EMERGY for Papua New Guinea (Doherty, 1990), 1050×10^{20} sej/yr.

[2] Metabolism of population:

$$(3.5 \times 10^{6}\ \text{people})(2500\ \text{kcal/day})(4186\ \text{J/kcal})(365\ \text{days/yr}) = 1.33 \times 10^{16}\ \text{J/yr}$$

[3] 10% of the human metabolism for social interation and learned information.

[4] 33-yr turnover time (0.03/yr); human DNA from item 7:

$$(0.03)(2.1 \times 10^{12}\ \text{J DNA}) = 6.3 \times 10^{10}\ \text{J/yr}$$

[5] Energy storage in population:

$$(0.2\ \text{dry})(454\ \text{g/lb})(3.5 \times 10^{6}\ \text{people})(150\ \text{lb ea})(5\ \text{kcal/g})(4186\ \text{J/kcal}) = 9.9 \times 10^{14}\ \text{J}$$

EMERGY storage:

$$(1020 \times 10^{20}\ \text{sej/yr})(33\ \text{yr}) = 3.36 \times 10^{24}\ \text{sej}$$

[6] Storage of learned population information (culture) based on 10 generations with social information flux from item 3:

$$(1050 \times 10^{20}\ \text{sej/yr})(10\ \text{generations})(33\ \text{yr/generation}) = 3.46 \times 10^{25}\ \text{sej}$$

Information carrier storing information as 10% of biomass joules in item 5.

[7] Human DNA:

$$(2.1\ \text{mg DNA/g dry nt.})(0.2\ \text{dry})(454\ \text{g/lb})(3.5 \times 10^{6}\ \text{people})(150\ \text{lb ea})(0.001\ \text{g/mg})(5\ \text{kcal/g})(4186\ \text{J/kcal}) = 2.1 \times 10^{12}\ \text{J human DNA in population}$$

Genetic differences from precursor stocks generated in 300,000 years.

$$(1020 \times 10^{20}\ \text{sej/yr})(300{,}000\ \text{yr}) = 3.06 \times 10^{28}\ \text{sej}$$

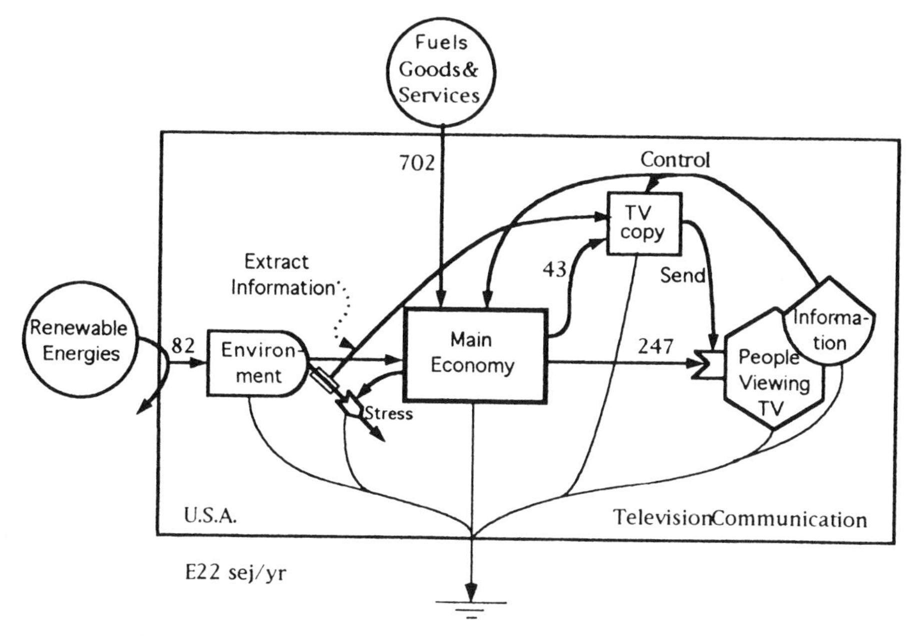

Figure 12.7. EMERGY flows in the U.S. television control system.

3.1, general shared information appears to have higher transformities than economic information and is capable of overriding control of markets and finance. As with ecosystems, growth appears to start with free competition, superseded later by the increasing role of the control played by shared information.

EMERGY AND INFORMATION BITS

In many branches of science, the "information" of components and configurations has been represented as the logarithm of the possibilities (possible combinations of units and their connections). (Here the word was redefined to refer to only one aspect of what that word may include in more general discourse.) This "information theory" index measures *complexity* on a logarithmic scale. The units obtained with these measures are in bits (the logarithm to the base 2) or nits (the natural logarithm, i.e., the logarithm to the base e). The log-of-possibilities measure gives the same value to useful information that has been selected through reinforcement during self-organization to operate systems as it does to useless complexity that will not operate anything.

Solar EMERGY per bit may be calculated as a measure of the information on the scale of the energy hierarchy. Useless complexity—in which little

TABLE 12.5. EMERGY Evaluation of U.S. Television

Note	Item	Solar EMERGY Flow ($\times 10^{22}$ sej/yr)
	TV Transmission	
1	Electricity	20.3
2	Assets cost	0.28
3	People	22.1
	Total	42.7
	Television Reception (1.62 × 10⁸ sets)†	
4	Electricity	4.5
5	Assets cost	9.5
6	People watching	263
7	Reception EMERGY per set yr	7.1×10^{14} sej/set/yr

* TV transmission using figures collected by R. Morton (1991).

[1] $(10.4 \times 10^{17}$ J/yr$)(2.0 \times 10^{4}$ sej/J $= 20.3 \times 10^{22}$ sej/yr.

[2] $(\$3.6 \times 10^{10})(0.05)(1.6 \times 10^{12}$ sej/\$$) = 0.28 \times 10^{22}$ sej/yr.

[3] $(3.87 \times 10^{5}$ people)(2500 kcal/day/person)(4186 J/kcal)(365 days/yr)$(150 \times 10^{6}$ sej/j$) = 22.1 \times 10^{22}$

† TV reception using figures collected by Morton (1991)

[4] (0.15 kWh/hr)(7.1 hr/day)(365 days/yr)$(3.6 \times 10^{6}$ J/kWh)$(160 \times 10^{6}$ sets)$(2 \times 10^{5}$ sej/J$) = 4.48 \times 10^{22}$ sej/yr.

[5] [(TV sets $(4.65 \times 10^{10}$ \$/yr) + cable $(1.28 \times 10^{10}$ \$/yr)] $(1.6 \times 10^{12}$ sej/\$$) = 9.5 \times 10^{22}$ sej/yr; \$287/set.

[6] Annual empower of USA, 900×10^{22} sej/yr, supporting a population each watching 7 hr/day. Empower share of people watching: (7 hr/24 hr)$(900 \times 10^{22}$ sej/yr$) = 263 \times 10^{22}$ sej/yr

[7] $4.5 \times 10^{22} + 9.5 \times 10^{22} = 11.5 \times 10^{22}$ sej/yr$/1.62 \times 10^{8}$ tv sets $= 7.1 \times 10^{14}$ sej/set/yr

work has been exerted to make it useful—has low EMERGY, whereas highly evolved and adapted complexity has high EMERGY because of the larger and longer flows of EMERGY required for its development and maintenance. Data on energy and information bits of systems may be readily unified by calculating solar EMERGY per bit. Complexity (i.e., the number of possible arrangements and connections) increases rapidly with number of units. If EMERGY of maintaining, copying, adapting, retrieving, and applying information goes up proportionately, then a resource limit to information use will be reached.

Tribus and McIrvine (1971) made the pioneering suggestion that higher-quality information-processing electronics use less energy per bit. Since higher-quality information is that which has had more transformation work, the energy remaining is less but the EMERGY is greater. Apparently, the higher the transformity, the more EMERGY there is that can be transmitted for the same effort, and the greater the effect.

At the chemical scale of molecular populations, the logarithm of possibilities is entropy (defined in another way in Table 2.1). Sometimes information measures on the larger scale of the environment are called *macroscopic entropy*. Because there are so many molecules, the number of bits in a molecular population is vastly greater than those in landscape units on an environmental scale. Information theory has had limited application because bits at one scale are not very comparable with those at another. However, evaluating the EMERGY/bit ratio puts these complexity measures on a common scale. The same number of bits on a small scale involves a smaller EMERGY than that same number of bits on a landscape scale.

Biodiversity

Traditional in ecology is the measurement of species variety, with several indices for comparing ecosystems. One graph plots the number of species found as a function of the area searched. For many areas, increasing the area means increasing the EMERGY flow, which can be evaluated from the environmental inputs (see Chapter 7). Thus, species-area curves can be used to evaluate the EMERGY required for levels of diversity. For example, in Figure 12.8 the tree species biodiversity is related to area, and thus to EMERGY flow, for the Luquillo Forest in Puerto Rico; the usual x- and y-coordinates have been swapped to show the very high EMERGY requirements of higher biodiversities.

Widely used in ecology is the *Shannon-Weaver-Wiener diversity index,* which calculates the bits per individual of a set of units or pathways. For species,

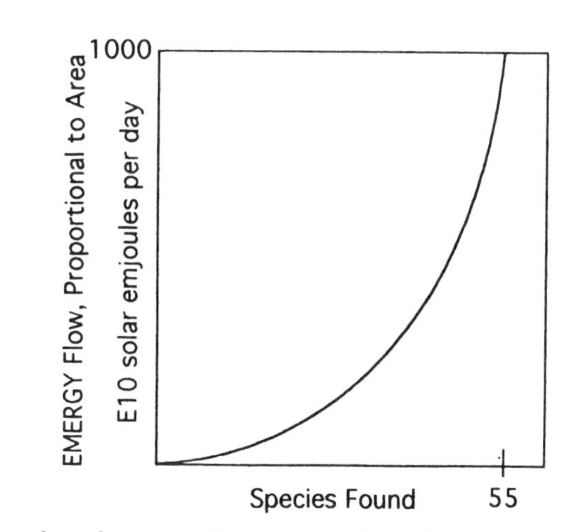

Figure 12.8. Area-based EMERGY flow as a function of the tree-species biodiversity in the Luquillo Forest of Puerto Rico (Keitt, 1991, using data of R. F. Smith, 1970).

values range from a less than 1 to 7 bits per individual in some rainforests. The EMERGY per bit for small units such as microbes is much smaller than that for large units such as trees, even though the number of bits per individual are the same.

Keitt (1991), using satellite tapes of the Luquillo Forest, considered both species diversity and satellite pixel diversity. The EMERGY per bit of diversity increased with area. Diversities of pixels (each 30 m^2) involved more EMERGY than do species diversities. In other words, he found solar EMERGY per bit increasing with scale.

Figure 12.9 shows the same complexity at three levels of size and energy hierarchy. The bits of complexity are the same, but the EMERGY per bit is larger for the larger realm of resource use. Solar EMERGY per bit is determined by the EMERGY flow required to support the configuration. It can be used to compare complexity of different scales.

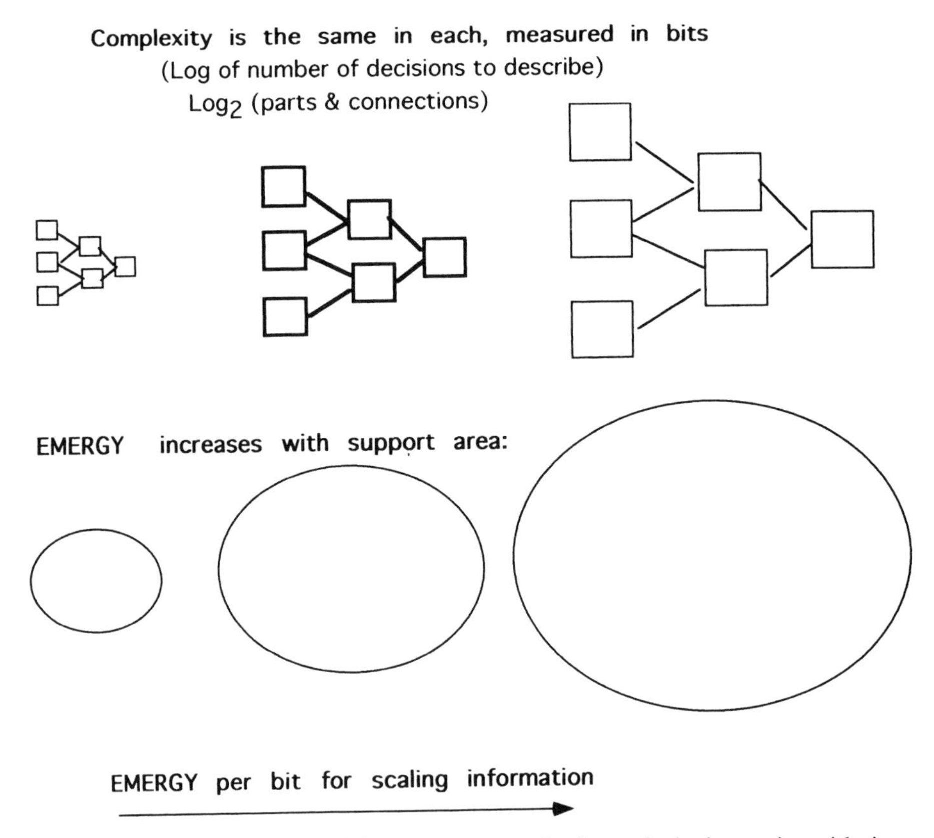

Figure 12.9. Demonstration of the EMERGY per unit of complexity increasing with size. Diagram shows the same information bits on three scales, each with a larger EMERGY storage and requirement.

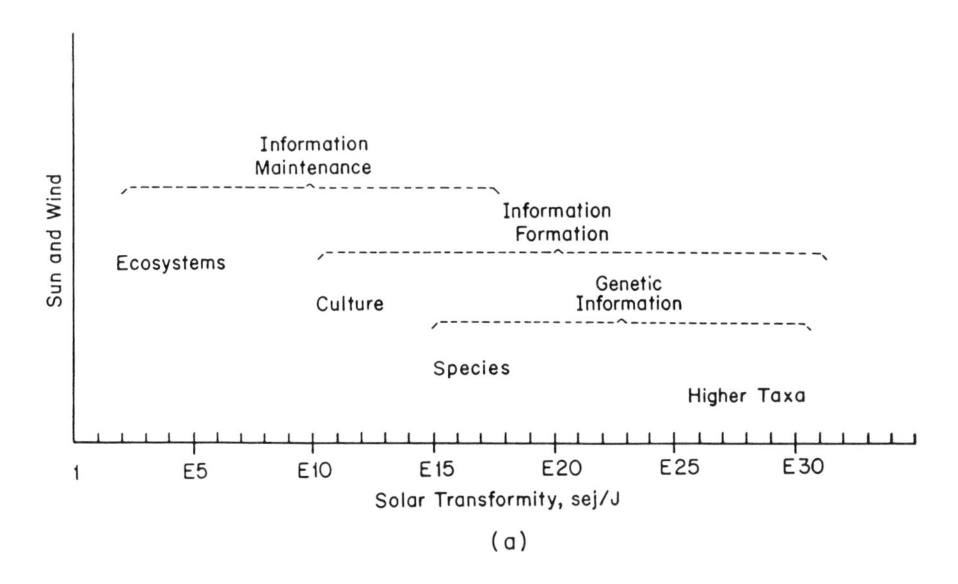

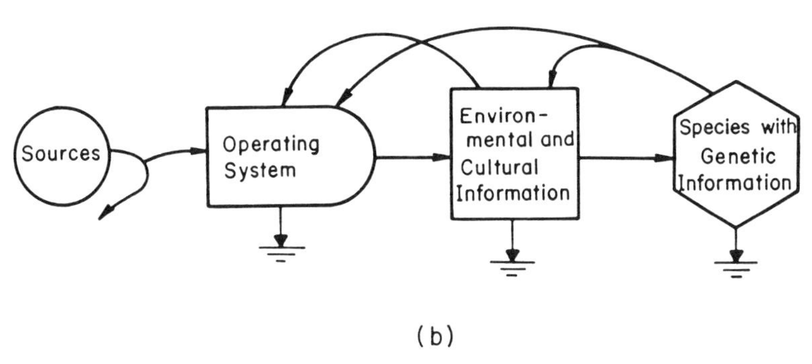

(b)

Figure 12.10. Classes of information on a generalized spectral diagram of quantity and solar transformity.

SUMMARY

EMERGY accounting was used to define and evaluate information in nature and in human affairs by considering genetic inheritance, life cycles, learning, biodiversity, evolution, information bits, television, and global sharing. Several ways were given for estimating the EMERGY flow of human services. Information has a wide range of transformities, as summarized in Figure 12.10.

CHAPTER 13

EMERGY OVER TIME

Because systems of each scale of size and time pulsate, EMERGY flows, storages, and indices also oscillate. Small-scale systems pulsate more frequently than systems of larger scale. For the environmental-economic system of the global scale, the pulse is that of the modern economy. It grows on a cycle ranging from decades to centuries. If pulsing designs maximize performance in the long run, then the baseline of thermodynamically maximum empower may be cyclic. In this chapter, we consider the implications of the pulsing paradigm for EMERGY accounting, varying transformities, and appropriate policies for each stage in oscillation. Study of EMERGY changing with time is facilitated with computer simulation models.

Models of natural systems have often been used as general paradigms for the global system that includes society. The older ecological paradigm of successional growth to a steady state and sustainable climax has to be replaced with the concept of recurring pulses in which production and consumption alternate. In natural systems of many kinds, pulses occur in a repeating cycle.

THEORY OF PULSING PARADIGM

Explained in Chapter 2 was the theory that alternating production and consumption makes a greater EMERGY contribution in the long run. Maximum performance comes with an alternation between times when there is slow net storage of products and times of rapid net consumption of these products. The pulse of transformations by consumers generates higher-quality realms of symbols, art, knowledge, sports, and military capacity.

Growth and succession in oscillating ecological and/or economic systems can be visualized as a short segment within their longer continuing oscillations (Figure 13.1). The growth and leveling portions of the curve somewhat represent the old ecological concepts of succession and climax, but now it is necessary to consider the stage of coming down, after which there is regrowth again. In the simplest of pulsing models (Figure 13.2) there are three storages: the dispersed materials, the gradually assembled resource products, and the storage of temporary assets that are part of the consumer frenzy.

According to the pulsing paradigm, there is a period when net production of primary products is positive, with some growth of consumers. Then, when product storage and consumer storage reach a threshold that accelerates autocatalytic and higher-order pathways, the system shifts to a temporary net consumption of products as the consumer assets make a temporary surge. In the case of the global economy, the accumulated products are the environmental resources, and the consumer storages are our economic, informational, and cultural assets.

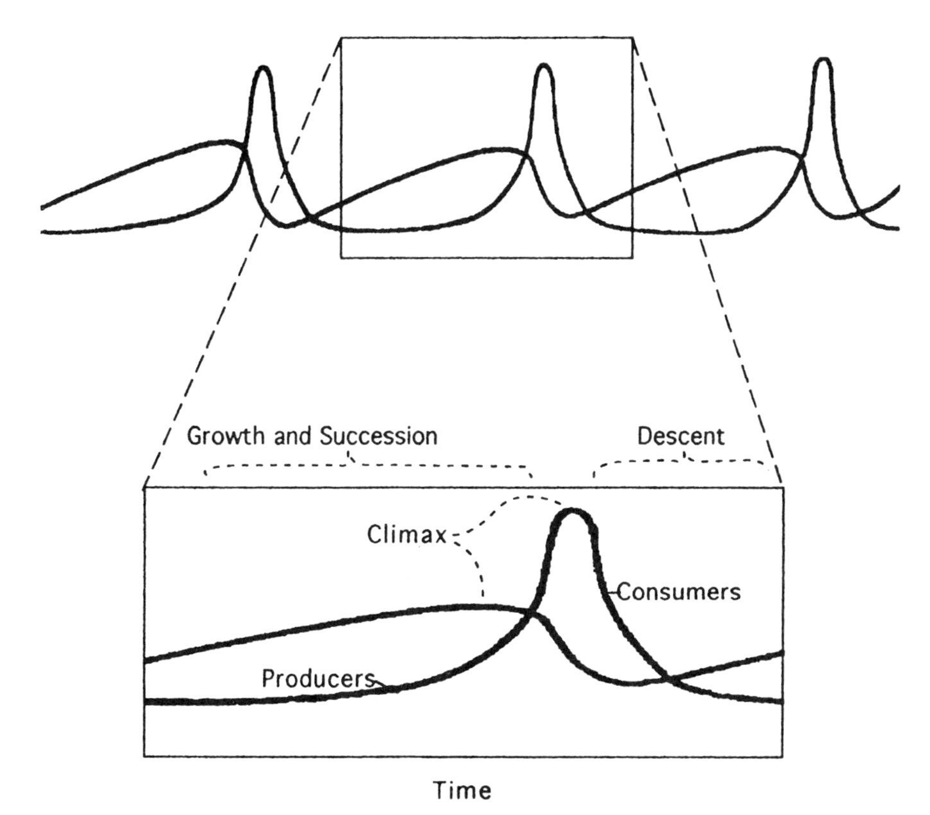

Figure 13.1. Growth, climax, and descent as part of a longer-range sustainable oscillation (H. T. Odum, 1983b, 1994b; reprinted with permission of J. Wiley & Sons, Inc.).

If pulsing maximizes power and performance, then each level of the hierarchy may have an optimum frequency for the alternating periods of gradual production and frenzied consumption. EMERGY, transformity, pulsing time, and interpulse period may be good measures. In the past we have sought constant, power-maximizing, thermodynamic limits for efficiencies, transformities, investment ratios, and so forth. If power is maximized by oscillating systems with storages filling and discharging along the levels of hierarchy, then we have to recognize that energy-transformation parameters have a varying baseline, with different values appropriate at different stages in the oscillatory cycle. Figure 13.2, for example, shows the minimodel, and Figure 13.1 the oscillation of environmental and economic storages that result from simulating its program.

PERSPECTIVES FROM THE MINIMODEL SIMULATION

Minimodels, as aggregated simplifications, can be given their mathematical equations and simulated to help us visualize the consequences in time of network configurations. When dynamic models are simulated, the equations for the state variables are calibrated in terms of energy, mass, or monetary units as may be appropriate for each. After the model is running in the usual way, equations that calculate the EMERGY and transformity of each storage can be added and these results included in tabular or graphical output.

Computer simulations of two simple minimodels are presented here to show the difference between steady sustainability and pulsing sustainability.

Steady Sustainability

Growth and succession in ecological and/or economic systems was often mentally visualized as growth and leveling of one storage representing all structure and diversity. The BASIC language program EMTANK was used to simulate the simple storage of natural capital in Figure 1.6. The program also graphed the solar EMERGY and transformity. During growth, transformity of the storage changed somewhat with size and turnover time (Figure 1.6b,c). The graph in Figure 1.6 reached a steady state after a short period of growth. In economics, the equivalent to the ecological-climax concept for the global economy is "sustainability." However, the steady-state type of sustainability may not be possible because short-term advantage favors consumers that use up accumulated reserves. What is generally observed is pulsing, with small oscillations nested in time and space within larger ones.

Pulsing Sustainability

To express the effect of pulsing on EMERGY flows, we simulated the model shown in Figure 13.2 (Alexander, 1978a). It includes recycling of materials,

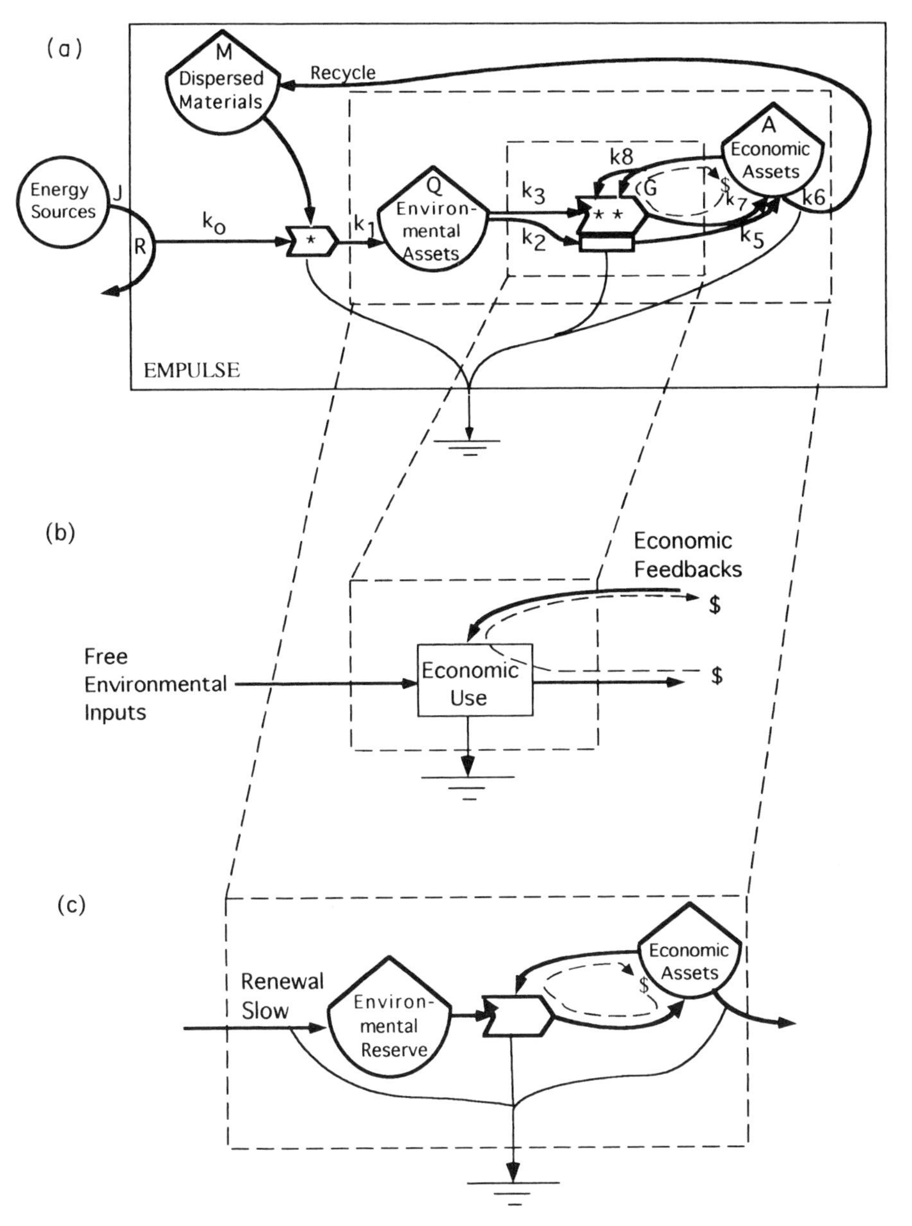

Figure 13.2. A model involving three levels of energy-transformation hierarchy with a design that sustains the pulsing pattern over time shown in Figure 13.1. Simulation program EMPULSE (see Appendix F, Table F.2): Energy sources, J with unused fraction R; total materials, k; dispersed materials by difference, M; environmental storage, Q; economic assets, A; constant gross economic product, G. Equations: $R = J/(1 + ko * M)$; $M = k - 0.1 * Q$; $DQ = k1 * R * M - k2 * Q - k3 * Q * A * A$; $DA = k7 * Q * A * A + k5 * Q - k8 * Q * A * A - k6 * A$. **(a)** Energy systems diagram. **(b)** Smaller window for the model of the environmental-economic interface within the pulsing model. **(c)** relationships in an intermediate-sized window for global models based on renewable and nonrenewable resources. See Figure 13.11.

competition of consumer pathways of higher mathematical exponents (Q, Q^2, Q^3, and so on), autocatalytic reinforcement, and interactive feedbacks from storages of transformed energy.

In this model, environmental assets are accumulated from a productive interaction between dispersed materials and energy pumped from a renewable environmental source. The lower pathway (linear) makes economic assets in proportion to the available environmental assets. The second pathway generates economic assets from environmental assets with a process that is quadratically autocatalytic, as is appropriate for high levels of energy processing. The upper pathway exiting the economic assets storage represents the loop of dispersing materials whose recycle is necessary for the next cycle. The results of the simulation (Figure 13.3) show the alternating rise and fall of economic assets, dispersed raw materials, and environmental assets.

Richardson and Odum (1981) and Richardson (1988) studied the power-maximizing properties of this design. Energy flows rise and fall as the model alternately fills and drains the storages of natural reserves and economic assets (Figure 13.3, top) and materials (Figure 13.3, bottom). Each pulse has its stages of growth, a climax, and a period of descent. Each pulse is part of oscillations that may be quasi-steady state when considered over a longer period. Figure 13.1 shows the way the older paradigm can be visualized as a part of a repeating oscillation (H. T. Odum, 1983b, 1994b).

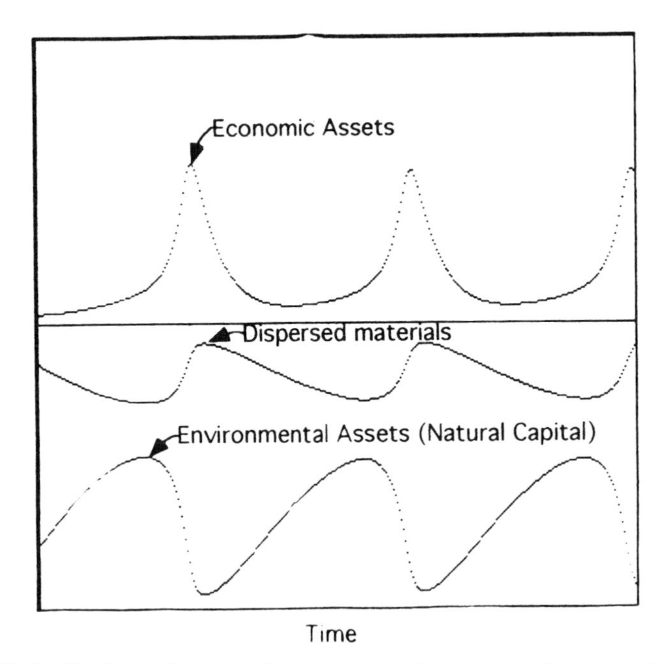

Figure 13.3. Oscillations of economic assets, materials, and environmental reserve in the pulsing minimodel shown in Figure 13.2.

Indices over Time for the Pulsing Environmental-Economic Interface

With the help of the inner, dashed-line box in Figure 13.2a, the window of our attention to the pulsing model may be narrowed to the environmental-economic interface of "economic use in Figure 13.2b. Economic-EMERGY indices of the simulation are graphed in Figure 13.4.

The first graph of prices (Figure 13.4a) was obtained by holding the money supply constant. Costs and prices of products are at a minimum when the rate of use of resources is largest, that is, when production of wealth is greatest. The property of decreasing cost per unit resource was misinterpreted by Simon and Kahn (1984) as evidence that resource shortages are not important to economic vitality. The simulation shows the way resources determine the curves produced by those authors. The declining-cost part of the curve occurs when the resources are being transformed into assets that help process more. The declining cost is the action of resources feeding back. Some of this is resource EMERGY in the form of technology increasing efficiency (as said by Kahn and Simon), but these are resource based (not permanent without resource use) and depreciate away when their EMERGY support lessens.

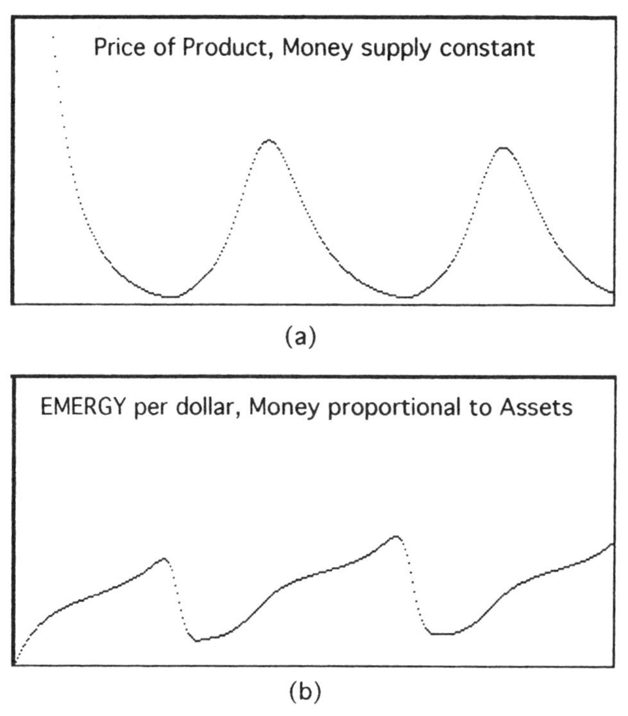

Figure 13.4. Oscillating economic properties of the pulsing minimodel shown in Figure 13.2. **(a)** Product prices, where money supply is held constant. **(b)** EMERGY/money ratio, where money is kept proportional to economic assets.

The EMERGY/money ratio is graphed in Figure 13.4b. Money was added in proportion to assets, so that there was no inflation. Toward the end of the rapid growth period the EMERGY/money ratio falls, as has been observed in highly developed countries (see Figures 10.4 and 10.5).

The EMERGY investment ratio (the ratio of the purchased EMERGY feeding back from the economy to the free contributions of EMERGY from the environment) during the pulsing simulation is graphed in Figure 13.5. It measures the intensity of economic development relative to environmental use. With the development of economic assets in the period of frenzied growth, the environmental loading increases. During the 20th century much natural capital has been used up to develop growth of economic assets, but in the next stage there may be a rebuilding of environmental resources with a decline of developed assets.

VARIATION IN TRANSFORMITY

If we evaluate the transformity of a newly initiated process, which has not been running long enough to develop its maximum efficiencies for full sustainable production, then a higher transformity (i.e., a lower efficiency) may be found at that point than is found later after the system is operating at highest output possible. For example, the transformity of steam-engine work calculated from steam engines in 1910 was higher than the transformity for the better engines that had been developed by 1940.

Eventually, through trial, error, and selection, transformities reach the theoretical lower limit (corresponding to the most efficient possible system for maximum power), reflecting the inherent thermodynamic limits of energy transformation in open systems. For example, to convert dilute solar energy falling on the earth into a concentrated fuel requires that much of the energy be degraded as part of the process of transforming a small part. If the green plants after 1 billion yr of evolution have achieved the maximum-possible

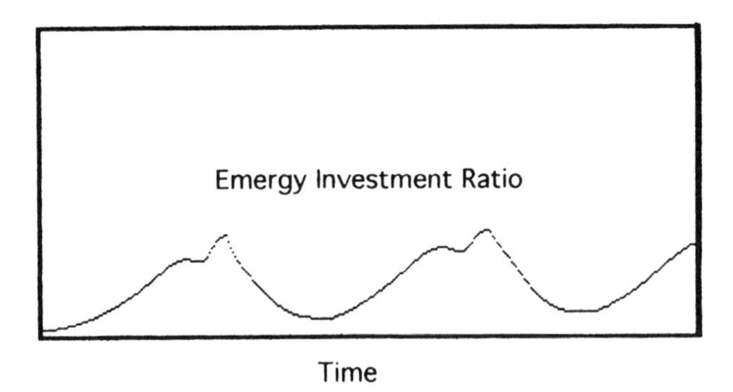

Figure 13.5. EMERGY investment ratio for development of economic assets during the oscillation of the pulsing minimodel shown in Figure 13.2.

sustainable output of organic production, then the transformity we obtain from evaluating biomass conversions from solar energy may not be improved by new technology.

However, efficiencies vary during pulsing cycles. The rise and fall of the consumer assets causes some oscillation in the transformities because of the different feedback loadings on the production process. The efficiency is the reciprocal of the transformity. If the general paradigm is a sustainable but pulsing pattern, some shifting of the efficiency may be competitive. As assets build up in a developed economy, the efficiency decreases (i.e., transformity increases). In other words, there is a trade-off between yield and efficiency. Both alternate during the sustainable oscillation that represents maximum sustainable empower in the longer run.

For the same conversion, higher transformity (lower efficiency) may maximize processes during periods of rapid growth. The best solar transformity possible is less efficient in times of accelerated growth. Later, as the high point of development of consumer assets is reached, efficiency of the same process may improve as the variable most important in maximizing empower. This discussion of efficiency variation for a single process should not be confused with the higher EMERGY and transformities of assets as a whole at climax.

In Figure 13.6 the transformities for penaeid shrimp production are plotted as a function of the investment ratio. Higher, less-efficient transformities go with more development (higher investment ratios). Efficiency is greater when the matching between purchased feedbacks and free environmental resources is more even (e.g., Figure 9.2a).

For those doing EMERGY accounting, the question is which transformity to use. The answer depends on what kind of analysis is being made. In some situations, the analysis is directed at finding out what is actually required in a particular stage of the pulsing cycle. In that case, the transformity determined for that stage of growth or efficiency should be used. On the other hand, some analyses seek to answer the question, what is the most efficient conversion possible? In that situation the lowest (most efficient) transformity known should be used.

Each new EMERGY analysis generates more transformities. As more and more analyses are done, and we obtain many independent values for the same kinds of products, it eventually becomes clear which are the consistent lower values. Then, when EMERGY accounting is applied to a new situation, we can judge better whether a process is efficient or inefficient, and which method is appropriately competitive, considering the phase of timing within the oscillating cycle.

EMERGY EVALUATION IN DYNAMIC SIMULATIONS WITH EXTEND

EXTEND[1] (for Macintosh and PC) is a remarkably flexible simulation program in which connecting icons on the computer screen set up the equations

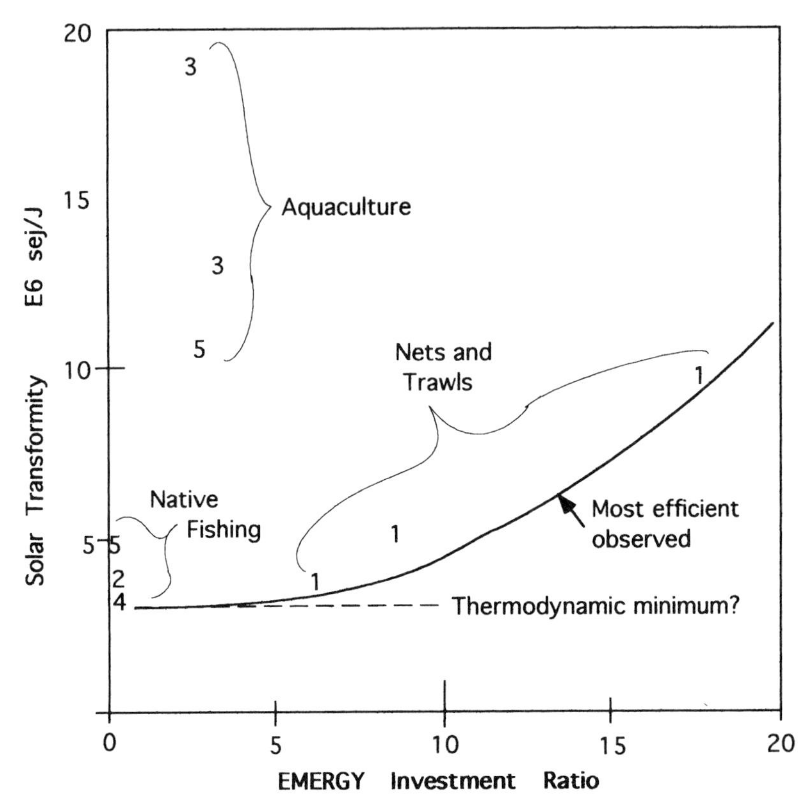

Figure 13.6. Results of EMERGY evaluations of shrimp production plotted in coordinates of solar transformity and EMERGY investment ratio. Dashed line indicates minimum transformities. Data points: 1, Gulf of Mexico, small, medium, and large vessels (Fonyo, 1983); 2, Sea of Cortez, Mexico (M. T. Brown et al., 1991); 3, aquaculture in Ecuador with (upper point) and without supplemental feed (Odum and Arding, 1991); 4, native fishery in southern Brazil (modified from Philomena, 1991); 5, Nayarit, Mexico (M. T. Brown et al., 1992).

and arrange computations for simulating models. We programmed blocks for each energy systems symbol so that an energy systems diagram that is arranged on the screen and given some calibration values automatically simulates the model (Odum and Peterson, 1995). These blocks are on the disk accompanying this text as an *energy systems library* usable by EXTEND. Figure 13.7 shows the screen obtained when the icons for all the blocks in the library are called up but not connected. Quantities or flows are graphed by connecting the block to be graphed to a plotter icon (quantity plotter or flow plotter). Plotter blocks are also programmed to calculate empower, EMERGY storage, and transformities. To graph these properties, connect systems blocks to plotter icons (the empower EMERGY storage, and transformity plotter icons, respectively).

For example, Figures 13.8 and 13.9 show the environmental-economic interface system model ENVUSE. (This program is one available on disk

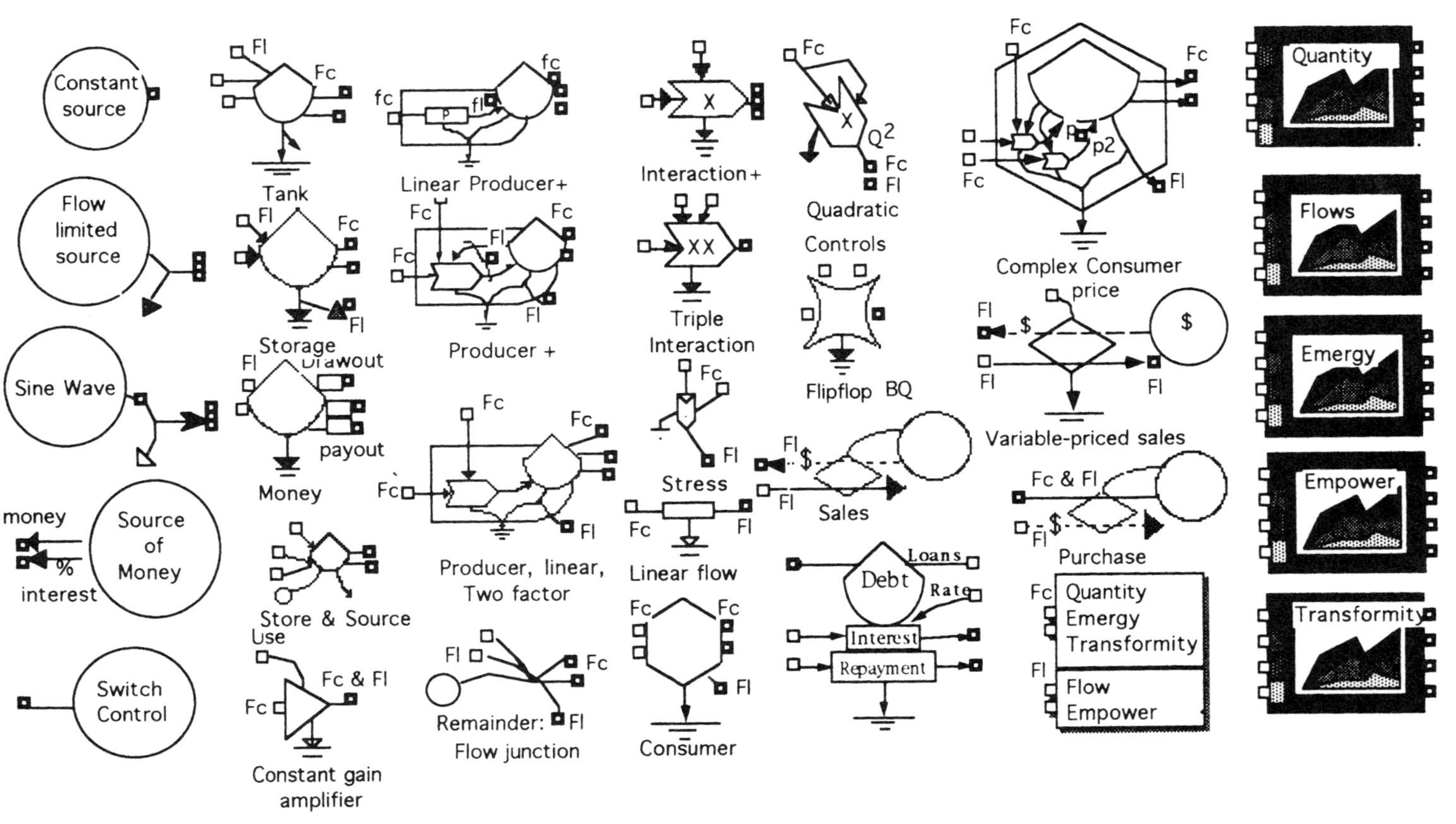

Figure 13.7. Macintosh screen showing icons of the energy systems library for EXTEND.

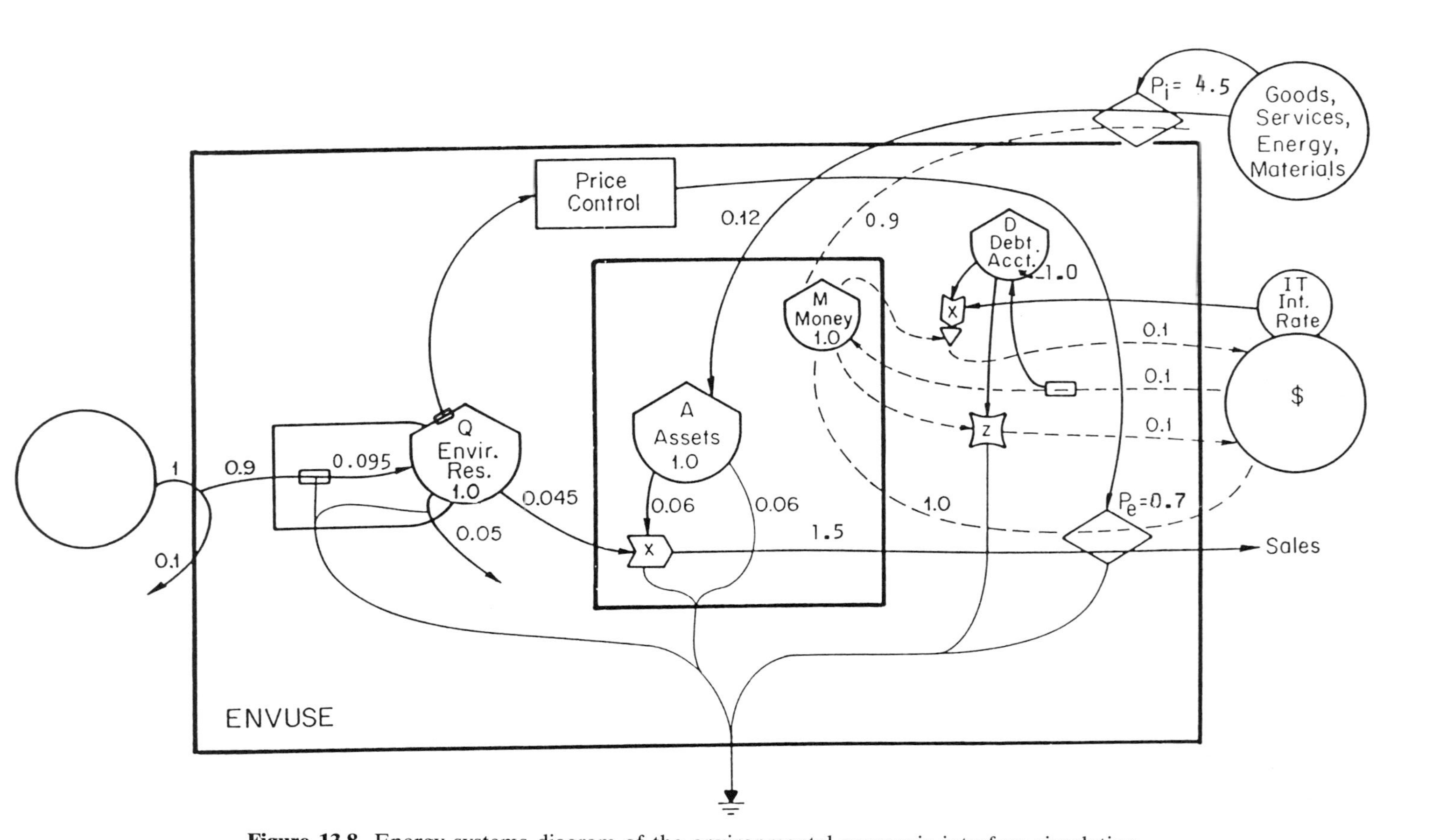

Figure 13.8. Energy systems diagram of the environmental-economic interface simulation program ENVUSE (Table F.3), including calibration data.

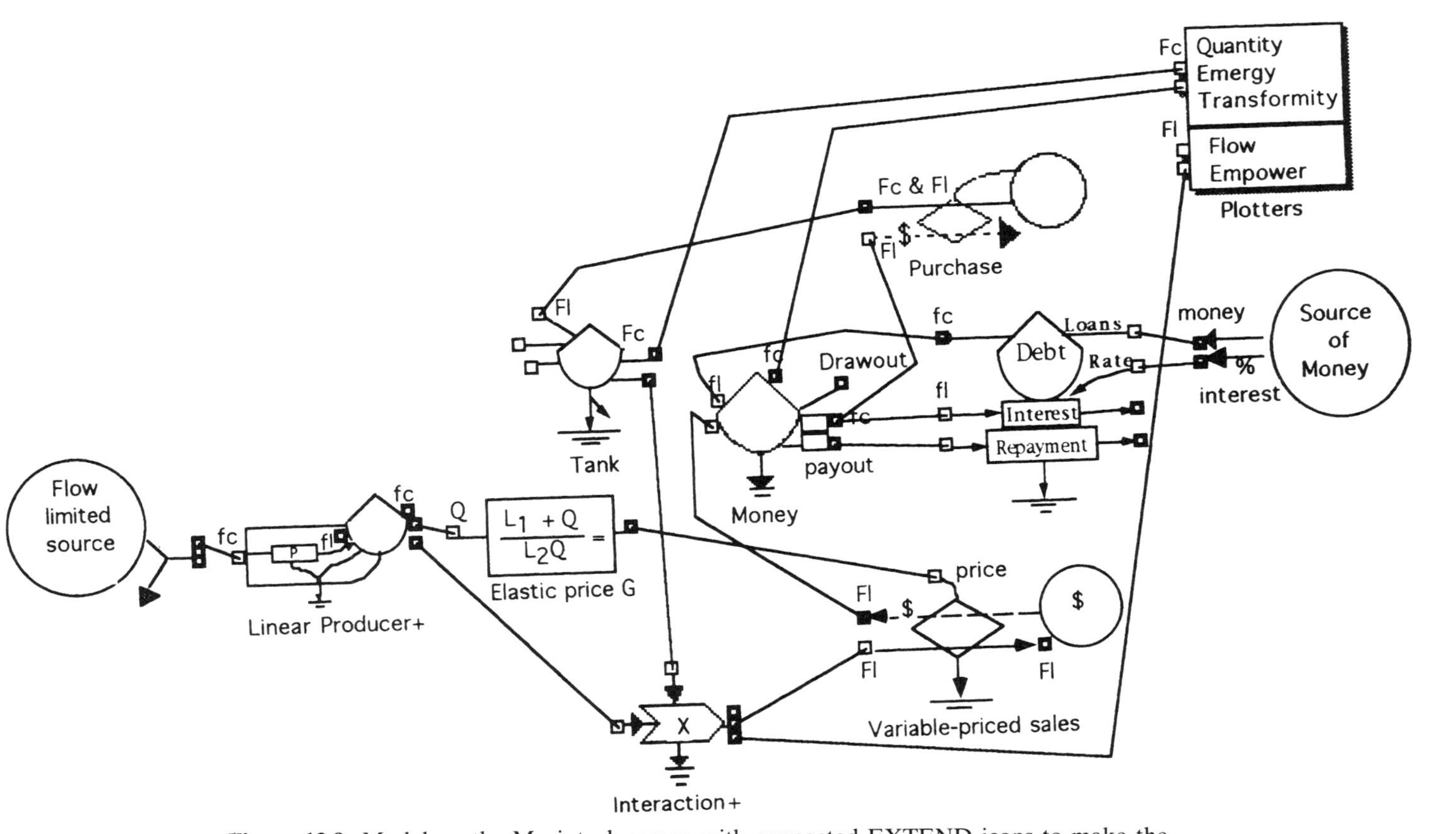

Figure 13.9. Model on the Macintosh screen with connected EXTEND icons to make the simulation program ENVUSE. On each block fc and fl identify connector inputs and outputs as forces and flows.

and in Appendix F. This model was simulated in BASIC and EXTEND. Table 13.1 lists the equations, and Table 13.2 shows the spreadsheet used to determine the coefficients from calibration data. Figure 13.8 is an energy systems diagram with calibration data, while Figure 13.9 is the screen view of the model on EXTEND with the symbol blocks connected. Graphs of storages and EMERGY characteristics resulting from the EXTEND simulation are plotted in Figure 13.10. Starting with small initial storages for economic assets, the assets grow, show oscillatory transients, and level off in a steady state.

The systems window that examines the environmental-economic interface was shown in Figure 13.2b to be a small part of the larger energy-transformation hierarchy that generates the sustainable pulsing. Interface models like this one can be simulated for more realistic time-space considerations by adding as external sources the pulses that come from the larger window.

TABLE 13.1. Equations for the Simulation Model ENVUSE.BAS (See Figures 13.8 and 13.9)

Storages (state variables):
- Q = stock of environmental produce
- A = assets of the economic user system
- M = money on hand
- D = debt account

Sources (outside influences):
- I = environmental sources, sun, rain, etc.
- JL = annual loans to the economic activity
- IT = interest rate on loans from outside
- PI = price of purchased inputs

Equations and Change Equations

Unused environmental input	R = I − k0 * R and therefore:
	R = I/(1 + k0)
Environmental Resource Storage	DQ = k1 * R − k2 * Q − k3 * A * Q
Economic production	= k4 * A * Q
Price of commodity	Pe = (L1 + Q)/(L2 * Q) a limit at 3
User assets	DA = k5 * M/Pi − k6 * A − k7 * A * Q
Money on hand	DM = Pe * k4 * A * Q + JL − k9 * M − L2 * M − k5 * M − IT * L − Z * k9 * M
Debt account	DL = JL − Z * K9 * M where repayment is when Z = 1
Rising input price	= Pi increase = 0.2/yr when switch W = 1

TABLE 13.2. Simulation Program ENVUSE Spreadsheet for Calculating Coefficients

Name of Item	Expression		Value		Coefficients
	Storages for Calibration				
Assets (economic)	A	=	1.00 E + 00		
Stock, unused steady state	Q	=	1.00 E + 00		
Stock, in use steady state	Q	=	1.00 E + 01		
Money	M	=	1.00 E + 00		
Debt (loan account)	D	=	5.00 E − 01		
	Sources for Calibration				
Environmental resources	I	=	1.00 E + 00		
Unused environmental resources	R	=	1.00 E − 01		
Loan rate	JL	=	1.00 E − 01		
Interest rate	IT	=	1.00 E − 01		
Price of purchased inputs	PI	=	4.50 E + 00		
	Flows and Their Coefficients				
Environmental input use	K0 * R	=	9.00 E − 01	Therefore	K0 = 9.00 E + 00
Stock production	K1 * R	=	9.50 E − 02	Therefore	K1 = 9.50 E − 01
Depreciation rate	K2 * Q	=	5.00 E − 02	Therefore	K2 = 5.00 E − 02
Rate of use of stock	K3 * A * Q	=	4.50 E − 02	Therefore	K3 = 4.50 E − 01
Commodity production	K4 * A * Q	=	1.50 E + 00	Therefore	K4 = 1.50 E + 01
Payment for services	K5 * M	=	9.00 E − 01	Therefore	K5 = 9.00 E − 01
Depreciation rate	K6 * A	=	6.00 E − 02	Therefore	K6 = 6.00 E − 02
Use by production	K7 * A * Q	=	6.00 E − 02	Therefore	K7 = 6.00 E − 01
Loan repayments	K9 * Z * M	=	1.00 E − 01	Therefore	K9 = 1.00 E − 01
Abundance price	1/L2	=	5.00 E − 01	Therefore	L2 = 2.00 E + 00
Scarcity price	(L1 + Q)/(L2 * Q)	=	7.00 E − 01	Therefore	L1 = 4.00 E − 01

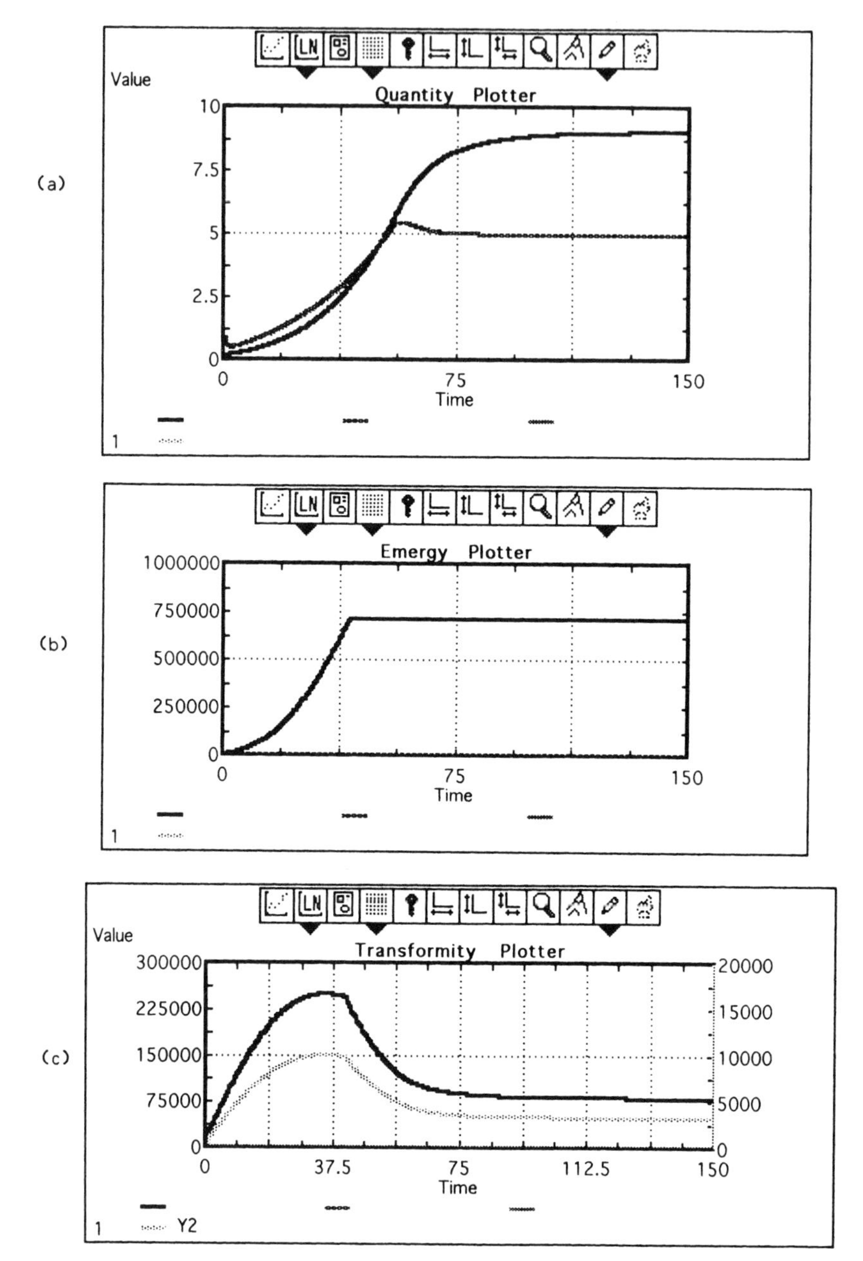

Figure 13.10. Graphs resulting from using EXTEND for simulating the ENVUSE program shown in Figure 13.8, 13.9. **(a)** Quantity plotter. **(b)** EMERGY plotter. **(c)** Transformity plotter. **(d)** Flow plotter. **(e)** Empower plotter.

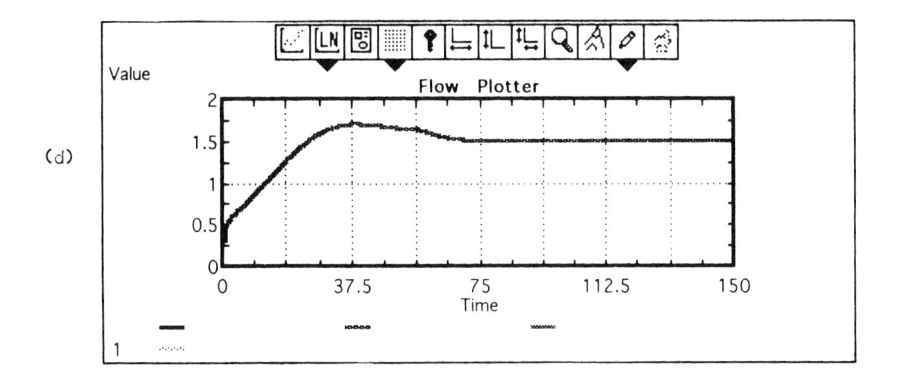

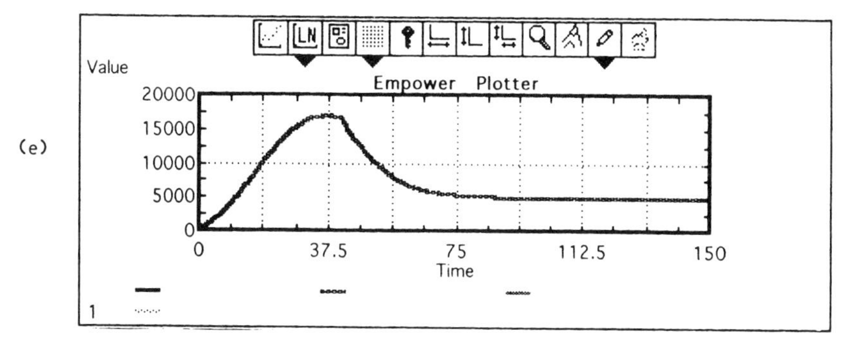

Figure 13.10. (*Continued*)

RENEWABLE-NONRENEWABLE MODELS OF ENVIRONMENT AND SOCIETY

Whatever the computer program language that is used for simulation, the essence of resource-based overview models of our global future is a very characteristic shape of growth and descent that results from interactive use of renewable and nonrenewable sources. Figure 13.11a shows one of this class of models. When this model is simulated with BASIC program RNRPRICE, graphs in Figure 13.11b result. Growth passes through a peak as the nonrenewable (slowly renewable) resources are reduced. The maximum production rate occurs before assets peak. With the money supply held constant, prices are least at the time of maximum productivity. The pulse of growth based on frenzied consumption (Figure 13.11b) is not unlike the famous pulse of fuel use calculated by M. K. Hubbert (1949) to show dramatically the temporary nature of a fuel-based civilization.

However, when viewed over a longer scale of time, the unsustainable climax of the renewable-nonrenewable scenario (the model of Figure 13.11) can be visualized as part of a continuing oscillation (Figure 13.2a). While helping us to understand growth and decline to a lower level in the short run, the smaller-

(a)

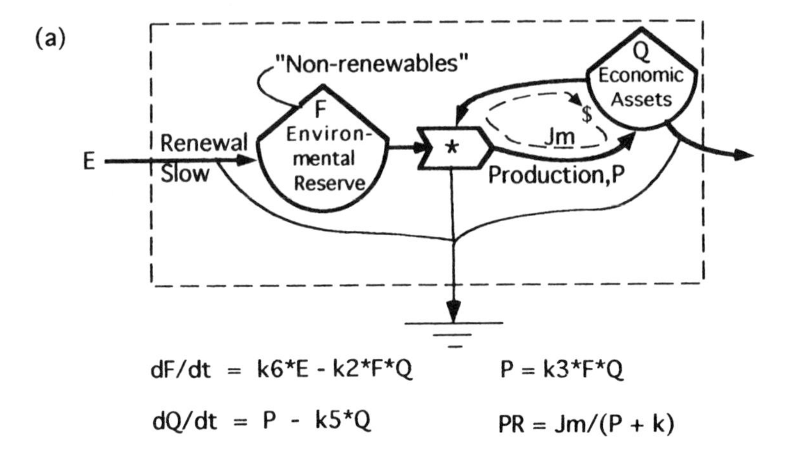

(b)

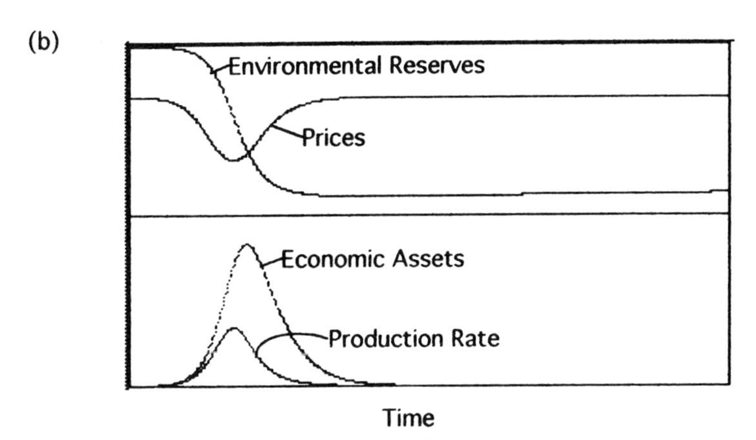

Figure 13.11. Example of a resource-based renewable-nonrenewable model of the global future. **(a)** Energy systems diagram. **(b)** Typical simulation with BASIC program RNRPRICE (Table F.4) results—nonrenewables, economic assets, productivity, and prices.

window model (Figure 13.2c, 13.11a) does not include enough of the longer-term, larger-scale processes to show the repetitive pulsing that may be characteristic of all systems in the long run.

For a wider perspective, we relate the renewable-nonrenewable systems window frame (Figure 13.2c) to the larger energy-transformation hierarchy (Figure 13.2a). As the window of attention is expanded, the nonrenewable resource is found to be pulsing but sustainable in the long run. Recall that simulation of production and consumption on the larger scale generates a steady-state alternation (Figure 13.1).

Models of this class build up high EMERGY in the storage of assets. If shared information is accumulated with growth, assets may develop a higher transformity along with a lowered turnover time and depreciation rate. In other words, the peak of stored EMERGY will be higher and longer than that produced by the simple minimodel shown in Figure 13.11.

NIGHT LIGHTS AND THE PULSING OF NATIONS

From satellite, the centers of human activity can be seen by the night lights, which are indicators of high-transformity EMERGY flows. Night lights in cities are a high-transformity means for increasing information processing over longer periods. The concentration of lights show the centers of hierarchy: small towns, cities, and national urban centers. As cultures and nations surge in activity, power, and influence, the rise and decline of civilized centers would look like the flashing lights of a Christmas tree to someone watching from satellite at night. In the 20th century, the civilization of the whole world has surged in unison because of the shared information generated with the vast EMERGY of the fossil fuels.

SUMMARY

Oscillations are observed in systems on every scale. Storages at lower levels of hierarchy are used up in transformations to fill storages at higher levels. Simulations of energy-constrained minimodels were used to help understand the stages in cycles of growth and descent. In the course of an oscillation, indices of environmental-economic interfaces, such as the investment ratio and efficiency, vary with the cycle. If oscillating systems are the most productive in the long run for thermodynamic reasons, then policies to maximize resource production and use should fit the stage of the cycle. EMERGY accounting indices may aid in the selection of alternative plans and projects that are appropriate for reinforcement at each phase of the cycle.

NOTES

[1] EXTEND is obtained from Imagine that, Inc., 6830 Via Del Oro, Suite 230, San Jose, California 95119-1353; Phone 408-365-0305. The General Systems Library with blocks on disk (Figure 13.7) for simulating with EXTEND can be obtained from the author.

CHAPTER 14

COMPARISON OF METHODS

The quest for measures of value has a long history, using many procedures involving money, energy, and other quantities. Martinez-Alier (1987) traced energy-based accounting ideas back into 19th century. Some of these are summarized with Table 14.1. In this chapter other methods of evaluation are compared with EMERGY accounting, which may be applied better if its relationship to related measures is understood.

MEASURES OF VALUE

Market value, defined as *what people are willing to pay,* is the principal method of accounting used in economics. It is a value determined by the human receiver according to short-range needs and expected benefits. As is well known, market prices go up and down with scarcity or abundance and other factors. The human economic behavior to pay more in response to scarcity helps the processes of smaller scales self-organize humans and their economy. As explained in Chapter 4, market prices are not helpful for direct evaluation of contributions from the environment and often respond inversely (prices are least when contributions are largest).

Real wealth needs a donor-determined value, a measure of what was required to make an item or generate a service. EMERGY is a donor-type value. Figure 14.1 shows donor- and receiver-type values. One type should not be confused with or used for the other.

TABLE 14.1. Historical Record of Ecological Economics*

Date	Author	Content
1750–1840	"Physiocrats"	Qualitative: resource basis of economics
1840	J. Joule J. Mayer	Energy defined as heat equivalents (first law)
1850	R. Clausius	Energy degradation (second law); evaluation of energy sources for economies
1867	K. Marx	Labor basis for value
1880	S. Podalinsky	Energy budgets of agriculture; energy theory of value; energy return on human labor
1881	E. Sacher	Energy determinism in history; energy budgets for regional economies
1885	P. Geddes	Energy-material requirements for production; energy basis for cities; social energetics
1886	L. Boltzman	Darwinian competition for energy; entropy and states
1902	L. Pfaundler	Energy carrying capacity
1909	W. Ostwald	Cultural advancement with new energy technology; energetic imperative; basis for culture; mental energy
1921	F. Soddy	Interest requires growth; financial capital not real capital; depreciation of capital assets; energy relation to inflation
1922	A. Lotka	Maximum power principle as fourth law; biogeochemical kinetics; population ecology

* See review by Martinez-Alier (1987).

Evaluating by Substitution of Effects on Production

As diagrammed in Figure 14.2a, production in ecosystems and in economic systems is usually proportional to the product of two or more necessary inputs (A and B in Figure 14.2a). The same production can be obtained with more of one input and less of the other. In this sense there is substitution. Figure 14.2b is a graph of the two input factors for constant production. In other words, if production $P = kAB$, then for constant production P_k, $A = (P_k/k)/B$; A is thus an inverse function of B, and the plotted graph of constant production is a hyperbola (isoquant in Figure 14.2b). Systems self-organize for maximum production and optimum efficiency, which requires that the quantity of each input to production be matched by the other with equal marginal effect (i.e., equal derivative of production as a function of input; no input more limiting than another). The dashed lines in Figure 14.2b show the point where A and B have similar effect on production.

In Figure 14.2c one of the inputs to production (A) is environmentally free (from the left), and one (B) (from above) is paid. For example, a fisherman's catch depends on the free contributions of fish from the ecosystem and the

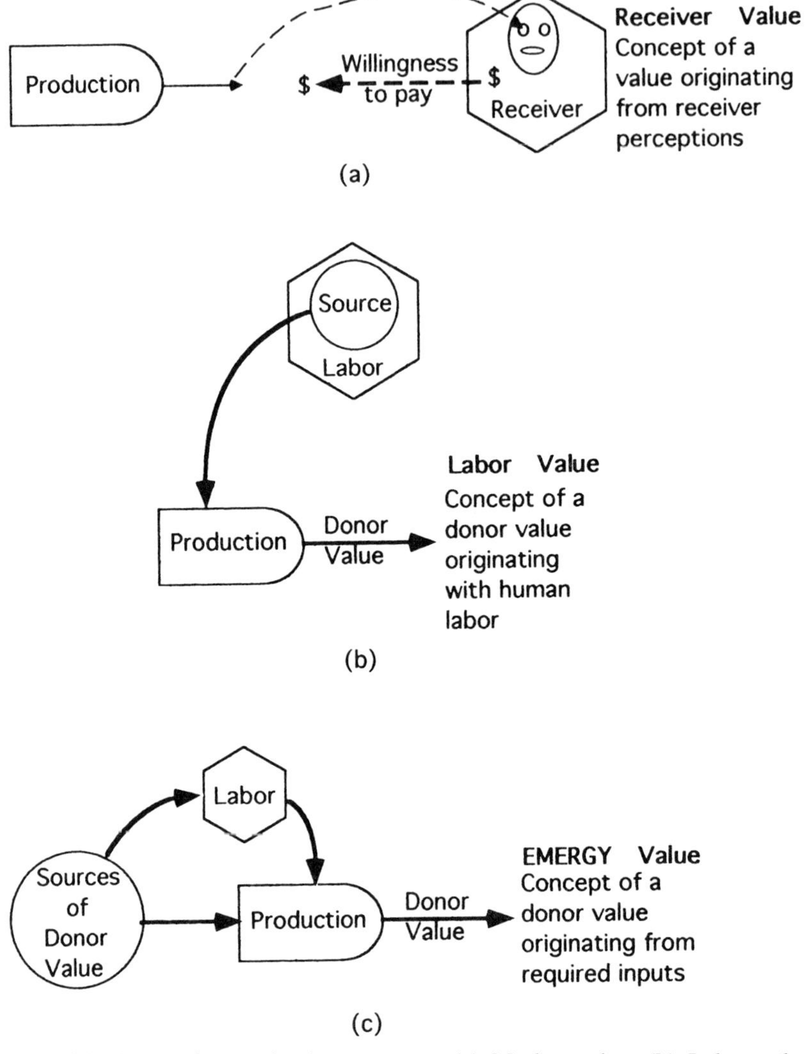

Figure 14.1. Comparisons of value concepts. **(a)** Market value. **(b)** Labor value of Marx. **(c)** EMERGY value.

paid inputs of fisherman services. The free inputs are evaluated by giving them a dollar value according to their substitution for paid inputs in Figure 14.2b. In other words, fish are evaluated according to their substitution for fisherman services. This is because the effect on the receiver of the product is the concept of value used.

However, as already discussed in Chapter 4, money flow only measures the human services of inputs and gives no clue as to how much total wealth is represented by the free environmental contribution (or the many paid

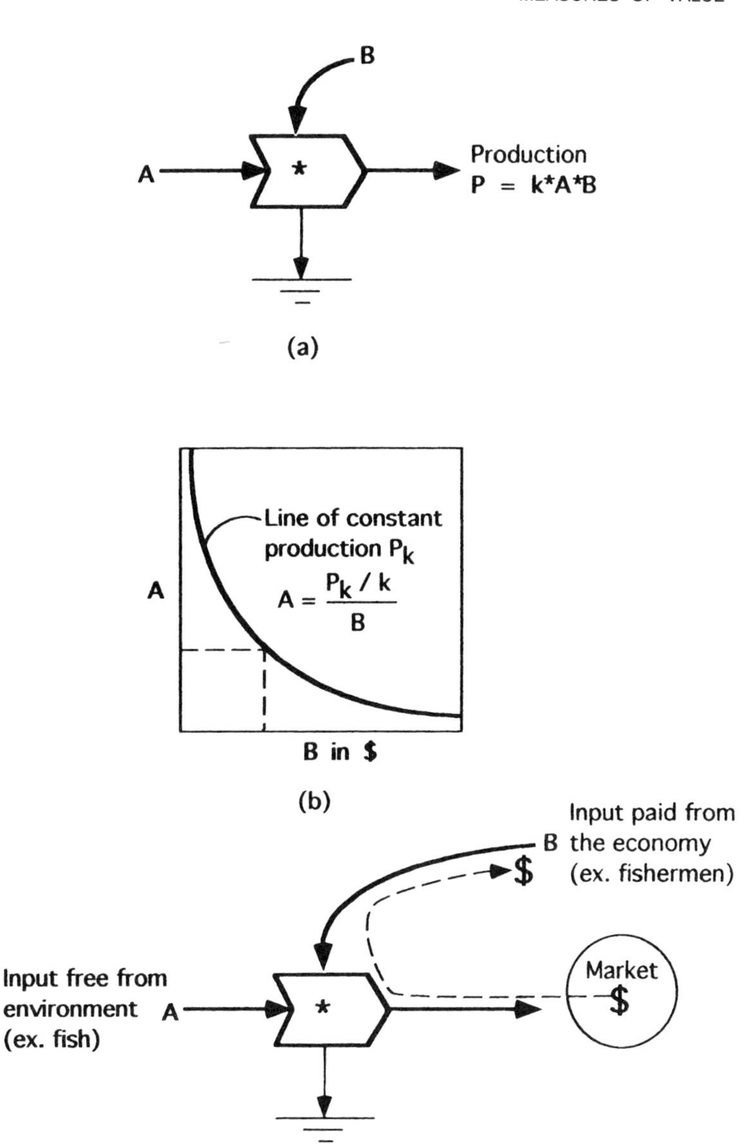

Figure 14.2. Production and the substitution of inputs. **(a)** energy systems diagram, **(b)** input values for constant production, **(c)** economic use interface.

inputs). Inputs to production are not equivalent but have different transformities. To truly substitute for something in a production process means to find an alternative with equivalent transformity and EMERGY contribution.

Using isoquant arguments, ridiculous suggestions have been made that fisherman can almost completely substitute for fish. A market economy tends to substitute fishermen for fish as the fish get scarce, as prices rise, and as productivity declines. Without fish stock the ecosystem cannot reproduce fish and the fishery collapses.

Labor Value

Karl Marx offered a labor concept of donor value in 1867. As sketched in Figure 14.1b, value is generated in proportion to the human services contributed. Value is measured in hours of labor or their gold equivalents. Marx's labor value is a measure of the previous contribution to the product and, like EMERGY, is a donor value carried forward through the economic network with products to users. Values generated from labor are considered to be used up (disappearing) at the point where the pathways return to intersect their source flow again (to labor). Thus, some aspects of labor value accounting resemble EMERGY accounting. Lonergan (1988) found similarities in numerical comparisons. However, Marx's labor value starts with humans as though value were a fountain that emerged within workers equally, without input requirements. Actually, the contributions by labor depend on the EMERGY inputs in resources, information, and the machines used by the human workers (evaluated in Chapter 12). For example, the EMERGY per hour of labor in the United States in the 20th century (Table 14.2) is vastly greater than that in 19th century (Woithe, 1994).

Technocrats and Energy Certificates

Starting in the 1930s an attempt was made to apply the energy theory of value in action programs by a group of people called *Technocrats,* led by Howard Scott. Technocrats advocated doing away with money and paying people with energy certificates. In addition to the error of evaluating all kinds of energy equally, they confused market value with public policy value, which is sharply differentiated in this book. Both market values and EMERGY values are essential, with different roles. Since the dollar at any time represents the EMERGY buying power (see the EMERGY/money ratios in Table 10.7), a dollar is an EMERGY certificate (not an energy certificate). There is little in common between technocracy accounting and EMERGY evaluations.

Energy Availability Analysis

In many approaches, available energy has been used as the measure of contribution (without the use of transformities to relate one form of energy to another). The word *exergy* is used for the sum of available energies of various kinds. In 1974 a workshop sponsored by the International Federation of Institutes for Advanced Study (IFIAS) was held to "standardize energy analysis." The resulting report (IFIAS, 1974) defined "Gibbs free energy" (chemical potential energy) as the measure to use (see later section, "Exergy and Process

TABLE 14.2. Energy, Embodied Energy, and EMERGY Assigned to Human Service**

Procedure and Type of Measure	Numerical Value and Units
Energy per Person	
Body metabolism, food energy units	2500 kcal/day 10.47×10^6 J/day
Embodied Energy, Fuel Equivalents per Person	
Fuel use per capita, USA, fuel units*	231,000 kcal/day 9.7×10^8 J/day
Input-output embodied fuel energy per person	229,000 kcal/day 9.58×10^8 J/day
EMERGY Use per Person	
Solar EMERGY‡	2.23×10^{10} solar emkcal/day 9.35×10^{13} sej/day
Coal EMERGY§	557,500 coal emkcal/day 23.3×10^8 coal emj/day

* U.S. national fuel use for 1993 (8.86×10^{19} J/yr from Table D1) divided by population of 250 million people, by 4186 kcal/J, and by 365 days.
† Energy inputs per dollar obtained with input-output methods and multiplied by income per capita in the U.S.A.:

$$(11{,}287 \text{ fuel kcal/1975 \$})(\$1.6 \times 10^{12} \text{ GNP})/[216 \times 10^6 \text{ people})(365 \text{ days/yr})]$$

‡ Solar empower use obtained by dividing total annual solar EMERGY use of the USA ($U = 7.85 \times 10^{24}$ sej/yr) by 365 days and by 230×10^6 people.
§ Solar empower use divided by solar transformity of coal (40,000 sej/J).
** For explanation of kilocalorie, emkcal, and related units, see footnote in Table 1.1.

Analysis). Some of the ratios and indices used in that report and by others are similar to those in this book except that IFIAS uses the concept of energy (not EMERGY). Figure 14.3 summarizes and compares some of these and their terminology. Unfortunately, there is an enormous range of EMERGY values for items having the same free energy. For example, 1 g of glucose has more free energy than 1 g of dynamite, but the EMERGY of dynamite is much larger because the transformity is higher, as we might expect with something that requires more work to prepare and has a commensurately larger effect. The error in using energy instead of EMERGY increases with the range of transformities compared.

Embodied-Energy Evaluations

Several concepts of embodied energy (sequestered energy) have been developed; *embodied energy* is *energy previously processed to make a product.*

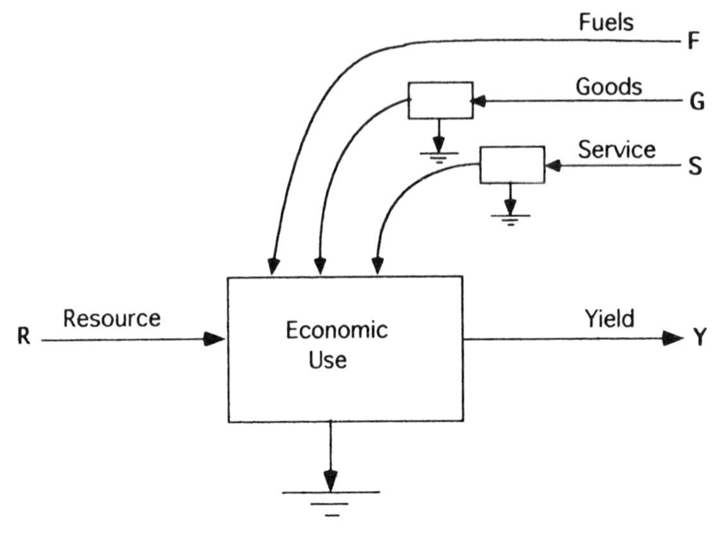

Figure 14.3. Comparison of terminology for indices. Definitions from IFIAS (1974) where flows are of energy:

Gross energy requirements:	GER = $R + F + G + S$
Net energy requirement:	NER = $R + F + G + S - Y$
ENERGY requirement of energy:	ERE = $(R + F + G + S - Y)/Y$

Definitions from Hall et al., (1986) where flows are of energy:

Energy return on investment	EROI = $Y/(F + G + S)$

Definition from Sedlik (1978):

Net energy gain:	$Y/(F + G + S)$

Definitions in this book where flows are of solar EMERGY:

Net EMERGY	$Y - (F + G + S)$
EMERGY yield ratio	$Y/(F + G + S)$

Energy is still in a product, but the embodied energy of that product has passed through. What is now called EMERGY was earlier called embodied energy. As described below, there are several, quite different concepts under the name "embodied energy," so it was necessary to coin the terms "EMERGY" and "transformity" and rigorously define them to eliminate confusion.

Twentieth-century agriculture, although involving new technology, is based on massive contributions of embodied energy. The energetics of ancient and modern agriculture have been extensively evaluated (H. T. Odum, 1967b, 1971a; Heichel, 1973; Pimentel et al., 1973; Pimentel and Pimentel, 1979; Fluck and Baird, 1980; Stanhill, 1984; Smil, 1991); however, these authors have generally been cautious about expressing the data on materials, energy, labor requirements, and yields on any common basis.

Many energy analysts, while recognizing that energy quality differences exist, have used energy types of similar quality interchangeably, while leaving out those like sunlight and human labor that are very different in quality. The extensive literature tabulating inputs of energy to various products and processes is very valuable, since these numbers, when multiplied by appropriate transformities to obtain EMERGY, can be used eventually to show which processes are best. See the useful tabulation of energy inputs to agriculture by Pimentel (1980), for example.

In my first evaluations (Odum, 1967b), available energy was given embodied-energy values in units of organic kilocalories, treating biomass and fossil fuels as being equivalent. See also the extensive books and papers by C. Hall, C. Cleveland, and R. K. Kaufman (some of which are cited in the References). Details on conventions and transformities used in our earlier publications are given by date in Appendix E so that these results can be updated in terms of solar EMERGY. Formal evaluations of methods at earlier stages were published as part of government contracts. Methods used in the Mississippi Evaluation study (Bayley et al., 1977) was evaluated by Meyers (1977). Criticisms and rebuttals of the evaluation of the power plant sites study were both included in the publication by the Nuclear Regulatory Commission (Odum et al., 1983).

A milestone in the energy analysis of a whole system of human society and environment was the Swedish study of the island of Gotland in the Baltic Sea (Jansson and Zucchetto, 1978a,b; Jansson, 1985, Zucchetto and Jansson, 1985). Economic, energy, and material flows were evaluated and related. For example, ratios of money flow to energy flow were given.

Human Services in Energy Analysis

Human service contributions have been handled in different ways. If evaluation was in terms of the human body's metabolic energy (without multiplying by a transformity), the values were very small compared to other sources; some authors simply omitted it. Other researchers evaluated human inputs using the results of the input-output method using monetary flows (see below). Some (i.e., Fluck and Baird, 1980) counted human labor for the hours they worked, but did not count the time during the human's life activity even though it was necessary as indirect support of the work. Table 14.2 contains three ways of evaluating human labor: metabolic energy, national fuel share per person, and national EMERGY share per person.

Exergy and Process Analysis

In recent years the available potential energy of sources of mechanical energy quality and related energy types has been given the name *exergy* and used as a measure of work. Exergy includes potential chemical energy (Gibbs or Helmholtz free energy) of gases and solutions measured as mechanical work by integrating the product of pressure and volume due to changes in molecular concentrations. Since available energy is what is measured in process analysis, it is sometimes called *availability analysis* or *exergy analysis* (Ahern, 1980; Sussman, 1980; Moran, 1989).

Evaluation of available energy in industrial processes is called *process analysis.* Joules of *exergy* have been used as a measure of work without multiplying by a transformity except for the case of electric power, where sometimes fuels and electric power are compared by first multiplying by 4 (the coal transformity of electricity). The forms of energy usually included in exergy are those in a middle range of the energy transformation hierarchy (not including low-spectrum energy such as sunlight, or high-spectrum energy such as human service). Process analysis shows the energy required for processes, but since different transformity energies are used as if they are interchangeable, and since many low-energy–high-transformity inputs are omitted, the results are incomplete for EMERGY evaluations. All that is necessary to complete the evaluation is to multiply each exergy item by its transformity.

Within the chemical substances listed in a handbook of chemistry, there is a vast range of transformities. The chemical potential energy of these substances (exergy) gives little indication of the orders-of-magnitude range of energy requirements for their formation. A major project by a large organization such as the American Chemical Society or the Bureau of Standards is needed to set up tables of transformities of chemical substances. By dividing EMERGY by exergy, the transformity of that exergy is obtained, thus showing for each kind of exergy its position in the energy hierarchy and its contributions to systems.

Giampietro and Pimentel (1991) used the product of land area times time as a rough measure of the contribution of embodied energy (see Chapter 7). Folke (1988, 1990), in considering aquaculture production, used the area of the Baltic Sea capable of equivalent productivity as a common measure.

INPUT OUTPUT CONCEPTS OF VALUE

Assigning Embodied Energy with Input-Output Methods

Another method of energy accounting uses input-output data. Energy inputs to a network are assigned to the pathways according to the flow of some quantity that is conserved as it circulates (money, materials, or energy). We can call the circulating quantity the carrier. Energy inputs assigned in this way, according to the distribution of the carrier flows on the network, are called the *input-output embodied energy.*

As already shown in Figures 6.10–6.12, the flows of any network can be represented either by numbers of the pathways of the systems diagram or in an input-output table. For example, flows of energy are shown both ways in Figure 6.10a, and material flows are shown both ways in Figure 6.11a. Money circulation is a countercurrent to the flows of the commodities and is represented by dashed lines in Figure 6.12a, with numbers on the pathways and in the equivalent input-output table. Whereas pathway values of energy flow are largest on the left of an energy systems diagram (the lower-transformity side), money flows are concentrated in pathways to the right (where higher-quality human service is concentrated).

Where embodied energy is calculated for the economic system, various input flows of fuel energies are usually considered equivalent. Environmental energies are not usually included. A matrix procedure is used to assign the input energies to the flows of something else (the carrier) on the pathways (Figures 14.4 and 14.5). For each sector the embodied energy assigned to its output production pathway (pathway X_j in Figure 14.5b) divided by the flow of that pathway is called its "energy intensity."

Input-Output Embodied Energy Assigned to Monetary Flows

Since money is paid only to people, embodied energy assigned according to the circulation of money can be used to evaluate the embodied energy in human services in a network (Bullard, et al., 1976). For example, Figure 14.4a shows a circulation of money in a three-sector network, and Figure 14.4b has the values in an input-output table. When the matrix procedure for assigning embodied energy from outside is applied (Figure 14.5), energy intensities are calculated (vector e), multiplied by sector outputs (x) as a result embodied energy is assigned to the pathways as drawn in Figure 14.4c.

In summarizing the energy analysis of the 1970s and 1980s, Spreng (1988) describes as useful the practice of adding the embodied-energy estimates of human-service contributions from the input-output method to the energy of other inputs determined from process analysis (availability analysis).

Krenz (1976) describes an input-output method that evaluates the direct and indirect contributions of one sector to another. This procedure has been used to calculate the contributions of an energy-processing sector, such as coal, to the final demand sector (human consumers). The procedure first uses the input-output table of dollar flows to calculate the number of dollars of final demand traceable to the energy-processing sector. Then the energy/money ratio for that sector (coal) is multiplied by the number of dollars to the final demand to obtain the energy responsible for that much final demand. However, no justification has been given for assigning nonhuman energy transformations (physical, chemical, and biological) according to human money transactions.

The procedure of allocating the embodied energy of human service to dollar pathways using the matrix procedure is not very different from the simpler procedure that we have used to assign EMERGY by multiplying the dollars paid by the EMERGY/money ratio (see Chapter 7). Both procedures assume that an average relationship between embodied energy and wages is appropriate for the gross estimation of human inputs. When the embodied energies of human service are assigned to dollar pathways, human labor is treated as a pool of uniform EMERGY flow that is being split. These procedures

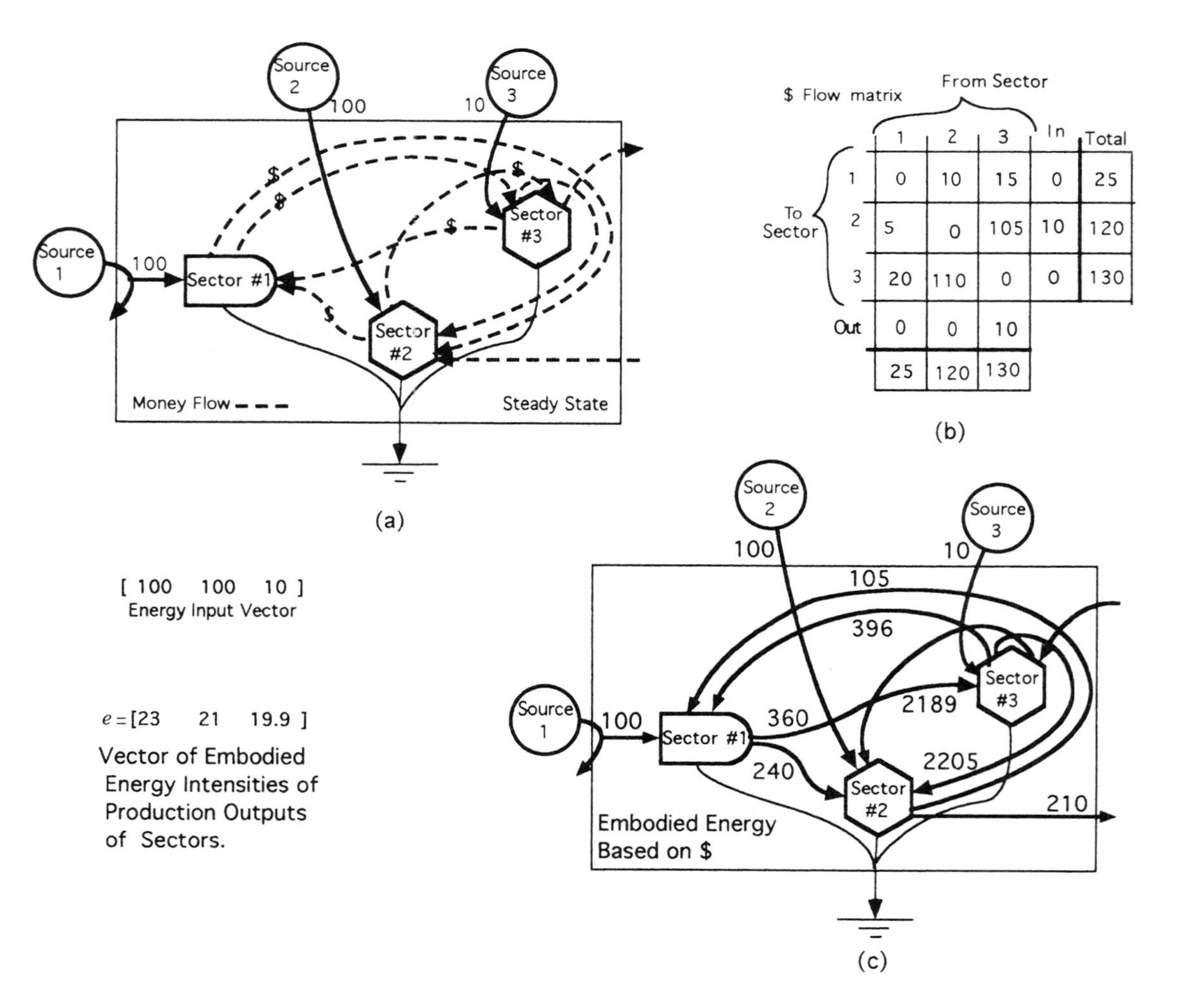

$ Flow matrix

To Sector \ From Sector	1	2	3	In	Total
1	0	10	15	0	25
2	5	0	105	10	120
3	20	110	0	0	130
Out	0	0	10		
	25	120	130		

(b)

are valid to whatever extent human service is flexible and transferable and can be represented by average wages. Karl Marx found that the pool of labor of his time was paid by the economic system according to an average wage. However, modern society's pay scales for higher-quality services are not similar and may not be proportionate to the embodied energy flows. A better method when data are available (Chapter 12) assigns EMERGY to human labor according to transformity of each service.

Input-Output Embodied Energy Assigned to Materials or Energy Flows

Much more controversial are procedures assigning embodied energy according to the input-output data on circulating materials or energy flow as the carrier. Apparently, no reason has ever been given for why embodied energy should have a network distribution according to the flow of a material cycle. Embodied energy is not in constant proportion to any material. For example, much more energy is embodied in 1 g of carbon dioxide from a power-plant exhaust or from a human being than from a typical soil surface. Some materials (heavy metals) may be concentrated in the high-transformity part of a network, and the matrix procedure would assign more embodied energy at that end. Yet, other materials in the same system are concentrated at the lower end, so the input-output procedure (Figure 14.5) would give opposite results.

To assign inputs of available energy to embody pathways according to their energy flow as the carrier, the heat-sink flow of unavailable energy is connected back to the sources to make a fictitious closed circulation (dashed line in Figure 14.6b. Energy is made to be conserved in a closed loop. Available energy is assigned to energy flow using the first law, conservation of energy. This is contrary to the second law. No reason has apparently been given for why available energy should be allocated according to an energy-flow network where most of the flow is degraded heat. Energy and embodied energy have different algebras.

Input-output procedures require that the flows in a network be aggregated so that pathway branches represent splits of equivalent kind and, as a result, the embodied energy is branched in proportion to the carrier flows. In most energy systems diagrams the flows are aggregated by necessary sector, with each sector's flows necessary to the others (see Figure 6.10). Branches are coproduct flows of different kinds, and different transformities, and each branch carries all the embodied energy of that sector undivided among the branches. Illustrating this point, Figure 6.10b shows the EMERGY flow for the

Figure 14.4. Evaluation of input-output embodied energy based on a three-sector network of circulating money. **(a)** Systems diagram. **(b)** Input-output table of money flows. **(c)** Embodied energy flows evaluated using the procedure in Figure 14.5.

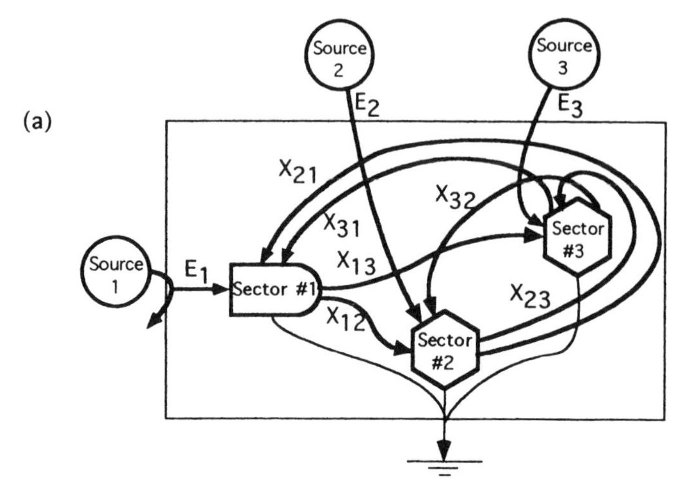

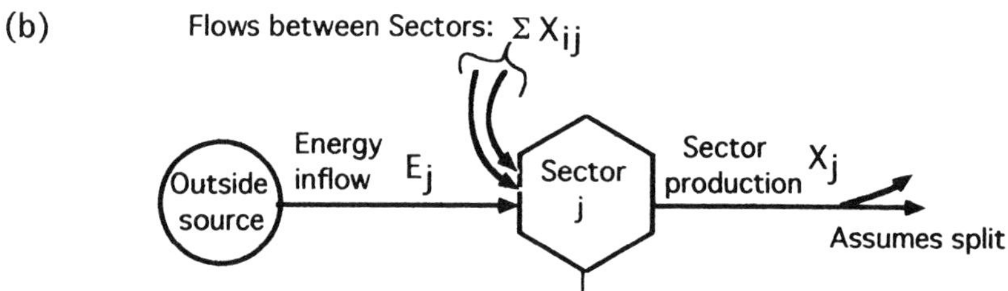

(c) Calculations:

Embodied energy equation for one sector: $E_j + \Sigma e_i X_{ij} = e_j X_j$

Where: e_j = energy intensity of productive output of sector j

And X's are flows of carrier quantity ($, Materials, etc)

$\underline{E} + \underline{e}\,\underline{\underline{X}} = \underline{e}\,\hat{\underline{\underline{X}}}$ where $\underline{\underline{X}}$ is a matrix of intersector flows

$\hat{\underline{\underline{X}}}$ is a diagonal matrix of sector outputs

$\underline{e}$ Is a row vector of energy intensities of output productivities

$\underline{E} = \underline{e}\,(\hat{\underline{\underline{X}}} - \underline{\underline{X}})$ and therefore $\underline{e} = \underline{E}\,(\hat{\underline{\underline{X}}} - \underline{\underline{X}})^{-1}$

Embodied energy of sector outputs = $\underline{e}\,\hat{\underline{\underline{X}}}$ (example: Figure 14.4c)

Figure 14.5. Procedure used for the input-output calculation of embodied energy. **(a)** Three-sector network with letters designating flows. **(b)** Equation for one sector. **(c)** Equations and matrix operations used to calculate embodied energy flows and intensities.

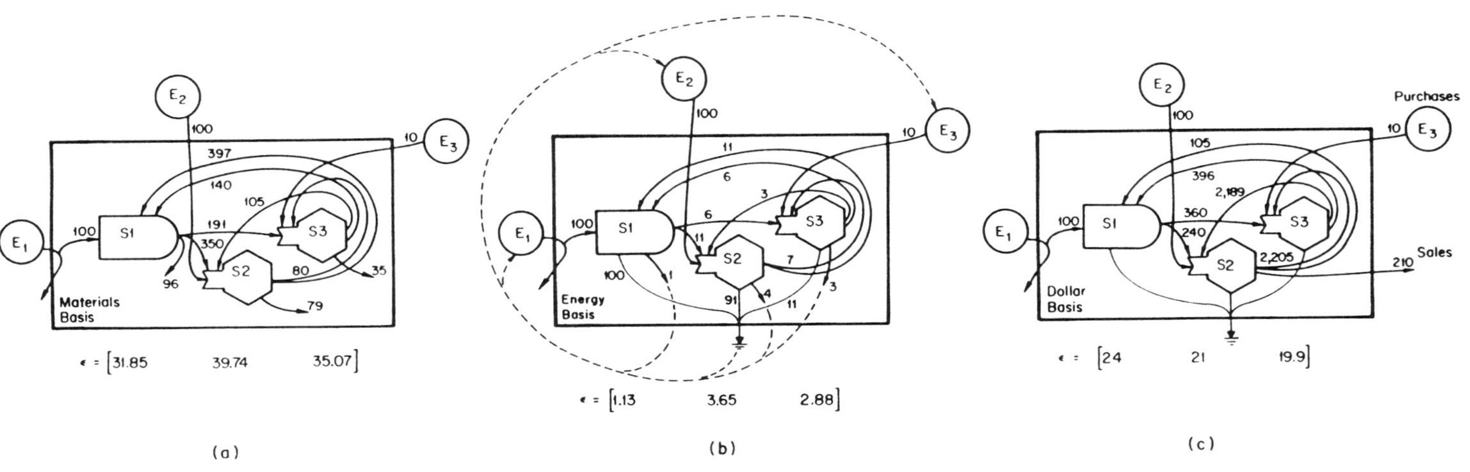

Figure 14.6. Comparison of input-output embodied energy assigned for different carriers in the same system. **(a)** Circulating materials. **(b)** Energy flows. **(c)** Circulating money (see Figure 14.4). Epsilon is a vector of embodied energy intensities.

closed loops of the network in Figure 6.10a. Since there is no energy availability in heat-sink flows, the EMERGY of the degraded heat is zero.

Figure 14.6 shows the results of assigning input-output embodied energy to one system in three different ways—using its energy flow, its material flow, and dollar flow as the carrier. The same embodied energy assigned by input-output procedures for the same system can differ by orders of magnitude, depending on which carrier of the many quantities circulating on the pathways is used.

Much of the confusion in ecological economics over evaluation methodologies comes from different aggregation principles used in developing models. If energies of the same kind are separated and recombined, as in the example of river water passing around both sides of an island, then they are additive. Their EMERGY values are split and recombined additively. Both flows are the same product and have the same transformity. We call these flows splits. Network models aggregated to have only this kind of branching are readily handled with input-output methods and matrix operations.

Problems arise when additive methods appropriate to split models are applied to models in which flows are aggregated according to major sectors that produce different kinds of outputs (coproducts) that are used elsewhere in the system in multiplicative interactions. As explained in Chapter 6, such models are aggregated to contain mutually necessary sectors that produce by-products (coproducts). These outputs are different, have different uses in the systems, and have different transformities, and each coproduct of a unit carries the whole EMERGY of their formation. However, if the diverging coproducts are later recombined, the separate EMERGY flows are not recombined, as this would be double counting. EMERGY is used up when its feedback pathway intersects its inflow (i.e., when it crosses its own tail). See Chapter 6.

Since branching of products in the real world has both the splitting of product flows of one kind and the branching of by-products of different kinds, neither of the two oversimplified ways of modeling may be sufficient alone to represent real systems. The energy systems language has provision for showing which kind of branching is involved so there can be clarity about what an author's model contains.

Matrix Operations for Tracking Energy in Networks

Network energy data in tabular form can be processed through various matrix operations to track energy and generate new insights, concepts, and indices; this is the subject of a theoretical field of network analysis developed by Patten (1978, 1992), Patten et al. (1990), Ulanowitz (1986), Higashi et al. (1989, 1992), and Wulf et al. (1989), to name a few.

Since there are feedbacks of energy, materials, and services in typical self-organized networks, some energy is recycled as it is transformed and dispersed. In one approach, the available energy still in the system (not yet dispersed into the heat sinks) is tracked by repetitive calculations. In some of the ecosystems studied, less than 5% of the original energy remains after seven passages, but some small amounts continue for many more internode passages. However, the EMERGY concept regards transformed energy as being different (having a changed transformity), so available energy is never recycled.

In one method (Higashi et al., 1989), the energy pathways, including those that are cyclic and looping, are unfolded to form a straight chain that represents the reaggregation of energy flow according to the number of steps beyond the input sources. A matrix procedure puts energy values on these. The result is an *unfolded ecosystem* model. For the traditional view that available energy of any kind is equally capable of doing work, then the unfolded energy chain represents a hierarchy of energy use.

However, this procedure aggregates energy flows which have different transformities into the same level in the new model. The procedure traces the contributions of EMERGY through the changed network. Patten (1995) uses the tracking procedures to divide the EMERGY inflow by the downstream energy flows passing through each step of the straightened energy chain until all the energy has been dispersed in the heat sinks. The quotients of input-available energy divided by straight-chain energy flows have the dimensions of transformity (EMERGY per unit of energy), and increase in passing downstream in steps along the unfolded chains. Since the energy flow in each passage from one inflow pulse is only a part of the total energy flow in the pathways at steady state, it might be useful to call these *partial transformities.* They are different from the transformities of the original network.

Ascendancy

Two of the most important properties of systems to be measured are (1) the total function and (2) the complexity of the systems parts and connections. Ulanowicz (1986) combined these attributes into a single measure, *ascendancy,* in which the bits of information measuring complexity are multiplied by the energy flow in the pathways. In other words, the energy flows are used to give the weight of importance to pathways. However, according to the concepts in this book, the importance of a flow is not measured by energy but by EMERGY. Our procedure (Chapter 12) multiplies bits of complexity by EMERGY per bit to weight complexity measures.

Christiansen (1994) studied EMERGY-based ascendency. The tendency of energy-based ascendency to overemphasize the lower trophic levels was improved by using the concept of EMERGY. Ascendency, whether calculated with energy or EMERGY, decreases with aggregation when the number of units in the network is small (less than 10).

CRITICISMS OF EMERGY ACCOUNTING

Opposition to EMERGY accounting comes from several kinds of backgrounds. A number of authors have attacked what they understood to be EMERGY, but misused the concept.

Those used to market price evaluations don't consider that a donor value rather than a market value is required for evaluating real wealth for the larger scale of public benefit. Some authors with a background in humanism believe that people are at the center, with free democratic rights to judge value and resist any concepts of limits. Those authors who do not incorporate concepts of thermodynamics into their critiques may doubt that there are physical limits. However, having physical determinants toward which self-organization adapts in no way limits the creative initiatives of human society to find—whether rationally or by trial and error—those patterns that are thermodynamically maximal.

Some researchers believe that no one measure can represent everything, but they nonetheless try to use money, a single measure. A measure of everything has to be a function of a measureable property that is *in* everything, which is energy, not money.

Those in science and engineering, and thus used to the ideas of available energy as a measure of work and work as a measure of product, don't accept the principle of an energy hierarchy that requires all energy evaluations to be multiplied by transformities.

Those working in small-scale ecology, who do not accept the concept that there is controlling self-organization at larger scales of environment and economy according to energy design principles, see no scientific basis for measuring value beyond the scale of populations and species.

Some workers who are used to input-output accounting are confused by EMERGY; they don't realize that accounts based on a network aggregated so that pathways split the donor value are different from accounts based on networks aggregated so that pathways are coproducts with the same donor value. It is hoped that the details on EMERGY algebra in Chapter 6 and the comparisons in this chapter will clarify the differences.

Some authors, while accepting concepts, criticize transformities as being variable, not based on enough cases, and not accompanied by statistical confidence limits. The data used to calculate transformities are the same scientific and economic data that are available for any other purpose. They are heterogeneous (from many fields), may contain errors, and are uneven in quality. Correcting one transformity may cause changes in many others. With spreadsheet software, recalculation is no longer the staggering task of the earlier years. Each year, as more and more analyses are made, more transformities are generated. Our earlier proposals to government agencies to make more transformity calculations were not funded, possibly because the concepts were new and the importance unrecognized. One of the purposes of this book is to establish the present state of knowledge about the energy hierarchy and EMERGY accounting after our 25-year effort, so that some larger organization can take over the big task of developing transformity tables.

Not everyone realizes that a transformity of a product has several uses. An empirical evaluation represents an objective evaluation of what went into a product and its conversion efficiencies. The transformity of a new

TABLE 14.3. Methods of Calculating Transformities

Methods	Location in This Book
Evaluating main energy flows of geobiosphere aggregated as necessary coproducts	Chapter 3
Evaluating energy flows of geobiosphere that are splits of the main flows	Chapter 3
Evaluating environmental economic production examples (subsystems of the geobiosphere)	Chapters 4, 5
Evaluating accumulations in a stored reserve	Chapter 5
Evaluating transformities by combining other transformities	Appendix C
Evaluating transformations in published energy network diagrams	Chapter 2
Tracking EMERGY of each source through network and combining	Chapter 5
Evaluating network energy flow data with micro-computer program (Tennenbaum, 1988).	Chapter 5
Evaluating energy distribution graphs	Chapter 2
Inferring transformity from hierarchical positions indicated by turnover time	Chapter 3

experimental, and likely inefficient, system is a guide regarding the use of that same process again. On the other hand, for estimating the best possible production, a transformity is sought from systems that are long tested, improved by selective choices, and more likely to be operated at the best efficiency consistent with maximum empower. In any case, when there is any doubt, we believe people should determine their own transformities with the help of procedures given in this book. Table 14.3 lists some of the methods.

Still others complain of the complexity and flood of numbers when an EMERGY evaluation table is done in the detail that is sometimes requested. As explained in Chapter 1, the initial model for appraising a system can be made in more or less detail to fit the needs of the decision makers or other users. Aggregation or disaggregation need not affect the total EMERGY or the conclusions so long as everything important is included.

In 1975 our initiatives through Senator M. Hatfield of Oregon caused a federal law to be introduced requiring "net energy analysis" of new projects. Because the words "energy" and "embodied energy" were not clearly defined, the implementation of the law became confused and its purpose of preventing wasteful projects was circumvented. While noting the illegal substitution of economic analysis for energy analysis, the U.S. General Accounting Office (GAO, 1982) reviewed energy analysis methods describing three approaches: process analysis; input-output analysis; and our approach, which they called "ecoenergetics." They wrote:

> **Ecoenergetics has broad appeal in its emphasis on the fullest possible measurement of the embodied energy of labor, environmental systems, and solar energy, but its analytical boundaries are more extensive than seems appropriate for the analysis of alternative energy technologies, as we explain at greater length [elsewhere]. Moreover, a set of consistent quantitative methods has yet to be developed for it. Therefore we chose not to use ecoenergetics.**

Although the last of our publications cited was 1976, Spreng (1988) drew a similar conclusion. In the 20 years of work since these judgments, over 100 refereed papers, books, theses, dissertations, and published reports have been completed by many authors using EMERGY. EMERGY and transformity were rigorously defined in 1983. In the process we hope that a consistent quantitative method is now operational as set out in this book.

SUMMARY

The many approaches to environmental evaluation include market value for the short-term measurement of human interaction, and longer-range values based on the sustainable work contributions of nature. From an EMERGY point of view, many of the other approaches mentioned failed to measure real wealth. However, much of the data in the extensive literature can be adapted to EMERGY accounting by multiplying energy data by transformities and economic data (services) by EMERGY/money ratios. The principal need is a large organizational effort for developing transformity tables.

CHAPTER 15

POLICY PERSPECTIVES

This book contains concepts, methods, and examples of EMERGY accounting. It suggests that policy on many scales be based on maximizing EMERGY production and use. EMERGY accounting may show in advance what combinations of environmental work and human effort contribute most real wealth now and in the future. With this new tool, perhaps, we are ready to substitute evaluations for wasteful, adversarial decision making. With more global organization, shared information may be replacing the trial-and-error ways of finding what policies work. EMERGY calculations can be part of this trend.

This chapter suggests areas for applying EMERGY to policy. Although there is not room here for the many unpublished results, preliminary student projects in our classes over the last decade show the feasibility of the EMERGY accounting approach to most if not all kinds of policy. Appendix E (Table E-1) gives calculation details used in older publications and the dates to aid updating older "embodied energy" accounts to EMERGY evaluations.

SUSTAINABLE DESIGNS

As explained in Chapter 1, approaches to policy begin with systems overviews using diagrams that include important inputs, components, and relationships. Comparative study of the models of science and social science at all scales and the experience from diagramming hundreds of systems show common designs. Table 15.1 is a summary of system characteristics that prevail. Experience with evaluations so far support the general hypothesis that these designs contribute to maximum empower.

TABLE 15.1. Systems Designs and Properties for Maximum Empower

All units have depreciation pathways according to the second law of thermodynamics.
Operation is at optimum efficiency for maximum power.
Optimum efficiency is less during growth than at maturity.
Materials recycle.
Autocatalytic feedbacks reinforce production.
Consumption prevails by reinforcing production.
Reinforcements link units of scales below and above.
Energy transformations form a hierarchical web.
Storages increase with scale.
Replacement time increases with scale.
Transformity increases with scale.
Different sources interact.
Items of different transformity interact multiplicatively.
Energy transformations converge spatially.
Production pathways generate storages.
Small-scale pulses are filtered by units of larger scale.
Territory of input and influence increases with scale.
Repetitious pulsation prevails over level steady states.
Pool of information diversity is stored for contingency use.

For example, Figure 15.1 shows two designs for consumer spending, with and without feedback. Whereas many economic measures rate any job as a contribution according to the money spent, only the feedback-loop design (Figure 15.1b) is reinforcing. To maximize performance and be sustainable, plans and policies should include these characteristics. Designs that maximize empower should be looked for in all systems and be included in systems diagrams.

EMERGY EVALUATION OF HISTORY

A start has been made on a new field, the *energy systems evaluation of history*. Past economic patterns in Florida were compared with systems diagrams that contrasted old and newer resource uses (H. T Odum 1978b; Sipe 1978a). These overviews showed the way Native American cultures were displaced by higher empower of Spanish colonization and later developments. Energy systems perspectives were given for 19th-century Ireland and the impact of its potato famine (Odum, 1986a). EMERGY from the forestry- charcoal- silver- copper- and iron-processing technologies of Sweden was found to be the basis of the dominance by the Swedish empire in northern Europe in 1650 (Sundberg, 1992; Sundberg et al., 1994a,b). M. T. Brown (1977b) evaluated the Vietnam War. Woithe (1994) evaluated the EMERGY of the U.S. Civil War of 1860 showing the role of slaves, native resources, foreign exchange, and the trans-

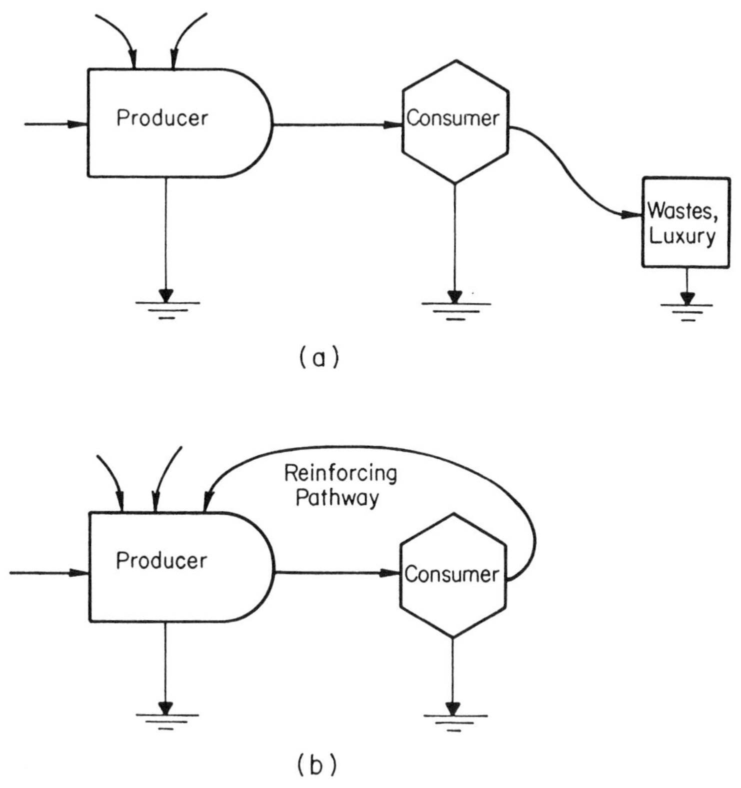

Figure 15.1. Comparison of consumers that reinforce production with those that do not. **(a)** Environmental production and consumption with consumer outputs of materials and services that do not reinforce production. **(b)** economic production and consumption with consumer output reinforcing production.

formities of key elements of battles. J. V. Boyles's (1975) dissertation in economics showed the use of embodied energy for historical calibration of modern monetary accounting. EMERGY accounting provides a quantitative measure to judge importance of causal factors and events in history.

ENVIRONMENTAL POLICY AND MANAGEMENT

By 1950 it was clear that economic development and environmental protection were on a collision course because there was no way then to put the two values on a common basis. EMERGY now provides that basis, and most of this book has dealt with EMERGY accounting for evaluating environmental resources and policies. If some Florida examples are any indication, EMERGY evaluation of environmental alternatives is much cheaper and quicker than economic evaluation of environment. For example, on an economic analysis of the alternatives for the Cross Florida Barge Canal, $500,000 was spent on questionnaires to find population preferences, whereas a much more rigorous emdollar evaluation possibly could have been done for $5000. Table 15.2 lists environmental policy applications not covered elsewhere in this book. As soon as additional transformity tables have been developed, EMERGY evaluation may be ready for assigning public values in courtrooms. EMERGY has been used in two environmental lawsuits settled out of court.

Environmental Management Based on Ecosystems

In the 1990s, at federal and state levels, public lands were declared to be managed according to ecosystems. An energy systems method of classifying ecosystems had been defined earlier (H. T. Odum et al., 1969, 1974a; Odum and Copeland, 1972), first used for coastal ecosystems of the United States. Unlike most classifications that pick one or two parameters as a basis for classifying everything, the energy systems method defines an ecosystem type according to whatever are the dominant sources of EMERGY. Classifying by dominant sources classifies the system as a whole because ecosystems develop their prominent characteristics (EMERGY signature) as part of self-organizing to maximize use of all sources and storages.

A methodology and application of ecosystem management was started in Florida with the "south Florida project" (Lugo et al., 1971; H. T. Odum and Brown, 1975) through the interest of undersecretary of the U.S. Department of the Interior, Nathaniel Reed, and the initiative of George Gardner. Much of Florida's lands and waters was mapped according to the ecosystem classification, and information on each was aggregated into overview models so as

TABLE 15.2. Applications of EMERGY Accounting for Environmental Policy

Proportion of land to leave in natural areas for seeding
Alternative allocations of water
Siting of power plants
Tax exemptions for environmental protection
Evaluation of plans to exchange land for debt
Indirect economic benefits of natural areas
Evaluation of toxicity impact
Alternative management of beaches
Evaluation of soil
Values of different ecosystems
Net benefit of antipollution technologies
Pesticide versus biodiversity means of preventing pest epidemics
Net benefits of dredge-and-fill projects
Net benefits of dams and fisheries
Values of rocks and minerals
Work of ecosystem succession
Net benefit of agricultural "subsidy"

to define what is important and organize quantitative data. Included in the classification of ecosystems of Florida were agroecosystems, underwater ecosystems, and zones of cities treated as urban ecosystems.

Table 15.3 summarizes the steps used for environmental management with the ecosystem method. The ultimate purposes were to understand the ecosystems, sensitivities, causalities and the means for managing for maximum EMERGY production and use.

AREA EVALUATIONS

Although the leaders of nations, states, counties, and cities have yet to take much notice, EMERGY evaluations of areas clearly show policies for improving economic vitality and making better use of the environment in order to further human and environmental prosperity.

EMERGY Evaluation of Nations

W. M. Kemp et al. (1981) evaluated the coal EMERGY budgets of 63 nations, combining fuel use and estimates of renewable environmental contributions to

TABLE 15.3. Procedure for Environmental Management by Ecosystems

1. Identify ecosystem types in the area using energy systems–based ecosystem classification.
2. Map ecosystems using aerial photographs and truth checking on the ground.
3. Develop overview systems diagrams to inventory and represent the important components, processes, and externally important factors. These summarizing models may include subsystem boxes, each of which may have its diagram.
4. From case histories identify the main sources of stress and impact to each ecosystem, the responses to the stress, and the recovery sequences and times.
5. From reports and publications, use the organization provided by the diagrams to assemble quantitative data on the properties and processes found to be important. In other words, obtain values for each unit and pathway flow (nutrients, energy biodiversity, etc.). Document these with tables, footnotes, and references.
6. Simulate aggregated ecosystem models with computers to help understand the time dimension for which the diagrams provide a connective dimension.
7. From systems diagrams write a verbal summary of the main parts, processes, sensitivities, and impacts as a guideline to field managers, permitters, and owners.
8. Determine EMERGY and emdollar values for flows, storages, and economic interactions with the economy.
9. Make decisions among alternatives that maximize the emdollar value of the ecosystem and its economic interactions. In other words, manage so as to make the economic uses of ecosystems maximize the contribution to the public good.

the economies (coal EMERGY evaluation). Their graphs suggest the hierarchical nature of the world with many low-EMERGY-using nations coupled to a fewer high-EMERGY-using nations. Starting with a comparison of 12 nations (H. T. Odum et al., 1983), a standard procedure for national EMERGY evaluation has been applied to many countries, including published details for the United States (H. T. Odum et al., 1987a), Switzerland (Pillet and Odum, 1984), Papua New Guinea (Doherty et al., 1993), Taiwan (Huang and Odum, 1991), China (Lan 1992, 1993), Mexico (M. T. Brown et al., 1992), Italy (Ulgiati et al., 1994), Sweden (Doherty et al., 1995), and Korea (S. M. Lee, 1994). These studies suggest changes in policies for resource use, trade, and internal investment to maximize economic prosperity. Unpublished student reports cover 60 other countries. See Chapter 10.

EMERGY Evaluation of Watersheds

EMERGY has been used to evaluate the environmental and economic inputs to regional watershed systems, their patterns of development, and alternatives. Systems diagramming and evaluation include the St. John's River, Florida (Bayley et al., 1977) and the Kissimmee-Okeechobee Basin, Florida (H. T. Odum and Brown, 1975). Successive evaluation of the Mississippi River suggested better ways of combining development with the original environmental contributions of waters, sediments and wetlands (Young et al., 1974; Bayley and Walker, 1976; Bayley et al., 1977; Zucchetto et al., 1980; Diamond, 1984; Odum et al., 1987c). EMERGY evaluation of the Amazon identified its resources and potentials (Christianson, 1984; H. T. Odum et al., 1985; M. T. Brown, 1986b; Browder et al., 1976).

EMERGY Evaluation of States

Counting unpublished analyses, a dozen states of the United States have been studied. Details are available for developed states of Texas (H. T. Odum et al., 1987a) and Florida (H. T. Odum et al., 1993) that contrast with Alaska (Woithe, 1992), an economy that resembles an exploited underdeveloped country in exporting rather than using its resources. With better resource policies than Alaska, Sweden has a larger, more balanced economy at similar latitude. Puerto Rico has a city-state pattern heavily based on imported resources and exported sales (Doherty et al., 1995). Other state evaluations were made in unpublished student reports.

EMERGY Evaluation of Counties

For EMERGY evaluations of whole Florida counties, a general energy systems methodology was developed for assisting environmental policy, identifying the emdollar contributions of the main resources, and addressing alternative development questions raised by local planners and county commissioners: Franklin County (Boynton, 1975), Volusia County (Hansen et al., 1982), Hendry County (DeBellevue, 1976a), Lee County (M. T. Brown 1975) and Sipe (1978b).

EMERGY Evaluation of Coastal Zones

Coastal zones and resources were evaluated for Texas (Odum et al., 1987a), the Nayarit coast of Mexico (M. T. Brown et al., 1992), the Sea of Cortez, Baja California (M. T. Brown et al., 1991), and Ecuador, (Odum and Arding, 1991). Very high emdollar values were generated by tides, rivers, discharges, and wave energies. An evaluation of the coastal zone of Holland included the giant marine dams constructed to block storm surge (H. T. Odum, 1984c).

EMERGY Evaluation of Cities

City evaluations began with energy-EMERGY analysis of Miami, Fla. (Zucchetto, 1975a), Hongkong (Stone and Wong, 1991), Jacksonville, Fla. (Whitfield, 1994), and Taipei, Taiwan (Huang, et al., 1993). New studies suggest that maps of empower density and transformity help identify the hierarchical zones of cities and indicate to planners appropriate locations for environmental amenity, housing, industry-utility-transport, and information centers.

EMERGY OF GOVERNMENT PROGRAMS

In public discussion of federal, state, and local budgets, the question is frequently asked, Which ongoing and proposed programs contribute most per dollar expended? EMERGY evaluations of alternative policies can show which programs contribute most real wealth or drain the least. Table 15.4 lists some areas where preliminary evaluations show policies amenable to EMERGY accounting.

Defense and War

A preliminary EMERGY evaluation by Dalton (1986) showed that the advantages to the United States from trade much exceeded the U.S. overseas defense effort. Because of heavy use of fuels and strategic materials, the emdollars of defense were about double the dollars spent directly on defense. EMERGY evaluations of inputs and destruction of war have been made for the U.S. Civil war (Woithe, 1994) and the Vietnam War (M. T. Brown, 1977); some preliminary evaluations to the Vietnam War from the United States have also been made of the Persian Gulf War. The EMERGY input was large compared

TABLE 15.4. Policies Amenable to EMERGY Accounting

Alternative methods of transportation
Spatial organization of cities
Alternative approaches to crime
Net benefits of health care
Alternatives approaches to unemployment
Distribution of income for maximum productivity
Educational priorities
Optimum population density
Selection of alternative programs for space
Allocation of research funds
Evaluation of equity in treaty provisions
Evaluation of cost and benefits of waste recycling
Net public benefit from growth
Net benefit from conservation measures
Evaluation of unpaid services
Alternative prepositions for information transmission

with the opposition's input, and the empower density destruction was very large. The EMERGY contributed and consumed in the Persian Gulf War was much larger than the annual oil flows in the region, but much less than the large oil reserves at risk.

The very high empower of globally shared information and concepts was described in Chapter 12. Facilitated by worldwide television and new information networks, a more united motivation may be developing in the world's population. The high EMERGY of shared visions for cooperative relationships may replace the old system of international organization by competitive defenses. As growth ceases, widely shared ideals regarding mutual reinforcement may limit the competitive excesses of profit capitalism, less useful in times of climax and descent.

EMERGY EVALUATION OF PRIVATE ENTERPRISE

In the short range, markets and what people are willing to pay guide private business and industry, but they give little indication of what people are going to want later when conditions are changed. In the long run, private industry survives if it achieves the maximum productivity possible with maximum compatible efficiency internally, while also making a better contribution to the economy of the environment and society outside. EMERGY evaluations of alternatives for private industry may be helpful for longer-range planning.

PROSPEROUS WAY DOWN

The world's rate of fuel consumption has apparently reached its maximum, and the renewable resources available are decreasing each year due to population increase and environmental encroachment. On an EMERGY basis the world's standard of living is already coming down. Already there are erratic contractions, arbitrary downsizings, and population-resource disasters. Much uncertainty and malaise can be avoided if EMERGY evaluations can be substituted for economic evaluation. If people can regain their commonsense view of real wealth, which EMERGY evaluation gives them, policies can be implemented for selective, slow, and deliberate, and prosperous descent. However, that is for another book.

SUMMARY

EMERGY evaluation provides a quantitative way to find what policies and patterns for humanity and nature are sustainable, with less trial and error, because they tend to anticipate self-organization for maximum prosperity.

AN EMERGY GLOSSARY

DAN CAMPBELL

Available energy Energy with the potential to do work (exergy).

Donor value A value of a product determined by the production process and not by what a person is willing to pay (e.g., the mass and energy of wood).

Emdollar value Dollars of gross economic product obtained by dividing the EMERGY of a product by the appropriate EMERGY/$ ratio; the dollars of gross economic product equivalent to the wealth measured in EMERGY.

EMERGY (spelled with an "M")—All the available energy that was used in the work of making a product and expressed in units of one type of energy.

Emjoule The unit of EMERGY that has the dimensions of the energy previously used (grams-centimeter squared per second squared, or $g\text{-}cm^2/sec^2$).

Energy A property that can be turned into heat and measured in heat units (kilocalories, British thermal units, or joules).

Energy hierarchy The convergence and transformation of energy of many small units into smaller amounts of higher-level types of energy with greater ability to interact with and control smaller units.

Energy systems language or energy circuit language A general systems language for representing units and connections for processing the materials, energy, and information of any system; a diagrammatic representation of systems with a set of symbols (Figure 1.2) that have precise mathematical and energetic meanings. See Appendix A.

Gross national product (GNP) The total market value of all final goods and services produced in an economy in 1 yr.

Investment ratio (EMERGY investment ratio) The ratio of EMERGY brought into an area from outside an economy, to the local, free environmental EMERGY used in the interaction.

Maximum power principle An explanation for the designs observed in self-organizing systems (energy transformations, hierarchical patterns, feedback controls, amplifier actions, and so on). Designs prevail because they draw in more available energy and use it with more efficiency than alternatives.

Net EMERGY The EMERGY yield from a resource after all the EMERGY used to process it has been subtracted.

EMERGY yield ratio The ratio of the EMERGY yield to that required for processing.

Next larger scale Larger territorial areas occupied by units with longer replacement times, which must be considered in determining system behavior because of the controls larger units exert on smaller-scale units and processes.

Reinforce The action of a unit or process to enhance production and survival of a contributing unit or process, thereby enhancing itself; a loop of mutually enhancing interactions.

Self-organization The process by which systems use energy to develop structure and organization.

Maximizing EMERGY The process by which the maximum power principle operates within a system to select from among the available components and interactions the combination that results in production of the most EMERGY.

Second law of thermodynamics The principle that energy concentrations disperse spontaneously and that any energy transformation has some of its available energy dispersed in the process.

Solar transformity Solar EMERGY per unit of energy, expressed in solar emjoules per joule (sej/J).

Sustainable use The resource use that can be continued by society in the long run because the use level and system design allow resources to be renewed by natural or human-aided processes.

Systems ecology The field that came from the union of systems theory and ecology and provides views on many scales for emergy analysis.

Transformity The EMERGY of one type required to make a unit of energy of another type. For example, since 3 coal emjoules (cej) of coal and 1 cej of services are required to generate 1 J of electricity, the coal transformity of electricity is 4 cej/J.

Turnover time or replacement time The time for a flow to replace a stored quantity. For example a flow of 10 gallons of water per day will replace a 1000-gallon tank of water in 100 days.

Wealth Usable products and services however produced.

APPENDIX A

USE OF ENERGY SYSTEMS SYMBOLS

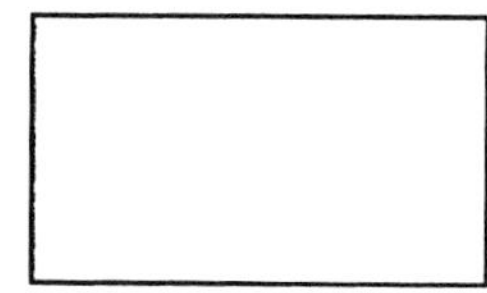

System Frame. A rectangular box is drawn to represent the boundaries that are selected. Boundaries selected must define a three dimensional prism around the system. For example, the analysis of a city would probably include its political boundaries for its lateral boundary, a plane below the ground surface (e.g., 10 m), and a plane above the city (e.g., 100 m).

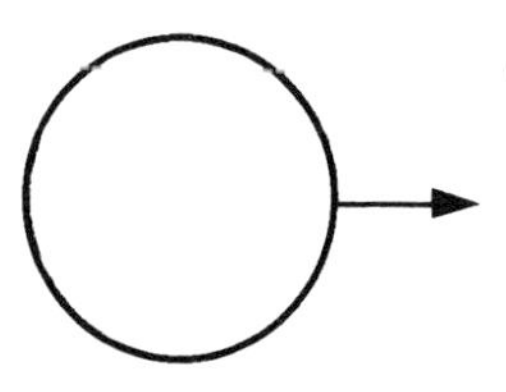

Source. Any input that crosses the boundary is a source, including pure energy flows, materials, information, genes, services, and inputs that are destructive. All of these inputs are given a circular symbol. Sources are arranged around the outside border from left to right in order of their solar transformity, starting with sunlight on the left and information and human services on the right. No source inflows are drawn into the lower side of the frame.

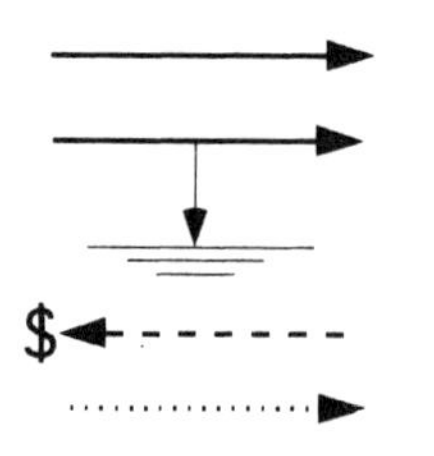

Pathway Line. Any flow is represented by a line, including pure energy, materials, and information. Money is shown with dashed lines. Where material flows of one kind are to be emphasized, use dotted lines (or color). Barbs (arrowheads) on the pathways mean that the flow is driven from behind the flow (donor driven) without appreciable backforce from the next entity. Lines without barbs flow in proportion to the difference between two forces and may flow in either direction.

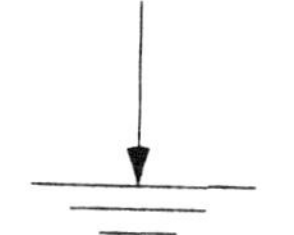

Heat Sink. The "heat-sink" symbol represents the dispersal of available energy (potential energy) into a degraded, used state, not capable of further work. Representing the second energy law, heat-sink pathways are required from every "transformation" symbol and every tank. At the start, one heat sink may be placed at the center bottom of the system frame. Then two lines at about 45° to the bottom frame border are drawn to collect heat-sink pathways. Using finer lines or yellow lines for heat sinks keeps these from dominating the diagram. No material, available energy, or usable information ever goes through heat sinks, only degraded energy.

Outflows. Any outflow that still has available potential, materials more concentrated than the environment, or usable information is shown as a pathway from either of the three upper system borders, but not out the bottom.

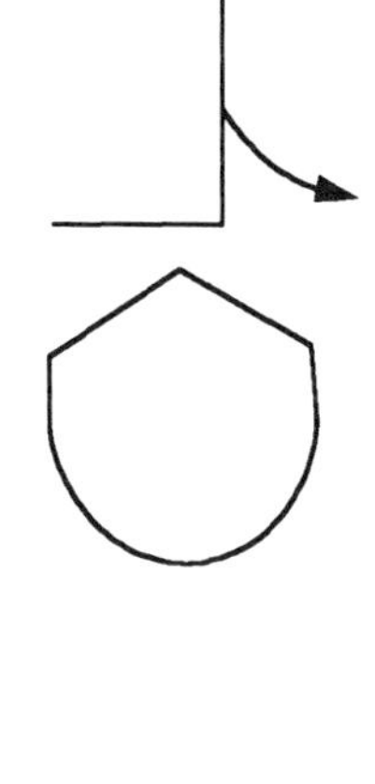

Storage Tank. Any quantity stored within the system is given a "tank" symbol, including materials, pure energy (energy without accompanying material), money, assets, information, image, and quantities that are harmful to others. Every flow in or out of a tank must be the same type of flow and measured in the same units. Sometimes a tank is shown overlapped by a symbol of which it is part. For example: wood storage, a part of the Radiata pine population in Figure 5.1.

Adding Pathways. Pathways add their flows when they join or when they go into the same tank. No pathways should join or enter a common tank if they are of different type or transformity or are measured in different units. A pathway that branches represents a split of flow into two of the same type (e.g., Figure 6.2a, 6.3a).

Interaction. Two or more flows that are different and both required for a process are connected to an "interaction" symbol. The flows to an interaction are drawn to the symbol from left to right in order of their transformity, with the lowest-quality one connecting to the notched left margin. The output of an interaction is an output of a production process, a flow of product. These should usually go to the right, since production is a quality-increasing transformation.

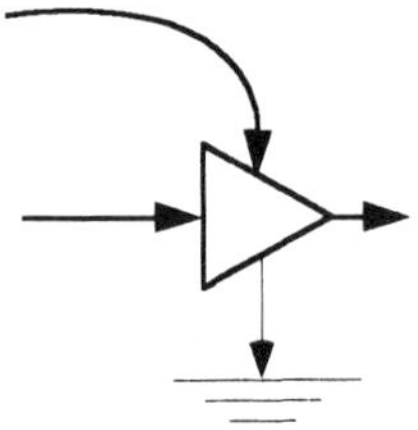

Constant Gain Amplifier. A special interaction symbol is used if the output is controlled by one input (entering the symbol from the left), but most of the energy is drawn from the other input (entering from the top).

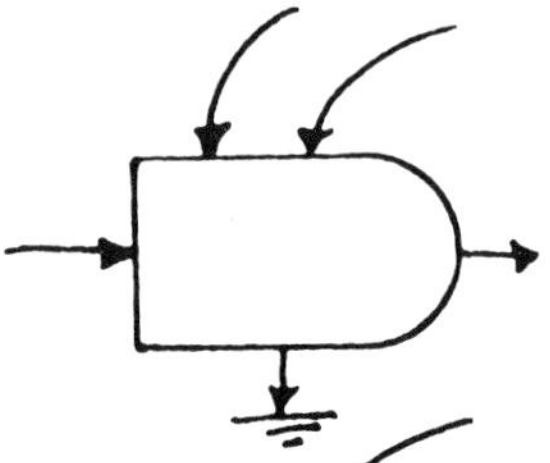

Producers. "Producer" symbols are used for units on the left side of the systems diagram that receive commodities and other inputs of different types interacting to generate products. The "producer" symbol implies that there are intersections and storages within. Sometimes it may be desirable to diagram the details of interactions and processes inside. Producers include biological producers, such as plants, and industrial production.

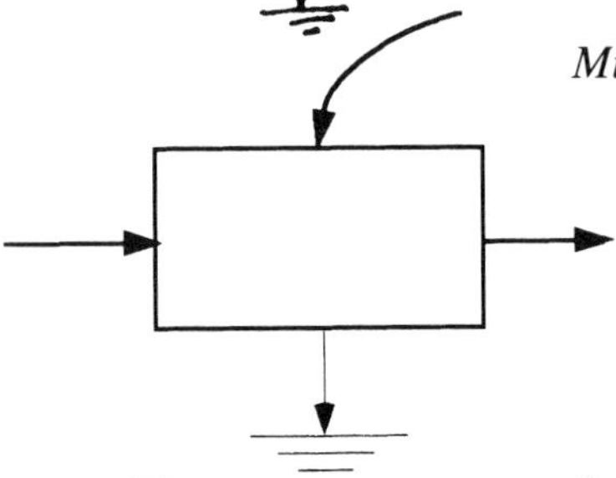

Miscellaneous Box. The rectangular box is used for any subsystem structure and/or function. Often these boxes are appropriate for representing economic sectors such as mining, power plants, commerce, and so on. The box can include interactions and storages with products emerging to the right. Details of what goes on within the consumer is not specified unless more details are described or diagrammed inside.

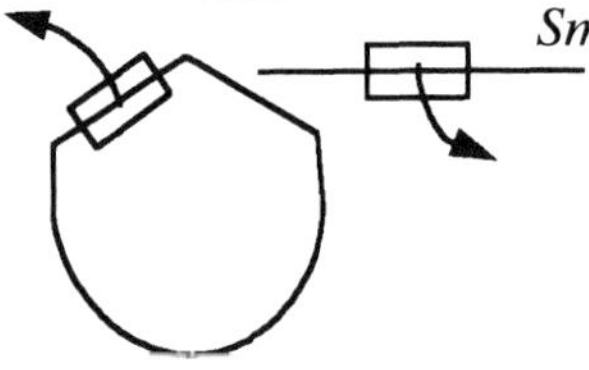

Small Box. A very small box on a pathway or on the side of a storage tank is used to initiate another circuit that is driven by "force" in proportion to the pathway or storage. This is sometimes called a "sensor" when it delivers its action without draining much energy from the original pathway or tank.

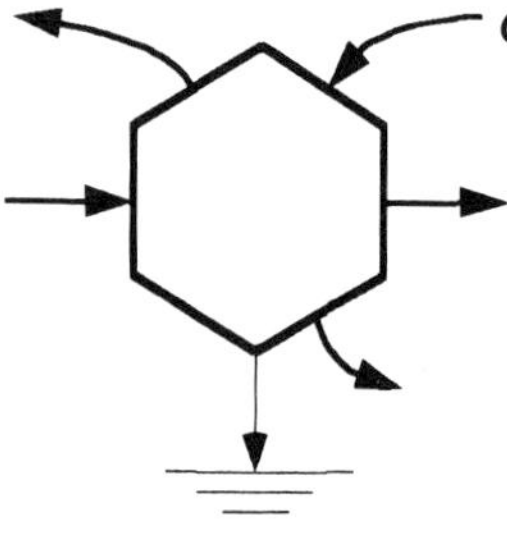

Consumers. "Consumer" symbols are used for units on the right side of the systems diagram that receive products and feedback services and materials. Consumers may be animal populations or sectors of society, such as the urban consumers. A "consumer" symbol usually implies autocatalytic interactions and storages within (e.g., Figure 2.3a). However, the "consumer" symbol is a class symbol (i.e., it refers to many similar but different units), and details of what goes on within the consumer are not specified exactly unless more details are diagrammed inside.

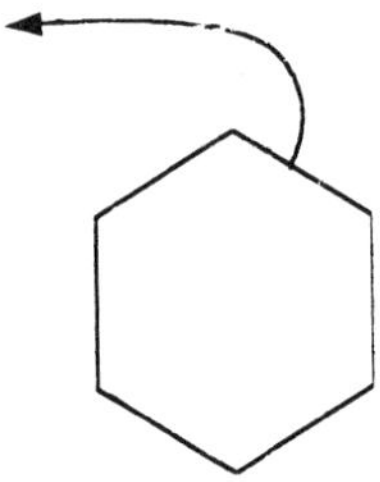

Counterclockwise Feedbacks. High-quality outputs from consumers, such as information, controls, and scarce materials, are fed back from right to left in the diagram. Feedbacks from right to left represent a diverging loss of concentration, the service usually being spread out to a larger area. These flows should be drawn with a counterclockwise pathway (up, around, and above the originating symbol—not under the symbol). These drawing procedures are not only conventions that prevent excess line crossing and make one person's diagrams the same as another's, but they make the diagrams a way of representing energy hierarchies.

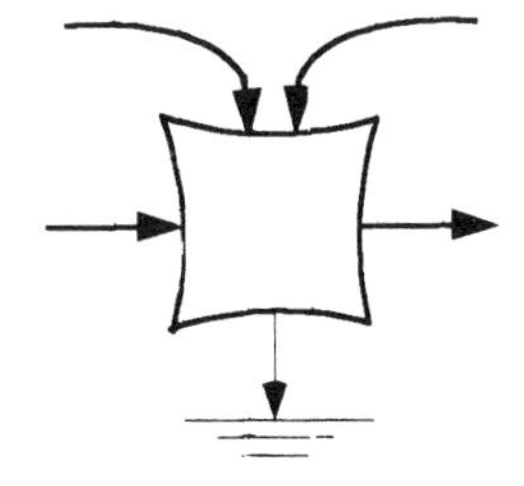

Switch. The concave-sided box represents switching processes, those that turn on and off. The flows that are controlled enter and leave from the sides. The pathways that control the switches are drawn entering from above to the top of the symbol. This includes thresholds and other information. Switching occurs in natural processes as well as with human controls. Examples are earthquakes, reproductive actions, and water overflows of a riverbank.

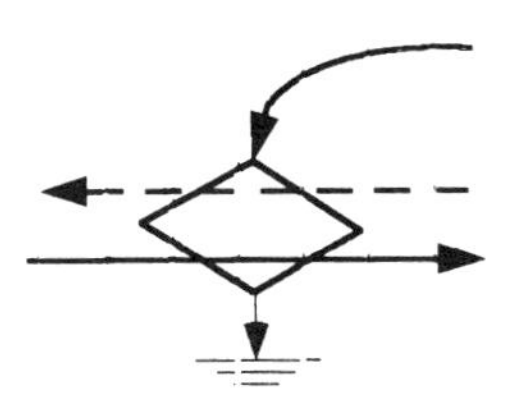

Exchange Transaction. Where quantities in one flow are exchange for those of another, the "transaction" symbol is used. Most often the exchange is a flow of commodities, goods, or services exchanged for money (drawn with dashed lines). Often the price that relates one flow to the other is an outside source of action representing world markets; it is shown with a pathway coming from above to the top of the symbol.

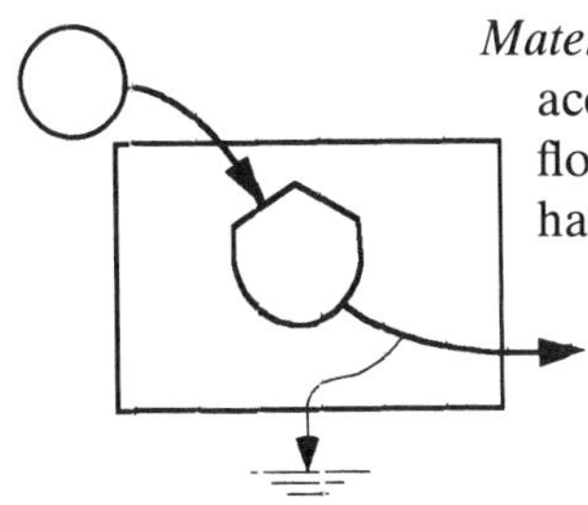

Material Balances. Since all inflowing materials either accumulate in system storages or flow out, each inflowing material, such as water or money, needs to have outflows drawn.

APPENDIX B

FORMULAS FOR ENERGY CALCULATIONS

I. FORMULAS FOR CALCULATING ENERGY FLOWS IN JOULES PER YEAR[1]

Direct Sunlight

$$(\text{area of country})(\text{average of insolation}) = (\text{— m}^2)(\text{— J/m}^2\text{/y}) = \text{— J/yr}$$

where the area of the country is the area of the land plus that of continental shelf.

Kinetic Energy of Wind Use at Surface

Kinetic energy of wind at 1000 m is multiplied by its height, density, eddy-diffusion coefficient, the wind gradient, and area of the system:

$$(\text{height})(\text{density})(\text{diffusion coefficient})(\text{wind gradient})(\text{area})$$

$$(1000\ \text{m})(1.23\ \text{kg/m}^3)(\text{— m}^3\text{/m/sec})(3.154 \times 10^7\ \text{sec/yr}) = (\text{— m/sec/m})^2(\text{— m}^2)$$

Typical values of eddy diffusion and vertical gradient coefficients are as follows:

[1] Where data are in kilocalories (kcal), multiply by 4186 J/kcal.

	Eddy diffusion ($m^3/m^2/sec$)		Vertical gradient (m/sec/m)	
	January	July	January	July
Flint, Mich.	40.2	8.3	8.0×10^{-3}	3.8×10^{-3}
Oakland, Calif.	8.4	1.0	4.3×10^{-3}	1.6×10^{-3}
Tampa, Fla.	2.8	1.7	2.3×10^{-3}	1.5×10^{-3}

For other data, see environmental evaluation manual (Odum et al., 1987).

Chemical Potential Energy in Rain

$$(\text{Area including shelf})(\text{rainfall})(G) =$$

where G is Gibbs free energy of rainwater relative to salt water within evapotranspiring plants or in seas receiving rain; $G = 4.94$ J/g.[2]

$$(\text{— m}^2)(\text{— m/yr})(4.94 \text{ J/g})(1 \times 10^6 \text{ g/m}^3) = \text{— J/yr}$$

Chemical Potential Energy in River

$$(\text{volume flow})(\text{density})(G) =$$

where G is Gibbs free energy of river water relative to seawater:

$$G = \frac{(8.33 \text{ J/mole/deg})(300°\text{C})}{(18 \text{ g/mole})} \ln\left(\frac{(1 \times 10^6 - S) \text{ ppm}}{965{,}000 \text{ ppm}}\right) \text{J/g}$$

where S is dissolved solids in parts per million.

[2] Gibbs free energy for 10 ppm rain relative to seawater salinity in evapotranspiring plants or to estuaries receiving fresh waters:

$$G = \frac{RT}{w} \ln\left(\frac{C_2}{C_1}\right) = \frac{(8.33 \text{ J/mole/deg})(300°\text{C})}{18 \text{ g/mole}} \ln\left(\frac{999{,}990}{965{,}000}\right)$$

Calculating the natural logarithim separately, we get:

$$2.3 \log_{10} \frac{999{,}990 \text{ ppm}}{965{,}000 \text{ ppm}} = 0.0355$$

Thus

$$G = 4.94 \text{ J/g} = 4.94 \times 10^6 \text{ J/m}^3 \text{ rainwater.}$$

$$G = (138.8 \text{ J/g}) \ln \left(\frac{(1 \times 10^6 - S) \text{ ppm}}{965{,}000 \text{ ppm}} \right.$$

$$(\text{— m}^3\text{/yr})(1 \times 10^6 \text{ g/m}^3)(G \text{ J/g}) = \text{— J/yr}$$

Chemical Potential Energy with Waters Used Within a Country

Combine chemical potential energies calculated for rain and rivers:

rain + inflowing rivers − outflowing rivers

If rivers reach the sea within the national boundary, combine:

rain + inflowing rivers

An alternative approach is to combine:

water evapotranspired + waters reaching sea within boundaries

Earth Cycle (Steady-State Uplift Balanced by Erosion

$$(\text{land area})(\text{heat flow per area}) = (\text{— m}^2)(\text{— J/m}^2\text{/yr})$$

Typical heat flows are: old, stable areas, 1×10^6 J/m²/yr; rapid orogeny, 3–10 $\times 10^6$ J/m²/yr.

Net Uplift

$$(\text{area})(\text{uplift rate})(\text{density})(0.5)(\text{uplift})(\text{gravity}) = (\text{— m}^2)(\text{— m/yr})$$
$$(\text{—} \times 10^3 \text{ kg/m}^3)(0.5)(\text{— m})(9.8 \text{ m/sec}^2) = \text{— J/yr}$$

Net Loss of Earth

Loss of clays from the area in river discharge or wind that is in excess of formation rate; a typical formation rate is: 31.2 g/m²/yr:

$$(\text{earth cycle rate})(\text{density}) =$$
$$(\text{—} \times 10^6 \text{ m/yr})(\text{—} \times 10^6 \text{ g/m}^3) = \text{— g/m}^2\text{/yr formation}$$
$$\text{erosion outflow} - \text{formation rate})(\text{area of country}) =$$
$$(\text{— g/yr}) - (\text{— g/m}^2\text{/yr})(\text{— m}^2) = \text{— g/yr}$$

Chemical Potential Energy in Imported and Exported Commodities Whose Value is Used in Reactions with Oxygen (Food, Fiber, Wood, etc.)

$$(\text{weight per year})(G) =$$

where G is the Gibbs free energy of oxidation with atmosphere. For organic substances with high free energies and small entropy changes of state in oxidation, G is practically equal to the bomb calorimetry values of heat of combustion (enthalpy changes). See tables of kilocalorie value in nutrition tables and handbooks. For carbohydrates, starch, wood, etc., assume about 4 kcal/g; for proteins, wool, etc., about 5 kcal/g dry; for fats and oils, about 7–9 kcal/g. Multiply by 4186 to get value in joules.

Where G (Gibbs free energy) is small, calculate its value from the chemistry of the reaction:

$$(\text{— Tn/yr})(G\ \text{J/g})(1 \times 10^6\ \text{g/Tn}) = \text{— J/yr}$$

where Tn is metric tons.

Net Loss of Topsoil

Topsoil erosion rates in excess of the profile formation rates are evaluated. See Appendix A18. Areas with mature vegetation are assumed to have little net gain or loss of topsoil.

A typical formation rate is: 1260 $g/m^2/yr$ or 8.54×10^5 $J/m^2/yr$ in areas showing natural vegetation succession.

Typical erosion rates of topsoils from farmed areas (Larson et al., 1983) from the United States are:

Area	$g/m^2/yr$	Area	$g/m^2/yr$
Pacific states	250	Cornbelt, delta area	1000
Mountain states	260	Southeastern states	850
Plains	500	Appalachian states	1250
Northeastern states	700		

Energy of net loss:

$$(\text{farmed area})(\text{erosion rate}) - (\text{successional area})(\text{formation rate}) =$$
$$(\text{— } m^2)(\text{— } g/m^2/yr) - (\text{— } m^2)(\text{— } g/m^2/yr) = g/yr$$

or

$$(\text{— Tn/m}^2\text{/yr})(\text{— m})(\text{— organic fraction})(5.4 \times 10^6 \text{ kcal/Tn})(4186 \text{ J/kcal}) = \text{— J/yr}$$

Geopotential in Inflowing Rivers

$$(\text{flow volume})(\text{density})(\text{height of river entry} - \text{river egress})(\text{gravity}) = (\text{— m}^3\text{/yr})(\text{— m})(1 \times 10^3 \text{ kg/m}^3)(9.8 \text{ m/sec}^2) = \text{— J/yr}$$

Geopotential in Rain Used

$$(\text{area})(\text{mean elevation}^{3})(\text{runoff})(\text{density})(\text{gravity}) = (\text{— m}^2)(\text{— m})(\text{— m/yr})(1 \times 10^3 \text{ kg/m}^3)(9.8 \text{ m/sec}^2) = \text{— J/yr}$$

Ocean Waves Absorbed at the Shore

$$(\text{shore length})(\tfrac{1}{8})(\text{density})(\text{gravity})(\text{height squared})(\text{velocity}) = (\text{— m})(\tfrac{1}{8})(1.025 \times 10^3 \text{ kg/m}^3)(9.8 \text{ m/sec}^2)(\text{— m})^2(\text{— m/sec}) = (3.154 \times 10^7 \text{ sec/yr}) = \text{— J/yr}$$

where velocity is: square root of (gravity)*(depth)

$$= [(9.8 \text{ m/sec}^2)(\text{— m})]^{1/2}$$

use depth of water where wave height is measured.

Tide Absorbed in Estuaries

The total energy of the tides absorbed in estuaries is:

$$(\text{area elevated})(0.5)(\text{tides/yr})(\text{height squared})(\text{density})(\text{gravity}) = (\text{— m}^2)(0.5)(706\text{/yr})(\text{— m})^2(1.0253 \times 10^3 \text{ kg/m}^3)(9.8 \text{ m/sec}^2) = = \text{— J/yr}$$

where $0.5 \times$ height = center of gravity.

[3] Elevation measured relative to the low point on the nation's border where rivers leave the country.

Tide Absorbed on Continental Shelves

The formula for this is the same as the estuary formula above multiplied by a factor ranging from 0.1 to 0.5.

Chemical Potential Energy in Imported and Exported Commodities Whose Value is in its Concentration[4]

$$(\text{weight per yr})(G) =$$

where G is the Gibbs free energy per unit weight relative to concentration of the commodity in the environment. For example, for iron ore, $G = 14.2$ (Gilliland and Eastman, 1981):

$$(\text{— Tn/yr})(G\ \text{J/g})(1 \times 10^6\ \text{g/Tn}) = \text{— J/yr}$$

Coal Flows or Outflows

$$(\text{weight/yr})(\text{energy/unit weight}) = (\text{— Tn/yr})(3.18 \times 10^{10}\ \text{J/Tn}) = \text{— J/yr}$$

Oil Inflows or Outflows

$$(\text{barrels/yr})(\text{energy/bbl}) = (\text{— bbl/yr})(6.28 \times 10^9\ \text{J/bbl}) = \text{— J/yr}$$

Natural-Gas Inflows or Outflows

$$(\text{volume of gas/yr})(\text{energy/unit volume}) =$$
$$(\text{— thousand cubic ft/yr})(1.1 \times 10^9\ \text{J/thousand cubic ft}) = \text{— J/yr}$$

or

$$(\text{— therms/yr})(1.055 \times 10^5\ \text{J/therm}) = \text{— J/yr}$$

[4] Effective concentration is that solution concentration in equilibrium with the solid. For solids it is the solution concentration in which they are used. Environmental concentration is the solution concentration of waters in the soils and surface waters of the nation. Molecular weight is the mean molecular weight of the effective components of the commodity.

$$G = \frac{(8.33\ \text{J/mole/deg})(300°\text{C})}{\text{effective molecular weight}} \ln\left(\frac{\text{effective concentration}}{\text{environmental concentration}}\right)\ \text{J/g}$$

Flow of Electric Power

(power units for a time)(energy per unit of power for a time) =
(— kWh/yr)(3.60×10^6 J/kWh) = — J/yr

or

(capacity of power plant)(% of capacity)(hours/yr)(energy/unit) =
(— kW)(— %/100)(1.40×10^9 J/kW/yr) = — J/yr

where kW = kilowatt; kWh = kilowatt-hour

Electrical Output of Nuclear Plants

Electricity delivered:

(— kWh)(3.6×10^6 J/kWh)

where kWh is kilowatt-hours.

Heat Production of Fission

(weight of uranium use per time)(fraction ^{235}U)(energy per unit ^{235}U)
(— Tn/yr U_3O_8)(0.007)(1×10^6 g/Tn)(7.95×10^{10} J/g^{235}U) = — J/yr

Imported or Exported Service (Data in Dollars Paid)

This yields the data in dollars per year from a particular economy, each of which has a different EMERGY/$ ratio.

II. FORMULAS USED FOR CALCULATING ENERGY STORAGES IN JOULES

Geothermal Heat Storage Potential

(reservoir volume)(density)(specific heat)(ΔT)($\Delta T/T$)(0.5)

where ΔT is Kelvin temperature difference between heat storage temperature (T) and background temperature, $\Delta T/T$ is the Carnot reversible efficiency, and efficiency at maximum power is half Carnot (0.5): gcal = small calorie.

(— m^3)(— $\times 10^3$ kg/m^3)(1×10^3 g/kg)(— gcal/g/deg)[(— − — deg)^2/
(— deg K)](0.5)(4.186 J/gcal) = — J

Potential Energy in Stored Organic Matter (Fuels, Soil, Peat, Wood, Etc.)

$$(\text{volume of material})(\text{density})(\text{organic fraction})(G) = \text{—}\ \text{J}$$

Typical soil, 5.4 kcal/g; fraction organic, 0.03 g/g; density, 1.47 g/m^3:

$$(1 \times 10^6\ \text{cm}^3/\text{m}^3)(1.47\ \text{g/cm}^3)(0.03)(5.4\ \text{kcal/g})(4186\ \text{J/kcal}) = 10.0 \times 10^8\ \text{J/m}^3$$

$$(\text{—}\ \text{m}^3)(10.8 \times 10^8\ \text{J/M}^3) = \text{—}\ \text{J}$$

or

$$(\text{weight})(\text{chemical potential energy per unit weight})$$
$$(\text{—}\ \text{Tn})(1 \times 10^6\ \text{g/Tn})(\text{—}\ \text{kcal/g})(4186\ \text{J/kcal}) = \text{—J}$$

or for fuel gas:

$$(\text{volume})(\text{chemical potential energy per volume}) = (\text{— thousand cft}^3)(1.05 \times 10^9\ \text{J/thousand cft}^3) = \text{—}\ \text{J}$$

Geopotential Energy of Elevated Materials (Water, Mountains, Rock)

$$(\text{volume})(\text{density})(\text{gravity})(\text{height of center of gravity of mass}) = (\text{—}\ \text{m}^3)(1 \times 10^3\ \text{kg/m}^3)(9.8\ \text{m/sec}^2)(\text{—}\ \text{m}) = \text{—}\ \text{J}$$

Chemical Potential Energy of Water and Groundwater Storages

$$(\text{water volume})(\text{density})(G) = \text{—}\ \text{J}$$

where G is Gibbs free energy of water relative to saltwater:

$$G = \frac{(8.33\ \text{J/mole/deg})(300°\text{C})}{19\ \text{g/mole}} \ln\left(\frac{(1 \times 10^6 - S)\ \text{ppm}}{965{,}000\ \text{ppm}}\right)\ \text{J/g}$$

where S is solutes in ppm.

To estimate the volume of groundwaters:

$$(\text{volume of land mass})(\text{porosity fraction}) = (\text{—}\ \text{m}^3)(\text{—}) = (\text{—}\ \text{m}^3) = \text{—}\ \text{m}^3\ \text{space containing water}$$

Typical porosities are:

Shale	0.10
Basalt	0.10

Sands	0.25
Granite	0.05
Gravels	0.40
Limestone	0.10

Chemical energy of water:

$$(\text{— m}^3)(1 \times 10^6 \text{ g/m}^3)G = \text{— J}$$

Chemical Potential Energy of Mineral Deposits

$$(\text{volume})(\text{density})(G) =$$

where G is Gibbs free energy of the mineral relative to the surrounding environment in which it is used, dispersed, or destroyed in chemical reactions. Typical values of G for common minerals are:

Mineral	J/g	Source
Phosphate deposits	58.3	Table 7.7, note 4
Copper ore	1.65	Gilliland (1978)
Bauxite (Al ore)	65.3	Van Moeske (1981)
Iron ore	14.2	Gilliland (1978)
Potassium (KCl)	702.0	Gibbs energy: 5×10^5 ppm/10 ppm
Nitrogen (NH_3)	2041.0	Anderson (1980)

Chemical Potential Energy of Bedrock

$$(\text{volume})(\text{density})(G) = \text{— J}$$

where G is the Gibbs free energy of the bedrock relative to states after weathering.

Typical values of G (Gilliland et al., 1978) are:

	Density (g/cm^3)	G (J/g)
Limestone	1.95	50
Granite	2.61	50
Shale	2.40	100
Basalt	2.79	172
Sand, sandstone	3.17	611

$$(\text{— m}^3)(1 \times 10^6\ \text{cm}^3/\text{m}^3)(\text{— g/cm}^3(G) = \text{— J}$$

Nuclear Energy

Heat equivalents from Schipper (1976):

$$\text{(weight of uranium ore)(fraction } ^{235}\text{U in ore)(heat per weight)} =$$
$$(\text{— Tn})(0.007)(1 \times 10^6\ \text{g/Tn})(7.95 \times 10^{10}\ \text{J/g } ^{235}\text{U}) = \text{— J}$$

APPENDIX C

TRANSFORMITIES

Ten methods are suggested for calculating solar transformities (Table 14.3). Main environmental energy flows are calculated from data on the hierarchical energy web of the geobiosphere (methods 1 and 2). Other transformities are calculated from analysis of subsystems of energy production and transformation (3). Solar transformities may also be calculated from storage development times (4), by combining other transformities (5), from data on energy flows in networks (6), by means of a computer-solved matrix evaluation (7), by source tracking in an energy network (8), from hierarchical distribution graphs (9), and turnover time (10).

This appendix contains solar transformities used in this book, indicating tables or references where they were calculated.

TABLE C.1. Solar Transformity of Electric Power

Note	System	Solar Empower (sej/yr)*	Electric Power (J/yr)	Transformity
1	Coal power plant	160,000	1	160,000
2	World stream geopotential	9.44×10^{24}	10×10^{20}	94,400
3	Hydroelectric power, Sweden	1.95×10^{24}	2.43×10^{17}	80,246
4	Wood power plant, Jari, Brazil	2.38×10^{20}	1.17×10^{15}	203,418
5	Solar voltaic grid, Austin, Tex.	7.5×10^{17}	1.8×10^{12}	416,666
6	Hydroelectric, Tucurui, Brazil	1.65×10^{22}	1.0×10^{17}	165,000
7	Wood power plant, Thailand	2.42×10^{14}	3.6×10^{9}	67,222
8	Oil power plant, Thailand	7.14×10^{14}	3.6×10^{9}	197,777
9	Coal power plant, Thailand	6.10×10^{14}	3.6×10^{9}	169,444
10	Lignite power plant, Thailand	5.47×10^{14}	3.6×10^{9}	151,944
11	Lignite power plant, Texas	5.4×10^{21}	2.65×10^{16}	204,384
	Mean			173,681

* 18% added to those EMERGY evaluations that were made before tide values were added to global solar EMERGY budget (items 6 and 11)

[1] Assuming 4 coal emj/J of electric power and 40,000 sej/J coal.

[2] Global calculation made with assumptions about the empower required for the mountain uplift, the carving of basins, and the construction of dams. Global solar empower, 9.44×10^{24} sej/yr, generates an average stream flow over land of 39.6×10^{3} km^3 runoff (Todd 1970) and maintains an average land elevation of 875 m (Ryabchikov, 1975). Average land and average streams were taken as by-products of shared empower.

Stream geopotential:

$$(39.6 \times 10^{12}\ m^3/yr)(875\ m)(1000\ kg/m^3)(9.8\ m/sec^2) = 3.39 \times 10^{20}\ J/yr.$$

Electric power potential = stream geopotential times efficiency of hydroelectric conversion taken as 80%:

$$(3.39 \times 10^{20})(0.8) = 2.7 \times 10^{20}\ J/yr\ \text{electrical}.$$

For a 25% feedback of empower from the economy for dam and operation, the net yield of electricity could be:

$$\tfrac{3}{4}(2.7 \times 10^{20}\ J/yr) = 2.0 \times 10^{20}\ J/yr.$$

If stream energy in the long run has to carve a basin half the time to allow for generation of electricity the other half, then the electric output is half, or 1×10^{20} J/yr.

[3] Realized electric power in 1988: 72 terawatt-hr (Sweden, 1990):

$$(72 \times 10^{9}\ kWh/yr)(860\ kcal/kWh)(4186\ J/kcal) = 2.60 \times 10^{17}\ J/yr$$

For 80% efficiency, input geopotential is:

$$2.6 \times 10^{17}/0.8 = 3.25 \times 10^{17}\ J/yr.$$

Time for erosion to make a basin may be assumed to be similar to the time for filling with sediment. Thus the dam in the long run operates for half the time as it fills with sediment, eroding for half using the same stream energy. Either consider the long-range electric yield as half, or consider the short-term operation as receiving the prorated EMERGY of the carved basin as equivalent to the input geopotential (2 times geopotential in use):

$$(2)*(3.25 \times 10^{17} \text{ J/yr}) = 6.5 \times 10^{17} \text{ J/yr input geopotential.}$$

For third-order streams, solar transformity from Figure 2.8 on the Mississippi River is 3×10^4 sej/J and therefore the input solar EMERGY is:

$$(3 \times 10^4 \text{ sej/J})(6.5 \times 10^{17} \text{ J/yr}) = 1.95 \times 10^{22} \text{ sej/yr}$$

Using $\frac{1}{4}$ of empower feedback for dam and operation, net electric yield is:

$$(3.25 \times 10^{17})(0.75) = 2.43 \times 10^{17} \text{ J/yr}$$

[4] Rainforest logs are supplied in a steady state from a 100-yr rotation requiring 2.324×10^9 m^2. Solar EMERGY from the main use of rain by trees and 3 mm transpiration, 4.94 J Gibbs free energy per gram of rainwater and solar transformity of rain, 1.82×10^4 sej/J:

$$(3 \text{ mm/day})(365 \text{ days/yr})(1 \times 10^{-3} \text{ m}^3\text{/mm})(1 \times 10^6 \text{ g/m}^3)(4.94 \text{ J/g water})(2.3 \times 10^9 \text{ m}^2)(1.82 \times 10^4 \text{ sej/J}) = 2.27 \times 10^{20} \text{ sej/yr}$$

plus solar EMERGY from fuels use (0.085×10^{20} sej/yr) and services used (0.025×10^{20} sej/yr). Electricity produced (1.67×10^{15} J/yr) minus electricity used in the processing: 0.032 J/yr debarking and chipping and 0.46×10^{15} J/yr in plant operations.

[5] Power grid evaluated by R. King and Schmandt (1991). See Table 8.2.

[6] Modified from M. T. Brown (1986b). Energy analysis of the hydroelectric dam near Tucurui, Brazil, pp. 82–91 in H. T. Odum et al. (1985). Electricity produced: 1.0×10^{17} J/yr based on 0.8 capacity factor and 4000 MW. Contribution to dam and operation from the economy: 4.25×10^{21} sej/yr. Contribution of geopotential of inflowing water and also the prorated contribution of the basin that was developed by the same streamflows earlier (see note 3):

$$(2.06 \times 10^{17} \text{ J/yr})(2.36 \times 10^4 \text{ sej/J}) = 4.87 \times 10^{21} \text{ sej/yr}$$

Total input includes this factor twice (present inflow + prorated basin emergy). In a full cycle of damming and allowing reerosion of basin, there is no net sediment diversion:

$$(4.87 + 4.87 + 4.25) \times 10^{21} \text{ sej/yr} = 13.95 \times 10^{21} \text{ sej/yr}$$
$$(13.95 \times 10^{21} \text{ sej/yr})(1.18 \text{ tidal correction}) = 1.646 \times 10^{22} \text{ sej/yr}$$

[7] Wood power plant (25 MW generating 173.5×10^3 kWh/yr) using eucalyptis plantation wood; evaluations by S. Doherty and Bo Hector (Doherty and Nilsson, 1992). Values estimated per megawatt-hour electric:

$$(1 \text{ mWh})(1000 \text{ kWh/mWh})(860 \text{ kcal/kWh})(4186 \text{ J/kcal}) = 3.59 \times 10^9 \text{ J/yr}$$

Solar EMERGY inputs in sej/mWh: Rain, 44×10^{12}; fertilizer, 6×10^{12}; labor, 7×10^{12}; plantation capital, 29×10^{12}; plant operational service, 96×10^{12}; power plant capital, 55×10^{12}; transmission, 6×10^{12}; total, 242×10^{12} sej/mWh.

[8] Oil-fired power plant; evaluations by S. Doherty and Bo Hector (Doherty and Nilsson, 1992). Values estimated per megawatt-hour electric:

$$(1 \text{ mWh})(1000 \text{ kWh/mWh})(860 \text{ kcal/kWh})(4186 \text{ J/kcal}) = 3.59 \times 10^{9} \text{ J/yr}$$

Solar EMERGY inputs in sej/mWh: oil, 402×10^{12}; oil services, 100×10^{12}; plant operational services, 131×10^{12}; capital, 40×10^{12}; transmission, 41×10^{12}; total, 714×10^{12} sej/mWh.

[9] Coal-powered plant; evaluations by S. Doherty and Bo Hector (Doherty and Nilsson, 1992). Values estimated per megawatt-hour electric:

$$(1 \text{ mWh})(1000 \text{ kWh/MWh})(860 \text{ kcal/kWh})4186 \text{ J/kcal}) = 3.59 \times 10^{9} \text{ J/yr}$$

Solar EMERGY inputs in sej/MWh: Coal, 380×10^{12}; oil services, 80×10^{12}; plant operational services, 109×10^{12}; capital, 58×10^{12}; transmission, 43×10^{12}; total, 610×10^{12} sej/mWh.

[10] Lignite power plant; evaluations by S. Doherty and Bo Hector (Doherty and Nilsson, 1992). Values estimated per megawatt-hour electric:

$$(1 \text{ mWh})(1000 \text{ kWh/MWh})(860 \text{ kcal/kWh})(4186 \text{ J/kcal}) = 3.59 \times 10^{9} \text{ J/yr}$$

Solar EMERGY inputs in sej/MWh: Lignite, 279×10^{12}; mining services, 93×10^{12}; plant operational services, 100×10^{12}; capital, 44×10^{12}; transmission, 30×10^{12}' total, 547×10^{12} sej/MWh.

[11] Big Brown lignite power plant, Texas (Odum et al., 1987a):

$$(7.27 \times 10^{13} \text{ J/day})(365 \text{ days/yr}) = 2.65 \times 10^{16} \text{ J/yr electric power produced}$$

Inputs evaluated in sej/day:
Mining inputs: Lignite mined for power plant, 73.7×10^{17}; topsoil lost, 3.1×10^{17}; fuel used, 0.032×10^{17}; electric power used, 0.49×10^{17}; equipment maintenance, 0.93×10^{17}; goods and services, 6.2×10^{17}; total, 84.45×10^{17} sej/day. Power-plant inputs: cooling water, 0.10×10^{17}; equipment maintenance, 1.24; goods and services, 40×10^{17}; total, 41.34×10^{17}; sej/day. Mining and power plant on a 1-year basis.

$$(365)(84.45 + 41.34) \times 10^{17} = 4.59 \times 10^{21} \text{ sej/yr}$$

Tidal correction to global transformities:

$$(4.59 \times 10^{21})(1.18) = 5.4 \times 10^{21} \text{ sej/yr}$$

TABLE C.2. Solar Transformity for Fuels

Note	Item	Solar Transformity ($\times 10^4$ sej/J)
1	Rainforest logs	3.2
2	Rainforest wood, transported and chipped	4.4
3	Liquid motor fuel	6.6
4	Crude oil	5.4
5	Natural gas	4.8
6	Coal	4.0
7	Peat	1.9
8	Lignite	3.7
9	Plantation pine	0.7
10	Charcoal	10.6

[1] Energy and EMERGY (Odum and Odum, 1983)

$$\frac{(8.3 \times 10^{12}\ \text{sej/m}^2\text{/100 yr})}{2.58 \times 10^6\ \text{J/m}^2\text{/100 yr}} = 3.23 \times 10^4\ \text{sej/J}$$

[2] Energy and EMERGY (Odum and Odum, 1983)

$$\frac{2.0 \times 10^5\ \text{sej/elect J}}{4.56\ \text{wood J/elect J}} = 4.38 \times 10^4\ \text{sej/J}$$

[3] 1.65 coal J/J liquid motor fuel (Slesser, 1978):

$$(4 \times 10^4\ \text{sej/J coal})(1.65\ \text{coal J/motor fuel J}) = 6.6 \times 10^4\ \text{sej/J motor fuel}$$

[4] 19% crude oil used in refining and transport (Cook, 1976)

$$\frac{6.6 \times 10^4\ \text{sej/J motor fuel}}{1.23\ \text{crude J/motor fuel J}} = 5.37\ \text{sej/J motor fuel}$$

[5] Natural gas is 20% more efficient in boilers than is coal (Cook, 1976):

$$(4 \times 10^4\ \text{sej/J coal})(1.2\ \text{coal J/natural gas J}) = 4.8 \times 10^4$$

[6] (1.7×10^5 sej/J elect power)/(4 coal J/J elect power) = 4.3×10^4 sej/coal J
From sedimentary cycle calculation, Table 3.5, 3.4×10^4 sej/coal J:

$$\text{Average } (4.3 \times 10^4 + 3.4 \times 10^4/2 = 3.9\ \text{sej/J}$$

Rounded to 4.0 as a temporary standard.
[7] Table 5.4.
[8] Lignite analysis (Odum et al., 1987a).
[9] Monterey pine (Table 5.2).
[10] Charcoal (Sundberg et al., 1991).

TABLE C.3. Solar Transformities and Mass EMERGY of Global Flows

Item	Transformity (sej/J)*	EMERGY/gram ($\times 10^9$ sej/g)*	Source
GLOBAL solar insolation	1	—	By definition
Surface wind	1,496		Table 3.2
Convective Earth Heat	6,055		Table 3.1
Oceanic rain, chemical potential	7,435		Note 1
Physical energy, rain on land	10,488		Table 3.2
Tidal energy absorbed	16,842		Table 3.1
Volcanic heat	18,000		Fig. 8.9
Chemical energy, rain on land	18,199		Table 3.2
Physical stream energy	27,874		Table 3.2
Waves absorbed on shores	30,550		Table 3.2
Continental earth cycle, heat flow	34,377		Table 3.2
Chemical stream energy	48,459		Table 3.2
Oceanic upwelling, inorganic carbon	7.8×10^5	0.18	Note 2
Oceanic upwelling, nitrate-nitrogen	2.6×10^6	1.05	Note 2
Oceanic upwelling, phosphate	3.8×10^7	9.5	Note 2

* sej/J = solar emjoules per joule; sej/g = solar emjoules per gram.

[1] $(9.44 \times 10^{24}$ sej/yr$)/[(2.57 \times 10^{14}$ m^3/yr$)(1 \times 10^6$ g/m$^3)(4.94$ J/g$)] = 7435$ sej/J.

[2] Phosporus upwelling flux from Garrels et al., (1975); nitrogen/phosphorus (9.0) and carbon/phosphorus (53.0) ratios from Redfield (1934); Gibbs free energies between deep water and the surface based on concentration ratios: phosphorus, 83/3; nitrate, 500/50; inorganic carbon, 600/20. The results are: 228 J/g C, 410 J/g N, 251 J/g P.

TABLE C.4. Transformity and EMERGY per Unit Mass in Earth Substances

Item	Transformity (sej/J)	EMERGY/gram ($\times 10^9$ sej/g)	Source
Oceal floor			
Oceanic basalt	0.15		Table 3.3
Pelagic and abyssal sediments	0.97		Table 3.3
Continents			
Granitic rocks	0.50		Table 3.3
Mountains on land	1.12		Table 3.3
Metamorphic rocks	1.45		Table 3.3
Continental sediment	1.88		Table 3.3
Volcanic extrusion at surface	4.5		Table 3.3
Global sedimentary cycle		1.0	Table 3.5
Shale	1.0×10^9	1.0	Table 3.5
Limestone	1.62×10^6	1.0	Table 3.5
Sandstone	2.0×10^7	1.0	Table 3.5
Evaporites	3.3×10^6	1.0	Table 3.5
Coal	4.0×10^4	1.0	Table C.2
Sedimentary Iron ore	6.2×10^7	1.0	Table 3.5
Bauxite (Aluminum ore)	1.5×10^7	1.0	Table 3.5
Soil clay from shale	—	2.0	Table 3.5
Top soil organic matter	7.4×10^4		Note 1
Clay from weathering	—	1.71	Note 2
Potassium fertilizer	3.0×10^6	1.1 (1.74/g K)	Note 3
Ammonia fertilizer	1.86×10^6	3.8 (4.6/g N)	Note 4
Phosphate fertilizer	1.01×10^7	3.9 (17.8/g)	Table 7.7

[1] Replacement time, 500 yr (Jenny, 1982); 3% organic content, 5.4 kcal/g dry in upper 0.45 m with density 1.4 g/ml.

[2] Earth formation and erosion rate: Uplift and erosion rate from Garrels et al., (1975); half of weathered uplift is clay (Siegel, 1974, after Krauskopf, 1967, and Goldick 1938):

$$(2.4 \times 10^{-5}\ \text{m/yr})(2.6 \times 10^6\ \text{g/m}^3\ \text{rock density})(0.5) = 31.2\ \text{g/m}^2\text{/yr}$$

$$(9.44 \times 10^{24}\ \text{sej/yr})/(31.2\ \text{g/m}^2\text{/yr})(1.5 \times 10^{14}\ \text{m}^2\ \text{continent area}) = 1.71 \times 10^9\ \text{sej/g}$$

[3] Potassium chloride from Dead Sea works in Israel (H. T. Odum and Odum, 1983, p. 477). EMERGY based on solar energy of evaporating water, energy in dry air, fresh water in processing, fuel, electricity, services, and hydrostatic head of water processing.

[4] Odum and Odum (1983).

TABLE C.5. Transformity and Solar EMERGY per Mass in Plant Products and Fuels

Item	Transformity (sej/J)	EMERGY/gram $\times 10^9$ sej/g	Source
Gross production, estuary	4.7×10^3		Note 10
Net production, estuary	9.0×10^3		Note 10
Plantation pine wood	6.7×10^3	0.10	Table 5.2
Estarine organic matter	1.1×10^4		Note 10
Peat	1.7×10^4	0.36	Table 5.4
Mulberry leaves	2.4×10^4		Table 4.2
Lignite	3.7×10^4		Note 1
Cornstalks	3.9×10^4		Note 8
Coal	4.0×10^4		Table C.2
Rainforest logs	4.1×10^4	0.39	Table C.2
Natural gas	4.8×10^4		Table C.2
Crude oil	5.4×10^4		Table C.2
Ethanol	6.0×10^4		Note 2
Liquid motor fuel	6.6×10^4		Table C.2
Corn	8.3×10^4	1.43	Note 8
Charcoal	1.07×10^5		Note 9
Electric power	2.0×10^5	—	Table C.1
Cotton	8.6×10^5		Note 1
Butter	1.3×10^6		Note 4
Smaller estuarine animals	1.5×10^6		Note 10
Caterpillar pupae	2.0×10^6		Table 4.2
Mutton	2.0×10^6		Note 6
Silk	3.4×10^6	72.	Table 4.2
Veal	4.0×10^6		Note 5
Wool	4.4×10^6		Note 7
Upper consumers, estuary	$30. \times 10^6$		Note 10
Aquaculture shrimp	13.0×10^6		Note 3

[1] Odum et al. (1987a).
[2] E. C. Odum and Odum (1984).
[3] H. T. Odum and Arding (1991).
[4] Energy analysis of Indian cattle by Mitchell (1979) using as solar EMERGY input that of the rain:

$$\frac{(4.10 \times 10^{10}\text{ solar emkcal/yr}}{3.25 \times 10^4\text{ kcal}} = 1.3 \times 10^6\text{ semkcal/kcal} = 1.3 \times 10^6\text{ sej/J}$$

[5] Data same as in note 4:

$$\frac{4.22 \times 10^{10}\text{ sekcal/yr}}{1.06 \times 10^4\text{ kcal/yr}} = 3.98 \times 10^6\text{ semkcal/kcal} = 3.98 \times 10^6\text{ sej/J}$$

[6] H. T. Odum and Odum (1983 p. 421); EMERGY inputs of rain, phosphate fertilizer, fuels, electricity, services, and government subsidy totaled 252×10^{13} sej/ha/yr (including tidal correction for world rain transformity) divided by annual production per hectare (1.27×10^9 J/yr).
[7] EMERGY data as in note 6:

$$\frac{252 \times 10^{13}\text{ sej/yr}}{5.68 \times 10^8\text{ J/yr wood production}} = 4.43 \times 10^6\text{ sej/J}$$

[8] Odum (1984b).
[9] Sundberg et al. (1991).
[10] Tennenbaum 1988

APPENDIX D

EMERGY/MONEY RATIOS

Average solar EMERGY/money ratios were calculated for countries (Table 10.7) by dividing the solar EMERGY used in that country from all sources in a year by the gross national product (GNP) for that year. The ratios used in this book are in solar EMERGY per U.S.$, but ratios of EMERGY (emjoules) per unit of local money can also be calculated for a given year using the currency market conversions for those data. EMERGY (in emjoules) per money ratios are useful for evaluating the EMERGY contribution in goods and services where data are given in dollars (Chapter 4). Or, vice versa: the average emdollar equivalents of EMERGY may be calculated to put EMERGY wealth in the economic terms familiar to most people.

Generally, EMERGY/money ratios decrease each year, partly due to inflation, partly due to economic development (which increases money circulation for the same resource use), and somewhat due to increasing efficiency in resource use. Although our Florida center has not had the funds to recalculate EMERGY/money ratios each year, an approximate calculation-interpolation can be made taking into account the two factors that change the most each year: the GNP and the fuel consumption.

Table D.1 for the United States for different years was made by adding the environmental EMERGY use evaluated for 1983 (Chapter 10) to the EMERGY of total fuel consumption for each year and dividing the sum by the GNP for that year.

TABLE D.1. Solar Emjoules/$ for United States for Different Years (Estimates Based on 1983 Analysis Changed According to Fuel Use and GNP)

Year	Fuel use		Solar EMERGY use‡	GNP	Solar EMERGY/$§
	$\times 10^{19}$ J/yr*	$\times 10^{24}$ sej/yr†	($\times 10^{24}$ sej/yr)	GNP ($\times 10^{9}$ $/yr)	($\times 10^{12}$ sej/$)
1947	3.47	1.87	5.87	231.3	25.4
1948	3.59	1.94	5.94	257.6	23.1
1949	3.33	1.80	5.84	256.5	22.6
1950	3.60	1.94	5.94	284.8	20.9
1951	3.89	2.10	6.10	328.4	18.6
1952	3.86	2.08	6.08	345.5	17.6
1953	3.97	2.14	6.14	364.6	16.8
1954	3.83	2.07	6.07	364.8	16.6
1955	4.22	2.28	6.28	398.0	15.8
1956	4.43	2.39	6.39	419.2	15.2
1957	4.42	2.39	6.39	441.1	14.5
1958	4.38	2.37	6.37	447.3	14.2
1959	4.58	2.47	6.47	483.7	13.4
1960	4.74	2.56	6.56	503.7	13.0
1961	4.82	2.60	6.60	520.1	12.7
1962	5.04	2.72	6.72	560.3	12.0
1963	5.24	2.83	6.83	590.5	11.6
1964	5.43	2.93	6.93	632.4	11.0
1965	5.69	3.07	7.07	684.9	10.3
1966	6.03	3.26	7.26	749.9	9.7
1967	6.14	3.32	7.32	793.9	9.2
1968	6.51	3.52	7.51	864.2	8.7
1969	6.85	3.70	7.70	930.3	8.3
1970	7.09	3.83	7.83	976.4	8.0
1971	7.25	3.92	7.91	1050.4	7.5

TABLE D.1. *(Continued)*

Year	Fuel use		Solar EMERGY use‡ ($\times 10^{24}$ sej/yr)	GNP ($\times 10^{9}$ \$/yr)	Solar EMERGY/\$§ ($\times 10^{12}$ sej/\$)
	$\times 10^{19}$ J/yr*	$\times 10^{24}$ sej/yr†			
1972	7.61	4.10	8.11	1151.8	7.0
1973	7.87	4.25	8.25	1306.6	6.3
1974	7.55	4.08	8.08	1412.9	5.7
1975	7.45	4.02	8.02	1328.8	6.0
1976	7.83	4.23	8.22	1700.1	4.8
1977	8.07	4.23	8.23	1887.2	4.4
1978	8.23	4.34	8.44	2107.6	4.0
1979	8.32	4.49	8.49	2417.8	3.5
1980	8.00	4.32	8.32	2633.1	3.2
1981	7.81	4.22	8.22	3053.	2.7
1982	7.47	4.04	8.03	3166.	2.5
1983	7.44	4.02	8.02	3305.	2.4
1984	7.82	4.22	8.22	3772.	2.2
1985	7.80	4.21	8.21	4015.	2.0
1986	7.82	4.22	8.22	4240.	1.9
1987	8.10	4.37	8.37	4527.	1.8
1988	8.43	4.45	8.55	4880.	1.75
1989	8.58	4.63	8.60	5267.	1.63
1990	8.58	4.63	8.60	5543.	1.55
1991	8.56	4.62	8.59	5737.	1.49
1992	8.69	4.69	8.66	6046.	1.43
1993	8.86	4.78	8.75	6378	1.37

* Fuels use in quads (1055 J/btu) $\times 10^{15}$ btu/yr, from U.S. Statistical Abstract (1990, 1994).
† Solar EMERGY of fuels used = fuel(joules) times 5.4×10^{4} sej/J (see Table C.2).
‡ Total EMERGY for each year based on sum of solar EMERGY in A (see below) plus B and C from 1983 study (Chapter 10); B and C were increased 1.165 times for later inclusion of tide in global EMERGY flux):

A. Fuels use in column 3
B. Renewables and soil use, 2.10×10^{24} sej/yr
C. Other continuing inputs, 1.87×10^{24} sej/yr

A. Fuels evaluated as oil-equivalents fuel (in joules) times the solar transformity of oil (5.4×10^{4} sej/J) (see Table C.2).
B. U.S. renewable EMERGY and soil loss for 1983 (Table 10.2):

$$(96.0 \times 10^{22} + 113.7 \times 10^{22} \text{ sej/yr} = 210 \times 10^{22} \text{ sej/yr}$$

C. Other EMERGY for 1983 calculated as the total solar EMERGY use by the U.S. (analyzed in 1983) minus the solar EMERGY due to renewables, soil loss, and fuels (for 1983); soil loss, 113.7×10^{22} sej/yr; fuel use in 1983, 4.45×10^{24} sej/yr (Table 10.2).
From U.S. sources: coal use, 93.1×10^{22}; natural gas, 88.8×10^{22}; oil use, 185.0×10^{22} sej/yr.
Imported sources: natural gas, 4.8×10^{22}; crude oil, 40.4×10^{22}; petroleum products, 26.1×10^{22} sej/yr:

$$(8.42 \times 10^{24} - 2.10 \times 10^{24} - 4.45 \times 10^{24} = 1.87 \times 10^{24} \text{ sej/yr}$$

§ Total solar EMERGY use divided by GNP for that year.

APPENDIX E

PARAMETERS FOR UPDATING EVALUATIONS

This appendix provides guidelines for revising older energy and EMERGY analyses. When better information becomes available, revise old EMERGY evaluation tables (embodied energy) substituting new data or new transformities. Before 1983 EMERGY was called "embodied energy" and transformity was called "energy transformation ratio" or "quality factor". After 1983, embodied joules were redefined as emjoules (H. I. Odum, 1986; Scienceman, 1987).

Several times between 1966 and 1994, the reference base of global EMERGY that we used for evaluating global environmental processes was increased to include additional input energies to the global system (solar energy over the ocean driving rain, wind, and waves to land; geologic energy; tides, etc.). The changes and dates are given in Table E.1 so that older work can be updated. Early on the solar transformity of fossil fuels was refined, which affected those analyses relating the environment to a fuel-based economy. The main effect of the later changes was to increase all the transformities by the same amount. For the most part, these additions increased the absolute EMERGY values without having much effect on ratios and conclusions.

In this book the baseline for annual global solar EMERGY use (solar empower) was set at 9.44×10^{24} sej/yr (solar emjoules per year) based on estimates in Table 3.1. So long as the same baseline is used, the exact value is not important to EMERGY evaluations, just as the exact height of sea level from the center of the earth is not important to measurements of energy in elevated water. To get a more exact baseline value (thus refining Table 3.1) may require a special project by specialists in geophysics to refine EMERGY evaluation tables for the earth processes. In the meantime, it may be desirable not to change the baseline further, since it does not affect the differences on which EMERGY accounting is based.

TABLE E.1. Chronology of EMERGY Conversions

1967–1971	Energy types of higher quality were expressed in units of organic matter (dry basis)—including wood, peat, coal, oil, natural gas, and living biomass—as if they were equivalent. This practice was used in *Environment, Power and Society* (H. T. Odum, 1971b) and in the report to the President's Science Advisory Council (Odum, 1967a,b). Direct sunlight equivalent to organic matter was taken as 1000 solar kilocalories per kilocalorie organic matter. Solar energy embodied in rains, winds, and geologic inputs was not included. Human services were evaluated using 10,000 organic kilocalories per 1969 $.
1973–1980	In a report to Congress (Odum et al., 1976b) and in first edition of *Energy Basis for Man and Nature* (H. T. Odum and Odum, 1976), energy qualities of plants, wood, and fossil fuels were differentiated. Calculations and comparisons were made in fossil-fuel equivalents. Direct sunlight equivalents of fossil fuels were taken as 2000 solar kilocalories per fuel kilocalorie. The energy/money ratio used was 25,000 fuel kilocalorie equivalents per 1973 $ in the first edition, and 11,000 coal kilocalorie equivalents per 1980 $ in the second edition.
1980–1982	With New Zealand analyses, a hierarchy of earth transformations was recognized so that the solar-energy-driven processes of the oceans and the atmospheric engine would converge rains, winds, and waves to the land; thus solar energy of the global processes could be embodied in these contributions to land productivity. In an energy analysis manual for the Nuclear Regulatory Commission (Odum et al., 1983), we used 6800 global solar equivalent kilocalorie and 19,600 fuel kilocalories per $ (also used in Table 14-1 in *Systems Ecology;* (Odum, 1983b).
1983–present	At IIASA in the summer of 1983 a solar equivalent of geologic heat sources was determined and included in the solar EMERGY basis of the global process (Odum and Odum, 1983). Also, calculations of solar equivalents of coal were estimated from a wood power plant and from rate of geologic cycling. Solar equivalents of coal were estimated as 40,000 solar kilocal/coal kilocal (40,000 sej/J) now defined as solar EMERGY. EMERGY/$ ratios were 2 trillion sej/$ or 11,944 coal emj/$ (1987 $). Calculations were standardized in units of solar EMERGY (rather than coal EMERGY). In this book a 15% increase was made in many transformities when tide was added to global solar EMERGY.

APPENDIX F

SIMULATION PROGRAMS FOR MACINTOSH USED IN THIS BOOK[1]

[1] To run these programs in BASIC on a PC computer, add lines for Screen and Color:

SCREEN 0,0: COLOR 0,1

TABLE F.1. Program EMTANK in BASIC for Simulating EMERGY Storage in Figure 1.7

```
  5  REM EMTANK (EMERGY and transformity of a Tank)
 10  REM If there are no exports from storage set Y = 0
 12  Y = 1
 15  REM The next line draws a box
 20  LINE (0,0) - (320,300),,B
 22  LINE (0,100) - (320,100),3
 24  LINE (0,200) - (320,200),3
 25  REM The given values for E and Q
 30  Q = 10
 35  EQ0 = 20
 37  Eq = 1
 40  E = 100
 41  Tq0 = .1
 42  DT = 1
 43  Q0 = 5
 44  t0 = 1
 45  REM The given coefficients values
 47  J = 10
 48  TJ = 1
 49  Tq = Eq/Q
 50  k1 = .02
 60  k2 = .02#
 65  REM The equation
 70  DQ = J - k1 * Q - k2 * Q
 90  IF DQ > 0 THEN x = 1
 93  IF DQ = 0 THEN DEQ = 0
 97  IF DQ < .01 THEN x = 0: GOTO 105
100  DEQ = x * TJ * J - Y * Tq * k2 * Q
103  GOTO 110
105  DEQ = Tq * DQ
110  Q = Q + DQ * Dt
120  Eq = Eq + DEQ * Dt
125  IF Eq < 0 THEN Eq = 0
130  Tq = Eq/Q
250  PSET (t/t0,200 - Tq/Tq0),3
250  PSET (t/t0,300 - Eq/EQ0),3
175  REM The time increments
180  t = t + Dt
185  REM Drawing a point on the graph
190  PSET(t/t0,100 - Q/Q0)
195  REM The loop
300  IF t < 320 GOTO 70
```

TABLE F.2. Simulation Program EMPULSE in BASIC for Figure 13.2

```
  2  REM MC
  3  REM EMPULSE
  4  CLS
  6  LINE (0,0) - (320,300),3,B
  7  LINE (0,150) - (320,150),3
  9  REM Levels Q and A normalized as 1; Tj * 1E6 going to Q; Tq * 1000
     going to A
 10  J = 100
 15  STJ = 1
 17  STQ = 1000
 19  stA = 100000!
 40  M = 1
 50  Q = 1
 60  a = .1
 70  K = 1
 73  EQ = 1
 75  EA = 100
 77  REM Coefficients
 80  k0 = 29
 90  K1 = .0015
100  K2 = .001
100  k3 = .005
130  K5 = .001
140  K6 = .01
142  k7 = .008
144  k8 = .0008
148  REM Scaling factors
150  Dt = 10
160  t0 = 10
170  M0 = .007
172  ED0 = 5E+07
175  q0 = .03
176  pr0 = 20
177  EQ0 = 1E+07
182  A0 = .03
184  sta0 = 1E+10
185  stq0 = 1E+07
186  NEYRO = .1
187  STAP0 = 2E+09
188  EIR0 = 1
190  EA0 = 1E+10
195  M = K - .1 * Q
200  R = J/(1 + k0 * M)
201  DQ = K1 * R * M - K2 * Q - k3 * Q * a * a
202  DA = K5 * Q - K6 * A + k7 * Q * A * A - k8 * Q * A * A
204  IFQ < .000001 THEN Q = .000001
```

TABLE F.2. *(Continued)*

```
206  a = a + Dt * DA
208  Q = Q + DQ * Dt
210  IFQ < .0000001 THEN Q = .0000001
215  IF (DQ/Q) > 0 THEN dEQ =
     STJ * 10000 * k0 * R * M - STQ * K2 * Q - STQ * k3 * Q * a * a
217  IF (DQ/Q) < .01 > 0 THEN dEQ = 0
220  IF (DQ/Q) < 0 THEN dEQ = STQ * DQ
222  STQ = EQ/(Q)
229  IF a < 1E-11 THEN a = 1E-11
230  IF (DA/a) > .001 THEN dEA = 1000 * STQ * k3 * Q * a * a + 1000 * STQ * K2 * Q
232  IF (DA/a) < .001 > 0 THEN dEA = 0
234  IF (DA/a) < 0 THEN dEA = stA * DA
236  EA = EA + dEA * Dt
237  stA = EA/a
239  IF EA < .000001 THEN EA = .000001
240  EQ = EQ + dEQ * Dt
241  IF EQ < .000001 THEN EQ = .000001
248  t = t + Dt
250  EAF = stA * k8 * Q * a * a: REM EMERGY flux of feedback
252  eap = (STQ * 1000 * k3 * Q * a * a + STQ * 1000 * K2 * Q + stA * k8 * Q * a * a):
     REM EMERGY flux of input from below
255  stap = eap/(k7 * Q * a * a + K5 * Q): REM Solar transformity of
     production flux of A
256  NEYR = (eap + STQ * 1000 * K2 * Q * a * a)/EAF
257  in = (STQ * 1000 * k3 * Q * a * a + STQ * 1000 * K2 * Q)
258  g = a: REM GNP
258  Pr = g/(k7 * Q * a * a): REM Price of product
259  ed = eap/g: REM EMERGY use per dollar circulated
260  EIR = EAF/in
261  GOTO 320
262  PSET (t/t0,148 - STQ/STQ0),3
263  PSET (t,t0,148 - STA/STA0),3
273  PSET (t,t0,148 - NEYR/NEYR0),3
278  PSET (t,t0,148 - EA/EA0),3
308  REM MATERIALS AT GLOBAL BACKGROUND CONCENTRATION
310  PSET (t/t0,300 - M/M0),3
320  REM RESERVES
322  PSET (t,t0, 148 - ed/ED0),3
323  GOTO 335
325  PSET (t/t0,148 - Pr/pr0),3
330  PSET (t/t0,148 - stap/STAP0),3
335  PSET (t/t0,300 - Q/Q0),3
340  REM URBAN ASSETS
350  PSET (t/t0,300 - a/A0),3
360  GOTO 400
370  PSET (t/t0,300 - EA/EA0),3
400  IF t/t0 < 320 GOTO 195
```

TABLE F.3. Computer Program ENVUSE in BASIC for Figure 13.8

```
 10 REM IBM PC
 20 REM ENVUSE.BAS
 30 CLS
 50 LINE (0,0) - (320,180),3,B
 60 LINE (0,90) - (320,90),1
100 REM Coefficients
105 K0 = 9
110 K1 = .5
115 K2 = .03
120 K3 = .045
125 K4 = 1.5
130 K5 = .8
135 K6 = .06
140 K7 = .06
150 K9 = .1
160 L1 = 1
170 L2 = 2
200 REM Scaling
205 Dt = .1
210 T0 = .3 : REM 96-year horizontal axis
215 Q0 = .03
220 A0 = .15
225 M0 = .1
230 DO = .05
235 REM Outside sources
240 IT = .1
245 JL = .1
255 PI = 6.6
265 I = 1
270 REM Starting conditions
275 Q = 1.6
280 M = .05
285 A = .1
290 D = 0
293 W = 0
295 Z = 1
300 REM Equations
315 R = I/(1 + K0)
320 IF W = 1 THEN PI = PI + .2 * DT
325 PE = (L1 + Q)/(L2 * Q)
327 IF PE > 3 THEN PE = 3
330 REM Change equations
332 IF D = 0 THEN Z = 0
333 IF D > 0 THEN Z = 1
335 DQ = K1 * R - K2 * Q - K3 * Q * A
340 DA = K5 * M/PI - K6 * A - K7 * Q * A
```

TABLE F.3. *(Continued)*

```
345  DM = PE * K4 * A * Q + JL - K5 * M - IT * L - Z * K9 * M
350  DD = JL - Z * K9 * M
355  A = A + DA * Dt
360  IF A < 0 THEN A = 0
365  Q = Q + DQ * Dt
370  IF Q < 0 THEN Q = 1E-10
375  M = M + DM * Dt
380  IF M < 0 THEN M = 0
385  D = D + DD * Dt
390  IF D < 0 THEN D = 0
400  REM Plotting graphs with time
405  PSET (T/T0,180 - Q/Q0),1
410  PSET (T/T0,180 - A/A0)),3
415  PSET (T/T0,90 - M/M0),3
420  PSET (T/T0,90 - D/D0),1
425  T = T + Dt
430  IF T/T0 < 320 GOTO 300
```

TABLE F.4. Simulation Program RNRPRICE in BASIC from Figure 13.11

```
  2  REM RNRPRICE (prices with source a slowly renewable storage)
  5  LINE (0,0) - (320,180),,B
  7  LINE (0,90) - (320,90)
 10  F = 89
 15  dt = 1
 20  J2 = 3.6
 30  J = 1
 40  PW = 0
 50  Q = .01
 55  E = 1
 60  K0 = .02
 70  K1 = .00041
 80  K2 = .002
 90  K3 = .7 * k2
100  K4 = .07
110  K5 = .05
115  k6 = .005
120  M = 2
130  I = 1
140  TO = 2
150  P = 1
160  Q0 = .2
165  P0 = .03
170  F0 = 1
175  PR0 = 2
180  P = 1
200  P = K3 * F * Q + PW
210  PR = 100/(P + .8)
230  DF = k6 * E-K2 * F * Q
240  DQ = P - K5 * Q
250  Q = Q + DQ * dt
260  F = F + DF * dt
265  PSET (T/T0,90 - PR/PR0)
270  PSET (T/T0,180 - P/P0)
280  PSET (T/T0,180 - Q/Q0)
300  PET (T/T0,90 - F/F0)
310  T = T + dt
320  IF T/T0 < 320 GOTO 200
330  END
```

REFERENCES

Cited references are marked with an asterisk(*). Uncited bibliography includes references concerned with energy, EMERGY evaluations, and related approaches.

*Ager, D. U. 1965. Principles of Paleontology. McGraw-Hill, New York.

*Ahern, J. E. 1980. The Exergy Method of Energy Systems Analysis. Wiley, New York. 295 pp.

Alekseev, G. N. 1986. Energy and Entropy, Translated from Russian by U. M. Taube. Mir Publishers, Moscow. 200 pp.

*Alexander, J. F. Jr. 1978a. Energy Basis of Disasters and the Cycles of Order and Disorder. Ph.D. Dissertation, Environmental Engineering Sciences, Univ. of Florida, Gainesville. 232 pp.

*Alexander, J. F. Jr. 1978b. Energy modelling in planning for disaster impact; Guatemala earthquake, 4 Feb., 1976. p. 105 in Systems in Sociocultural Development, ed. by K. Balkus. Proceedings of the Society for General Systems Research, Southeastern Meeting, Tallahassee, Fla.

Alexander, J. F. Jr. 1980. A global systems ecology model. pp. 443–456 in Changing Energy Futures, ed. by R. A. Fazzolare and C. B. Smith. Pergamon Press,

Alexander, J. F. Jr., J. Henslick, M. Lee, C. Palmer, R. Rongstad, N. Sipe, D. Swaney, and A. Whittman. 1980. An Energetics Approach to Assessing National Development Plans: A Case Study of Redwood National Park. Report to the National Park Service Office of Appropriate Technology. Center for Wetlands, Univ. of Florida, Gainesville (CFW-80-01). 201 pp.

Allesio, F. J. 1981. Energy analysis and the energy theory of value. Energy J. 2:61.

*Anderson, J. W. 1980. Bioenergetics of Autotrophs and Heterotrophs. Edward Arnold, London.

*Anonymous: Technocracy Mag. 1937. The energy certificate, A-10, 22 pp.

Anonymous: 1978. The utility of energy resource accounting and net energy analysis in energy technology assessment. Unpublished Report to U.S. Department of Energy, Development Sciences, Inc., Sagamore, Mass. 158 pp.

Ayres, R. U. 1978. Resources, Environment and Economics: Applications of the Materials/Energy Balance Principle. Wiley, New York. 207 pp.

*Bahr, C. M., J. W. Day, and J. H. Stone. 1982. Energy cost-accounting of Louisiana fishery production. Estuaries 5:209–215.

Bahr, L. M., R. Costanza, J. W. Day Jr., S. E. Bayley, C. Neill, S. G. Leibowitz, and J. Fruci. 1983. Ecological Characterization of the Mississippi Deltaic Plain Region: A Narrative with Management Recommendations. U.S. Fish and Wildlife Service, (FWS/OBS-82/69). 189 pp.

Baines, J. 1981. Indirect Energy of New Zealand's Foreign Trade 1969–78: An Energy Analysis of Imports and Exports. Report to N.Z. Energy Research and Development

Committee, Dept. of Chemical Engineering, Univ. of Canterbury, Christchurch, N.Z. 77 pp.

*Ballentine, T. 1976. A Net Energy Analysis of Surface Mined Coal from the Northern Great Plains. M.S. Thesis, Environmental Engineering Sciences, Univ. of Florida, Gainesville. 149 pp.

*Barbir, F. 1992. Analysis and Modelling of Environmental and Economic Impacts of the Solar Hydrogen Energy System. Ph.D. Dissertation, Dept. of Mechanical Engineering, Univ. of Miami, Miami, Fla. 176 pp.

Barney, G. O. 1980. The Global 2000 Report to the President's Council on Environmental Quality and Department of State, Washington, D.C. Report 2000. Vol. 1, 47 pp.; vol. 2, 766 pp.

Barr, T. D., and F. A. Dahlen. 1988. Thermodynamic efficiency of brittle frictional mountain building. Science 242:749–752.

*Bayley, S., and R. Walker. 1976. Energy Methodologies for Assessing the Regional Role of Transportation Systems and for Evaluating Environmental Economic Tradeoffs. Final Report to the Florida Dept. of Transportation Research. Dept. of Environmental Engineering Sciences, Univ. of Florida, Gainesville (CFW-76-40). 176 pp.

*Bayley, S., H. T. Odum, B. Hanley, and C. McDowell. 1977a. Example of an Energy Evaluation of a Transportation Project. Center for Wetlands, Univ. of Florida, Gainesville (CFW-77-01). 34 pp.

*Bayley, S. E., H. T. Odum, and W. M. Kemp. 1977b. Energy evaluation and management alternatives for Florida's east coast. pp. 87–104 in Transactions of the 41st North American Wildlife Conference. Wildlife Management Institute, Washington, D.C.

*Bayley, S., J. Zucchetto, L. Shapiro, D. Mau, and J. Nessel. 1977c. Energetics and Systems Modeling: a Framework Study for Energy Evaluation of Alternative Transportation Modes. Final Report to U.S. Army Engineer Institute for Water Resources, Kingman Building, Fort Belvoir, VA. IWR Contract Reports 77-6, 43 pp., and 77-10, 173 pp.

Best, G. R., P. M. Wallace, W. J. Dunn, and H. T. Odum. 1988. Enhanced Ecological Succession Following Phosphate Mining. (FIPR Publ. 03-048-54), Bartow. 160 pp.

Bhatt, R. 1986. Policy implications of Gold EMERGY. Unpublished report of policy research project. L.B.J. School of Public Affairs, Austin, Tex. 10 pp.

Biondi, P., G. Farina, and V. Panaro. 1987 L'analisi energetico in agricoltura. Riv. di Ing. Agr. 4:205–219.

Bockris, J. O., and Lee Handley. 1978. The alleged diluteness of solar energy. Energy Res. 295–296.

Bodger, P. S. 1988. Dynamics of an energy-economic system subject to an energy substitution sequence. Energy Syst. Policy 12:167–178.

Boggess, C. F. 1994. The Biogeoeconomics of Phosphorus in the Kissimmee Valley. Ph.D. dissertation, Environmental Engineering Sciences. Univ. of Florida, Gainesville. 234 pp.

*Boltzmann, L. 1886, 1905. The second law of thermodynamics. Address in English to Imperial Academy of Science in 1886. Populare Schriften. Essay 3; Selected Writings of L. Boltzmann. D. Reidel, Dordrecht, Holland.

*Boltzmann, L. 1886, 1905. Der zweite Haupsatz der mechanishen warme Theorie. Almanach K. Acad. Wiss. Mech. (Wien) 36:255–299.

Bosch, G. 1983. Energy flow in the West German processing of iron steel and end product subsystems. Appendix A13 in Energy Analysis Overview of Nations, ed. by H. T. Odum and E. C. Odum. Working Paper, International Institute of Applied Systems Analysis, (WP-83-82). 469 pp.

Boustead, I., and G. E. Hancock. 1979. Handbook of Industrial Energy Analysis, Wiley, New York.

*Boyles, J. V. 1975. Accounting for long-lived productive resources: the development and evaluation of general energy systems measurement procedures. Ph.D. Dissertation, Dept. of Accounting, Univ. of Florida, Gainesville. 198 pp.

Boyles, D. T. 1984. Bioenergy: Technology, Thermodynamics and Costs. Halstead, Wiley, New York. 158 pp.

*Boynton, W. R. 1975. Energy Basis of a Coastal Region: Franklin County and Apalachicola Bay, Florida. Ph.D. Dissertation, Environmental Engineering Sciences, Univ. of Florida, Gainesville. 389 pp.

Braat, L. C. 1992. Sustainable Multiple Use of Forest Ecosystems. Ph.D. Dissertation, Free University of Amsterdam, Netherlands. 196 pp.

Braat, L. C., and W. F. J. van Lierop. 1987. Economic-ecologic Modelling. North Holland, Amsterdam. 329 pp.

*Brady, N. C. 1974. The Nature and Properties of Soils. Macmillan, New York.

Brookes, L. G. 1972. Energy consumption and economic growth. Chem. Eng. 266:368–376.

Browder, J. A. 1976. Water, Wetlands and Wood Storks in Southwest Florida. Ph.D. Dissertation, Environmental Engineering Sciences, Univ. of Florida, Gainesville. 466 pp.

*Browder, J., C. Littlejohn, and D. Young. 1976. The South Florida Study: South Florida, Seeking a Balance of Man and Nature. Report to the Division of State Planning, Tallahassee, Fla. Center for Wetlands, Univ. of Florida, Gainesville (CFW 76-07). 128 pp.

*Brown, A. 1986. Research Information and High Technology Manufacturing. Student Policy Research Project Report, L.B.J. School of Public Affairs, Austin, Tex. 14 pp.

*Brown, M. T. 1975. Lee County: An Area of Recent Rapid Growth. Energy, Water and Land Use Analysis with Recommendations for Best Economic Vitality. Report to the Division of State Planning, Tallahassee, Fla. Center for Wetlands, Univ. of Florida, Gainesville (CFW 75-43). 64 pp.

*Brown, M. T. 1976. Lee County: An Area of Recent Rapid Growth. Center for Wetlands, Univ. of Florida, Gainesville; and the Bureau of Comprehensive Planning, Division of State Planning, Florida Dept. of Administration, Tallahasee, Fla. 57 pp.

*Brown, M. T. 1977a. Ordering and disordering in South Vietnam by energy calculation. pp. 165–191 in The Effects of Herbicides in South Viet Nam, Part B: National Academy of Science, Washington, D.C. Working Paper: Models of Herbicide, Mangroves, and War in Vietnam, ed. by H. T. Odum, M. Sell, M. Brown, J. Zucchetto, C. Swallows, J. Browder, T. Ahstrom and L. Peterson.

*Brown, M. T. 1977b. War, Peace and the Computer: Simulation of Disordering and Ordering Energies in South Vietman. pp. 393–418 in Ecosystem Modeling in Theory and Practice: An Introduction with Case Histories, ed. by C. A. S. Hall and J. W. Day. Wiley Interscience, New York.

*Brown, M. T. 1980. Energy Basis for Hierarachies in Urban and Regional Landscapes. Ph.D. Dissertation, Dept. of Environmental Engineering Sciences, Univ. of Florida, Gainesville. 359 pp.

Brown, M. T. 1981. Energy basis for hierarchies in urban and regional systems. pp. 517–534 in Energy and Ecological Modelling: Developments in Environmental Modelling 1. Proceedings of a Symposium Sponsored by the International Society for Ecological Modelling, ed. by W. J. Mitsch, R. W. Bosserman, and J. M. Klopatek. Elsevier Scientific, New York.

Brown, M. T. 1985. Assessing wetland values in landscapes dominated by humanity. Proceedings: National Wetland Assessment Symposium. Association of State Wetland Managers, Chester, Vt.

Brown, M. T. 1986a. Cumulative impacts in landscapes dominated by humanity. In Proceedings of the Conference: Managing Cumulative Effects in Florida Wetlands in Sarasota, Fla. ed. by E. D. Estevez, J. Miller, and R. Hamann. New College Environmental Studies Program (Publ. 37). Omnipress, Madison, Wis.

*Brown, M. T. 1986b. Energy analysis of the hydroelectric dam near Tucurui. pp. 82–91 in Energy Systems Overview of the Amazon Basin, ed. by H. T. Odum, W. T. Brown, and R. A. Christianson. Report to The Cousteau Foundation. Center for Wetlands, Univ. of Florida, Gainesville (Publ. 86-1). 190 pp.

Brown, M. T. 1987. The value of wetlands and other ecological systems. Proceedings of the National Wetland Symposium: Mitigation of Impacts and Losses. Association of State Wetland Managers, Chester, Vt.

Brown, M. T., and G. V. Genova. 1973. Energy indices in the urban pattern. Masters papers, Dept. of Architechture, Univ. of Florida, Gainesville.

*Brown, M. T., and T. R. McLanahan. 1992. EMERGY Analysis Perspectives of Thailand and Mekong River Dam Proposals. Report to The Cousteau Society, Center for Wetlands and Water Resources, Univ. of Florida, Gainesville. 60 pp.

Brown, M. T., and R. C. Murphy. 1995. EMERGY analysis perspectives on ecotourism, carrying capacity and sustainable development. In Maximum Power, ed. by C. A. S. Hall. Univ. Press of Colorado, Niwot, in press.

Brown, M. T., and M. F. Sullivan. 1987. The value of wetlands in low relief landscapes. p. 17 in The Ecology and Management of Wetlands, ed. by D. D. Hook. Croom Helm, Beckenham, England.

*Brown, M. T., S. Brown, R. Costanza, E. DeBellevue, K. C. Ewel, R. Gutierrez, M. Sell, W. J. Mitsch, D. E. Layland, and H. T. Odum. 1975. Natural Systems and Carrying Capacity of the Green Swamp. Report to Florida Dept. of Administration. Center for Wetlands, Univ. of Florida, Gainesville (CFW 75-24). 336 pp.

*Brown, M. T., S. Tennenbaum, and H. T. Odum. 1991. EMERGY analysis and policy perspectives for the Sea of Cortez, Mexico. Report to The Cousteau Society. CWF publ. 88-04. Center for Wetlands, Univ. of Florida, Gainesville. 58 pp.

*Brown, M. T., P. Green, A. Gonzalez, and J. Venegas. 1992. EMERGY Analysis Perspectives, Public Policy Options, and Development Guidelines for the Coastal Zone of Nayarit, Mexico. Center for Wetlands and Water Resources, Univ. of Florida, Gainesville. Vol. 1, 215 pp.; vol. 2, 145 pp. and 31 map inserts.

*Brown, M. T., R. D. Woithe, H. T. Odum, C. L. Montague, and E. C. Odum. 1993. EMERGY Analysis Perspectives on the EXXON *Valdez* Oil Spill in Prince William Sound, Alaska. Report to The Cousteau Society. Center for Wetlands and Water Resources, Univ. of Florida, Gainesville. 122 pp.

Brown, S., and A. E. Lugo. 1981. Management and Status of U.S. Commercial Marine Fisheries. Council on Environmental Quality, Washington, D.C.

*Budyko, M. I. 1974. Climate and Life. Academic Press, New York.

Bullard, C. W. III. 1975. Energy Costs, Benefits, and Net Energy. Center for Advanced Computation, Univ. of Illinois at Urbana-Champaign (CAC document 174), 70 pp.

Bullard, C. W. 1976. Energy costs, benefits, and net energy. Energy Syst. Policy 1(4):367–382.

Bullard, C., and R. Herendeen. 1975. The energy costs of goods and services. Energy Policy 3:268–278.

*Bullard, C. W., P. S. Penner, and D. A. Pilati. 1976. Energy Analysis: Handbook for Combining Process and Input-Output Analysis. Center for Advanced Computation, Univ. of Illinois at Urbana-Champaign, Ill. (CAC document 214).

Bulmer, D. W. 1985. Gold. pp. 17–24 in The Minerals Bureau of South Africa. South Africa's Mineral Industry (1984), Johannesburg.

Burnett, M. 1978. System efficiency and energy intensity. p. 65 in Systems in Sociocultural Development, ed. by K. Balkus. Proceedings of the Society for General Systems Research, Southeastern Meeting, Tallahassee, Fla.

Burnett, M. S. 1981a. Energy Analysis of Intermediate Technology Agricultural Systems. M.S. Thesis, Environmental Engineering Sciences, Univ. of Florida, Gainesville. 169 pp.

*Burnett, M. S. 1981b. A methodology for assessing net energy yield and carbon dioxide production of fossil fuel resources. pp. 711–714 in Energy and Ecological Modelling, ed. by W. A. Mitsch, R. W. Bosserman, and J. M. Kloptatek. Elsevier, Amsterdam. 839 pp.

*Burnett, M. S. 1981c. A methodology for assessing net energy and abundance of energy resources. pp. 703–710 in Energy and Ecological Modelling, ed by. W. A. Mitsch, R. W. Bosserman, and J. M. Kloptatek. Elsevier, Amsterdam. 839 pp.

Burns, T. P., M. Higashi, M. S. Wainwright, and B. C. Patten. 1991. Trophic unfolding of a continental shelf energy-flow network. Ecol. Model. 55:1–26.

Burr, D. Y. 1977. Value and Water Budget of Cypress Wetlands in Collier County, Florida. M.S. Thesis, Environmental Engineering Sciences, Univ. of Florida, Gainesville. 106 pp.

Cairncross, F. 1992. Costing the Earth. Harvard Business School Press, Boston, Mass. 341 pp.

Campbell, D. E. 1984. Energy Filter Properties of Ecosystems. Ph.D. Dissertation, Environmental Engineering Sciences, Univ. of Florida, Gainesville. 478 pp.

*Canoy, M. 1970. Desoxyribonucleic acid in two tropical forests. pp. G69–G70 in A Tropical Rainforest, ed. by H. T. Odum and R. F. Pigeon. Division of Technical Information, U.S. Atomic Energy Commission (TID2470). 1660 pp.

*Canoy, M. 1972. Desoxyribonucleic Acid in Ecosystems. Ph.D. Dissertation, Dept. of Zoology, Univ. of North Carolina, Chapel Hill.

*Capehart, B. L., and J. Blackburn. 1989. Florida's Enlightened Conservation Standards: $4 Billion Savings Are Projected. Strategic Planning and Energy Management, pp. 62–71.

Caplan, S. R. 1966. The degree of coupling and its relation to efficiency of energy conversion in multiple flow systems. J. Theoret. Biology. 10:209–235.

Chapman, P. O. 1974. The energy costs of producing copper from primary sources. Metals Mater. 8:107–11.

Charlier, R. H. 1993. Environmental, Economic and Technological Aspects of Alternative Power Sources. Oceanography Series vol. 56, Elsevier, Amsterdam. 556 pp.

*Christensen, V. 1994. EMERGY-based ascendency. Ecol. Model. 72:129–144.

Christianson, R. 1984. Interactions of Man and the Environment in Central America—an Energy Systems Perspective. Center for Environmental Policy. Unpublished report, 36 pp.

*Christianson, R. A. 1984. Energy Perspectives on a Tropical Forest/Planation System at Jari, Brazil. M.S. Thesis, Environmental Engineering Sciences, Univ. of Florida, Gainesville. 168 pp.

Cleveland, C. J., R. Costanza, C. A. S. Hall, and R. Kaufmann. 1984. Energy and the United States economy—a biophysical perspective. Science 225:890–897.

Cleveland, C. J. 1991. Natural resource scarcity and economic growth revisited—economic and biophysical perspectives. pp. 289–317 in Ecological Economics, ed. by R. Costanza. Columbia Univ. Press, New York.

Coker, A., and C. Richards. 1992. Valuing the Environment. Belhaven Press, London.

Colinvaux, P. A. 1980. The Fates of Nations: A Biological Theory of History. Simon and Shuster, New York. 383 pp.

Comar, V. 1993. An EMERGY Evaluation of the Central Amazon Town of Itacoatiara, Its Plywood and Veneer Industry, and the Floodplain of the Madeira River Basin. Masters Thesis, Ecology, Univ. of Amazonas, Manaus, Brazil. 147 pp.

*Cook, Earl. 1976. Man, Energy, and Society. Freeman, San Francisco.

Costanza, R. 1975. Spacial Distribution of Land Use, Incoming Energy and Energy Use in South Florida from 1900 to 1973. M.S. Thesis, Environmental Engineering Sciences, Univ. of Florida, Gainesville. 201 pp.

Costanza, R. 1978a. Energy costs of goods and services in 1967 including solar energy inputs and labor and government service feedbacks. Center for Advanced Computation, Univ. of Illinois at Urbana-Champaign, (ERG document 262). 46 pp.

Costanza, R. 1978b. Energy, Value and Exchange. p. 70 in Systems in Sociocultural Development, ed. by K. Balkus. Proceedings of the Society for General Systems Research, Southeastern Meeting, Tallahassee, Fla.

Costanza, R. 1979. Embodied energy basis for economic-ecologic systems. Ph.D. Dissertation, Environmental Engineering Sciences, Univ. of Florida, Gainesville, 254 pp.

Costanza, R. 1980. Embodied energy and economic evaluation. Science 210:1219–1224.

Costanza, R. 1981. Energy analysis and economics. pp. 119–146 in Energy, Economics, and the Environment, ed. by H. E. Daly and A. F. Umanak. Selected Symposia of the AAAS No. 64. Westview Press, Boulder, Colo.

Constanza, R., ed. 1991. Ecological Economics. Columbia Univ. Press, New York. 525 pp.

Costanza, R., and H. E. Daly. 1987. Toward an ecological economics. Ecol. Model. 38:1–7.

Costanza, R., and S. C. Farber. 1984. Theories and methods of valuation of natural systems: A comparison of willingness to pay and energy analysis based approaches. Man, Environ. Space Time 4(1):1–38.

Costanza, R., and R. A. Herendeen. 1984. Embodied energy and economic value in the United States Economy: 1963, 1967, and 1972. Resour. Energy 6:129–163.

Costanza, R., and C. Neill. 1984. Energy intensities, interdependence, and value in ecological systems: A Linear Programming Approach. J. Theor. Biol. 106:41–57.

Costanza, R., C. Neill, S. G. Leibowitz, J. R. Fruci, L. M. Bahr Jr., and J. W. Day Jr. 1983. Ecological Models of Mississippi Deltaic Plain Region: Data Collection and Presentation. U.S. Fish and Wildlife Service, (FWS/OBS-82/68), 339 pp.

Costanza, R., S. C. Farber, and J. Maxwell. 1989. The valuation and management of wetland ecosystems. Ecol. Econ. 1:335–361.

Cottrell, F. 1955. Energy and Society. McGraw-Hill, New York.

*Coultas, C. L., and F. C. Calhoun. 1976. Properties of some tidal marsh soils of Florida. Soil Sci. 40:73–75.

Coultas, C. L., and E. Gross. 1975. Distribution and properties of some tidal marsh soils of Apalachee Bay, Florida. Soil Sci. Soc. Am. Proc. 39(5):914–919.

*Curzon, F. I., and B. Ahlborn. 1975. Efficiency of a Carnot engine at maximum power output. Am. J. Phys., 43:22–24.

*Dalton, P. 1986. Defense Analysis. Student Policy Research Project report. L.B.J. School of Public Affairs, Austin, Tex. 14 pp.

*da Silva, J. G., G. E. Serra, J. R. Moreir, J. C. Concalves, and J. Goldemberg. 1978. Energy balance for ethyl alcohol production from crops. Science 201:903–906.

*DeBellevue, E. 1976a. Energy Basis for an Agricultural Region: Hendry County, Florida. M.S. Thesis, Environmental Engineering Sciences, Univ. of Florida, Gainesville. 205 pp.

*DeBellevue, E. 1976b. Hendry County, an Agricultural District in a Wetland Region—The South Florida Study. Center for Wetlands, Univ. of Florida and Division of State Planning, Florida Dept. of Administration, Tallahasee. 59 pp.

*DeBellevue, E., H. T. Odum, J. Browder, and G. Gardner. 1979. Energy Analysis of the Everglades National Park. pp. 31–43 in Proc. of the First Conference of Scientific Research in the National Parks, vol. 2. National Park Service, Washington, D.C.

De Vos, A. 1992. Endoreversible Thermodynamics of Solar Energy Conversion. Oxford Univ. Press, Oxford. 186 pp.

*Diamond, C. 1984. Energy Basis for the Regional Organization of the Mississippi River Basin. M.S. Thesis, Environmental Engineering Sciences, Univ. of Florida, Gainesville. 136 pp.

*Doherty, S. J. 1990. Policy Perspectives on Resource Utilization in Papua New Guinea Using Techniques of EMERGY Analysis. Masters Project Paper, Dept. of Urban and Regional Planning. Univ. of Florida, Gainesville. Center for Wetlands, 72 pp.

*Doherty, S. J., and P. O. Nilsson, eds. 1992. EMERGY analysis: A Biophysical Bridging between the Economies of Humanity and Nature. A report of initial studies to Vattenfall and the Royal Academy of Agricultural Sciences, Garpenberg, Sweden. Draft.

*Doherty, S. J., and M. T. Brown, with R. C. Murphy, H. T. Odum, and G. A. Smith. 1993. Emergy synthesis perspectives, sustainable development and public policy options for Papua New Guinea. Report to The Cousteau Society; Center for Wetlands and Water Resources, Univ. of Florida, Gainesville. 102 pp.

*Doherty, S. J., F. N. Scatena, and H. T. Odeum. 1994. EMERGY evaluation of the Luquillo Experimental Forest and Puerto Rico. Report of Cooperative Project 19-93-023 between International Institute of Tropical Forestry, Rio Piedras, P.R., and Center for Environmental Policy, Univ. of Florida, Gainesville. 90 pp.

*Doherty, S. J., H. T. Odum, and P. O. Nilsson. 1995. Systems Analysis of the Solar EMERGY Basis for Forest Alternatives in Sweden. Final Report to the Swedish State Power Board, College of Forestry, Garpenberg, Sweden, 112 pp. Draft.

Duckham, A. N., J. G. W. Joes, and E. H. Roberts, eds. 1976. Food Production and Consumption. North Holland Publ., Amsterdam.

Duff, P. M. D., A. Hallam, and E. K. Walton. 1967. Cyclic Sedimentation. Elsevier, New York.

*Eastern Research Group. 1993. A Review of Ecologial Assessment Studies from a Risk Assessment Perspective. Risk Assessment Forum, Environmental Protection Agency, Washington, D.C. (EPA/630/R-92/005) 442 pp.

*EIA/DOE. 1984a. Annual Energy Review 1984. Energy Information Administration, Dept. of Energy, Washington, D.C. 577 pp.

*EIA/DOE. 1984b. Electric Power Annual 1984. Energy Information Administration, Dept. of Energy, Washington, D.C. 577 pp.

*EIA/DOE. 1984c. Petroleum Supply Annual 1984. Energy Information Administration, Dept. of Energy, Washington, D.C. Vol. 1:

*EIA/DOE. 1985a. Natural Gas Monthly, June. Energy Information Administration, Dept. of Energy, Washington, D.C.

*EIA/DOE. 1985b. Petroleum Supply Monthly, June. Energy Information Administration, Dept. of Energy, Washington, D.C.

EIA/DOE. 1985c. Weekly Coal Production, August. Energy Information Administration, Dept. of Energy, Washington, D.C.

Ewel, K. C., and H. T. Odum, eds. 1985. Cypress Swamps. Univ. of Florida Press, Gainesville, 472 pp.

*Fairen, V., and J. Ross 1981. On the efficiency of thermal engines with power output. J. Chem. Phys. 75(11):5490–5496.

Fairen, V., M. D. Hatee, and J. Ross. 1982. Thermodynamic processes, time scales, and entropy production. J. Phys. Chem. 76:70–73.

Fishing Industry Board. 1978. Energy crisis looms for N.Z. Fishing Industry. Fishing Industry Board Bulletin 46 (September):1–5, Wellington, N.Z.

*Florida Bureau of Economic and Business Research. 1988. Florida Statistical Abstract, Univ. Press of Florida, Gainesville.

*Fluck, R. C., and D. C. Baird. 1980. Agricultural Energetics. AVI Publishing, Westport, Conn.

*Folke, C. 1988. Energy economy of salmon aquaculture in the Baltic sea. Environ. Manage. 12:525–537.

*Folke, C. 1990. Evaluation of ecosystem life-support in relation to salmon and wetland exploitation. Dept. of Systems Ecology, Stockholm Univ., Stockholm. 30 pp.

Folke, C., and T. Kaberger. 1991. Linking the Natural Environment and the Economy: Essays from the Eco-Eco Group. Kluwer Academic Publ., Boston, Mass. 305 pp.

Fontaine, T. D. 1981. A self-designing model for testing hypotheses of ecosystem development. pp. 281–291 in Progress in Ecological Engineering and Management by Mathematical Modelling 1. Second International Conference on the State of the Art in Ecological Modelling, Copenhagen, Denmark.

*Fonyo (Boggess), C. 1983. Embodied Energy Analysis of Shrimp Harvesting in Louisiana Waters. Unpublished manuscript. 24 pp.

Fowler, R. G. 1977. The longevity and searchworthiness of petroleum resources. Energy 2:189–195.

Frankena, F. 1978. Energy Analysis/Energy Accounting. Vance Bibliographies, Box 229, Monticello, Ill. 48 pp.

*Gardner, G. 1977. Comparison of Energy Analysis of Oil Shale. Masters Paper, Environmental Engineering Sciences, Univ. of Florida, Gainesville; pp. 196–232 in Energy Analysis Models of the United States, ed. by H. T. Odum and J. F. Alexander Jr. Report to Department of Energy, Center for Wetlands, Univ. of Florida, Gainesville.

*Garrels, R. M., F. T. Mackenzie, and C. Hunt. 1975. Chemical cycles and the Global Environment. William Kaufmann, Los Altos, Calif. 206 pp.

*General Accounting Office, U.S. Comptroller, General. 1982. DOE Funds New Energy Technologies without Estimating Potential Net Energy Yields. Report to Congress (GAO/IPE-82-1). 173 pp.

Genoni, G. P., and C. L. Montague. 1995. Influence of the energy relationships of trophic levels and of elements on bioaccumulation. Ecotox. and Environ. Saf. 30(2):203–218.

*Georgescu-Roegen, N. 1971. The entropy law and the economic process. Harvard Univ. Press, Cambridge, Mass. 455 pp.

*Georgescu-Roegen, N. 1977. The steady state and economic salvation: A thermodynamic analysis. Bioscience 27:266–270.

Georgescu-Roegen, N. 1979. Energy analysis and economic valuation. South. Econ. J. 45(4):1023–1058.

Giampietro, M., and D. Pimentel. 1990. Energy analysis models to study the biophysical limits for human exploitation of natural processes. pp. 139–184 in Ecological Physical Chemistry, ed. by C. Rossi and E. Tiezzi, Elsevier, London.

*Giampietro, M., and D. Pimentel. 1991. Energy analysis models to study the biophysical limits for human exploitation of natural processes. pp 139–184 in Ecological Physical Chemistry, edited by C. Rossi and E. Tiezzi Elsevier, Amsterdam, 651 pp.

Gibbons, D. C. 1986. The Economic Value of Water. Resources for the Future, Washington, D.C. 101 pp.

Gilliland, M. W. 1973. Man's impact on the phosphorus cycle in Florida. Ph.D. Dissertation, Environmental Engineering Sciences, Univ. of Florida, Gainesville. 269 pp.

*Gilliland, M. W. 1975. Energy analysis and public policy. Science 189(4208):1051–1056.

Gilliland, M. W. 1976. A geochemical model for evaluating theories on the genesis of Florida's sedimentry phosphate deposits. Math. Geol. 8:219–242.

*Gilliland, M. W. 1978. Energy Analysis, a New Public Policy Tool. Westview Press, Boulder, Colo. 110 pp.

*Gilliland, M. W., and M. Eastman. 1981. Free Energies for Iron, Aluminum, and Copper Ores. Contract report, Center for Environmental Research, Cornell University, Ithaca. 36 pp.

Gilliland, M. W., L. B. Fenner, and M. Eastman. 1978. Energy measures of rocks as environmental resources. Energy Policy Studies, East Paso, Tex. 68 pp.

Gilliland, M. W., J. M. Klopatek, and S. G. Hildebrand. 1981a. Net Energy: results for small-scale hydroelectric power and summary of existing analyses. Energy 6:1029–1040.

*Goldick, S. S. 1938. A study in rock weathering. J. Geol. 46:17–58.

Grassi, G. L. F. G., and H. E. Willaims. 1987. Biomass energy. Elsevier, N.Y.

Gregor, C. B., R. M. Garrels, F. T. Mackenzie, and J. B. Maynard 1988. Chemical Cycles in the Evolution of the Earth. Wiley, New York. 276 pp.

*Gunderson, L. 1989. EMERGY Analysis of Everglades National Park in 1989. Special class report. Center for Wetlands, Univ. of Florida, Gainesville. 14 pp.

*Gutenberg, B., and C. F. Richter. 1949. Seismicity of the Earth and Associated Phenomena. Princeton Univ. Press, Princeton, N.J. 273 pp.

Guthrie, D. P. 1979. Energetics analysis of resouce recovery systems for Hillsborough County, Florida. M.A. Thesis, Dept. of Urban and Regional Planning, University of Florida, Gainesville, 105 pp.

Gutierrez, R. J. 1978. Energy Analysis and Computer Simulations of Pastures in Florida. M.S. Thesis, Environmental Engineering Sciences, Univ. of Florida, Gainesville. 186 pp.

*Hall, C. A. S., ed. 1995. Maximum Power. The Ideas and Applications of H. T. Odum. Univ. Press of Colorado, Niwot. 454 pp.

Hall, C. A. S., and J. W. Day Jr. 1977. Ecosystem Modeling in Theory and Practice. Wiley, New York. 684 pp.

Hall, C. A. S., M. Lavine, and J. Sloane. 1979. Efficiency of energy delivery systems: part I. An economic and energy analysis. Environ. Manage. 3(6):493–504.

Hall, C. A. S., C. J. Cleveland, and M. Berger. 1981. Yield per effort as a function of time and effort for United States petroleum, uranium, and coal. pp. 715–724 in Energy and Ecological Modelling, ed. by W. J. Mitsch, R. W. Bosserman, and J. M. Klopatek. Elsevier Scientific, Amsterdam.

*Hall, C. A. S., C. J. Cleveland, and R. Kaufmann. 1986. Energy and Resource Quality, The Ecology of the Economic Process. Wiley, New York. 577 pp.

Hannon, B. 1973a. An energy standard of value. Ann. Am. Acad. Polit. Soc. Sci. 41:139–153.

Hannon, B. 1973b. Marginal product pricing in the ecosystem. J. Theor. Biol. 41:252–267.

Hannon, B. 1976. The structure of ecosystems. J. Theor. Biol. 56:535–546.

Hannon, B. 1979. Total energy costs in ecosystems. J. Theor. Biol. 80:271–293.

Hannon, B. 1982a. Analysis of the energy costs of economic activities: 1963–2000. Energy Syst. Policy J. 6:249–278.

Hannon, B. 1982b. Energy discouting. pp. 73–93 in Energetics and Systems, ed. by W. J. Mitsch, R. K. Ragade, R. W. Bosserman, and J. A. Dillon Jr. Ann Arbor Science, Ann Arbor, Mich.

Hannon, B. 1985. Linear dynamic ecosystems. J. Theor. Biol. 116:89–110.

Hannon, B. 1986. Ecosystem control theory. J. Theor. Biol. 121:417–437.

Hannon, B. 1991. Accounting in ecological systems. pp. 234–252 in Ecological Economics, ed. by R. Costanza. Columbia Univ. Press, New York.

Hannon, B., and C. Joiris. 1989. A seasonal analysis of the southern North Sea ecosystem. Ecology 70:1916–1934.

*Hansen, K. L., H. T. Odum, and M. T. Brown. 1982. Energy models for Volusia County, Florida. Q. J. Fla. Acad. Sci. 45(4):209–227.

Hayes, E. T. 1979. Energy resources available to the United States, 1985–2000. Science 203:233–239.

Healey, H. M. 1984. The economics of residential photovoltaic systems. Florida Solar Energy Center (FSEC-PF-64-84), pp. 1–6.

Healey, H. M., G. Atmaram, S. Kalaaghehy, and C. Maytrott. 1986. Design, installation, and preliminary operational results of a photovoltaic tracking research facility. Florida Solar Energy Center (FSEC-PF-112-86).

*Heichel, G. H. 1973. Comparative efficiency of energy use in crop production. Conn. Agric. Exp. Sth. Bull. 739. 26 pp.

Heichel, G. H. 1976. Agricultural production and energy resources. Am. Sci. 64:64–72.

Herendeen, R. A. 1973. Comparative efficiency of energy use in crop production. Center for Advanced Computation, Univ. of Illinois, Urbana. 22 pp.

Herendeen, R. A. 1974. Energy intensities in ecological and economic systems. J. Theor. Biol. 91:607–620.

Herendeen, R. A. 1981a. Net energy and true subsidies to new energy technology. pp. 681–687 in Energy and Ecological Modelling, ed. by W. A. Mitsch, R. W. Bosserman, and J. M. Kloptatek. Elsevier, Amsterdam. 839 pp.

Herendeen, R. A. 1981b. Energy intensities in ecological and economic systems. J. Theor. Biol. 91:607–620.

Herendeen, R. A., and C. W. Bullard. 1974. Energy Costs of Goods and Services, 1963 and 1967. Center for Advanced Computation, Univ. of Illinois, Urbana (CAC Document 140). 43 pp.

Herendeen, R. A., and C. W. Bullard. 1975. Energy costs of goods and services: 1963 and 1967. Energy Policy 3:268.

Herendeen, R. A., and C. W. Bullard. 1976. U.S. energy balance of trade, 1963–1967. In Energy Systems and Policy, vol. 1, no. 4, Crane, Russak.

Herendeen, R. A. 1978. Energy balance of trade in Norway 1973. Energy Systems and Policy 2:425–431.

Herendeen, R. A. 1989. Energy intensity, residence time, exergy, and ascendency in dynamic ecosystems. Ecological Modelling. 48:19–44.

Herendeen, R., and J. Tanaka. 1976. Energy Cost of Living. Energy 1:165–178.

*Higashi, M., T. P. Burns, and B. C. Patten. 1989. Food network unfolding: an extension of trophic dynamics for application to natural ecosystems. J. Theor. Biol. 140:243–261.

*Higashi, M., T. P. Burns, and B. C. Patten. 1992. Trophic niches of species and trophic structure of ecosystems: complementary perspectives through food network unfolding. J. Theor. Biol. 154:57–76.

Higashi, M., B. C. Patten, and T. P. Burns. 1993. Network trophic dynamics: the modes of energy utilization in ecosystems. Ecol. Model. 66:1–42.

Higashi, M., T. P. Burns, and B. C. Patten. 1993. Network trophic dynamics: the tempo of energy movement and availability in ecosystems. Ecol. Model. 66:43–64.

Hirata, H., and R. E. Ulanowicz. 1984. Information theoretical analysis of ecological networks. Int. J. Syst. Sci. 15:261–270.

Hirst, E. 1974. Food related energy requirements. Science 184:134–138.

Hirst, E., and J. C. Moyers. 1973. Efficiency of energy use in the United States. Science 179:1299–1304.

Hirst, E., R. Marlay, D. Greene, and R. Barnes. 1983. Recent changes in U.S. energy consumption: What happened and why. Annu. Rev. Energy 89:193–245.

Hittman Associates. 1975. Environmental Impacts, Efficiency and Cost of Energy Supply and End Use. HIT-593, vol. 2.

*Homer, M. 1977. Seasonal abundance, biomass, diversity, and trophic structure of fish in a salt marsh tidal creek affected by a coastal power plant. pp. 259–267 in Thermal Ecology, vol. 2, Division Technical Information, Energy Research and Development Administration, Oak Ridge, Tenn.

Hoon, V., and C. V. Seshadri. 1990. Energy studies of island communities with emphasis on time/energy availability for women's needs. Monograph series on the engineering of photosynthetic systems. 31:SHRI A.M.M. Murugappa Chettiar Research Centre, Tharamani, Madras, India. 69 pp.

*Hopkinson, C. S. Jr., and J. W. Day Jr. 1980. Net energy analysis of alcohol production from sugarcane. Science 207:302–304.

*Hornbeck, D. A. 1979. Metabolism of salt marshes and their role in the economy of a coastal community. MS Thesis (CFW-79-06). University of Florida, Gainesville. 174 pp.

Houghton, H. G. 1954. On the annual heat balance of the northern hemisphere. J. Meteorol. 11:1–9.

Howard, M. C., and J. E. King, ed. 1976. The Economics of Marx. Penguin Books, N.Y.

*Huang, S. L., and H. T. Odum. 1991. Ecology and economy: EMERGY synthesis and public policy in Taiwan. J. Environ. Manage. 32:313–333.

Huang, S. L., S. C. Wu, W. B. Chen. 1993. Ecological economic system and the environmental quality of the Taipei Metropolitan region. pp. 1–10 in Proceedings: Environmntal Quality Evaluation System of Metropolitan Areas, April 23–24, 1993. Dept. of Urban and Regional Planning, National Chung-Hsing University, Taipei, Taiwan.

*Hubbert, M. K. 1949. Energy from fossil fuels. Science 109:103–109.

Huettner, D. A. 1976. Net energy analysis: an economic assessment. Science 192:101–104.

Hyman, E. L. 1979. J. Environ. Syst. 9(4):313–324.

*IFIAS. 1974. Energy Analysis Workshop on Methodology and Conventions. International Federation of Institutes for Advanced Study, Stockholm, Sweden. 89 pp.

Inhaber, H. 1982. Energy Risk Assessment. Gordon and Breach, New York. 395 pp.

Institute of Gas Technology. 1978. Symposium Papers: Energy Modeling and Net Energy Analysis. Chicago. 795 pp.

Jahrbuch für Bergbau, Energie, Mineralol und Chemie (Yearbook for Mining, Energy, Crude Oil, and Chemistry). 1982–1983. Verlag Gluckauf Gmblt., Essen, Germany.

Jansson, A. M. The ecological economics of sustainable development—environmental conservation reconsidered. p. 31–36 in Perspectives of Sustainable Development. Stockholm Studies in Natural Resources Management No. 1. 115 pp.

*Jansson, A. M. 1985. Natural productivity and regional carrying capacity for human activities on the island of Gotland, Sweden. pp. 85–91 in Economics of Ecosystems Management, ed. by D. O. Hall, N. Myers, and N. S. Margaris. Dr. W. Junk, Dordrecht, Netherlands.

*Jansson, A. M., and J. Zucchetto. 1978a. Man, nature and energy flow on the island of Gotland, Sweden. Ambio 7(4):140–149.

*Jansson, A. M., and J. Zucchetto, eds. 1978b. Energy, economic and ecological relationships for Gotland, Sweden—A regional systems study. Ecological Bulletin No. 28. Swedish National Science Research Council, Stockholm, Sweden. 154 pp.

Jansson, B. O. 1972. Ecosystem approach to the Baltic Problem. Bulletin Ecological Research Committee, Swedish Natural Resource Council, Stockholm, 16:1–82.

*Jenny, H. 1982. The Soil Resource: Origin and Behavior. Springer-Verlag, New York.

Jerez, P. G. 1972. Energy flow estimates in cotton production in the tropics (Nicaragua). M.S. Thesis, Dept. of Botany, University of Florida, Gainesville, 71 pp.

Jiang Youxu, Xu Deying, Wang Yamhui, Nie Daoping, Lam Zaiping, Zhu Yangang, Ma Yuiping, Wang Fengyou, Wu Honggi, Ma Keping, Yang Gruoting, Dong Shilin, Lu Zhu, Dong Shilin, and Chang Jie. 1993. Chinese translation of H. T. Odum's Systems Ecology. Chinese Academy of Forestry, Beijing. 772 pp. English language original published in 1983, John Wiley. 544 pp.

Johansson, T. B., H. Kelly, A. K. N. Reddy, R. H. Wiliams, amd L. Burnham, eds. 1993. Renewable Energy Sources For Fuels And Electricity. Island Press, Washington, D.C. 1160 pp.

Johnson, R. W. 1990. The public trust doctrine and Alaska oil. pp. 1–22, No. 8.2 in Appendix M of Alaska Oil Spill Commission (Vol. 3). Sea Grant Legal Research Report by Bader, T. R. Johnson, Z. Plater, and A. Rieser. Alaska Sea Grant Program, Univ. of Alaska, Fairbannks.

Jorgensen, S. E. 1992. Exergy and ecology. Ecolo. Model. 63 (1992). 185–214.

Johnstone, I. M. 1978. In search of the Steady State. Thesis, School of Architecture, Univ. of Auckland, N.Z. 229 pp.

Kangas, P. C. 1983. Landforms. Succession and Reclamation. Ph.D. Dissertation. Dept. of Environmental Engineering Sciences, Univ. of Florida, Gainesville. 185 pp.

Kaufman, R. 1987. Biophysical and Marxist Economics: Learning from each other. Ecol. Modelling 39:91–105.

Kauffman, S. A. 1993. The Origins of Order. Oxford Univ. Press, Oxford, UK, 709 pp.

Kaufman, R. K., and C. A. S. Hall. 1981. The energy return on investment of imported petroleum. pp. 697–701 in Energy and Ecological Modeling, ed. by W. J. Mitsch, R. W. Bosserman, and J. M. Klopatek. Elsevier Scientific, Amsterdam.

*Keitt, T. H. 1991. Hierarchical organization of energy and information in a tropical rain forest ecosystem. M.S. Thesis, Environmental Engineering Sciences, Univ. of Florida, Gainesville. 72 pp.

Keller, P. A. 1992. Perspectives on interfacing paper mill wastewaters and wetlands. M.S. Thesis, Environmental Engineering Sciences, Univ. of Florida. 133 pp.

Kemp, M. 1975. Energy evaluation of cooling alternatives and regional impact of power plant at Crystal River, Florida. pp. 49–166 in Report to Florida Power Corporation. Environmental Engineering Sciences, Univ. of Florida, Gainesville.

Kemp, W. M. 1977. Energy Analysis and Ecological Evaluation of a Coastal Power Plant. Ph.D. Dissertation, Environmental Engineering Sciences, Univ. of Florida, Gainesville. 582 pp.

Kemp, W. M. 1981. The Influence of National Renewable Energies on National Economies of the World in Energy and Ecological Modeling. Elsevier Scientific, New York.

*Kemp, W. M., H. McKellar, amd M. Homer. 1975. Value of higher animals at Crystal River estimated with energy quality ratios. pp. 372–392 in Power Plants and Estuaries at Crystal River, Fla. Final Report to Florida Power Corporation, ed. by H. T. Odum ed. (Contract GEC-159-918-200-188.19). 540 pp.

*Kemp, W. M., W. H. B. Smith, H. M. McKellar, M. E. Lehman, M. Homer, D. L. Young, and H. T. Odum. 1977a. Energy cost-benefit analysis applied to power plants near Crystal River, Florida. pp. 507–543 in Ecosystem Modeling in Theory and Practice, C. A. S. Hall and J. Day, eds. Wiley-Interscience, New York.

*Kemp, W. M., F. V. Ramsey, and H. T. Odum. 1977b. Power Plant in Its Environment at Anclote, Florida. Report to the Florida Power Corporation. Center for Wetlands (CFW 77-24) Univ. of Florida, Gainesville. 215 pp.

*Kemp, W. M., W. R. Boynton, and K. Limburg. 1981. The influence of natural resources and demographic factors on the economic production of nations. pp. 827–839 in Energy and Ecological Modeling, ed. by W. J. Mitsch, R. W. Bosserman, and J. M. Klopatek. Elsevier Scientific, Amsterdam.

*King, R. J., and J. Schmandt. 1991. Ecological Economics of Alternative Transportation Fuels. Report to Texas State Dept. of Energy. Unpublished report, L.B.J. School of Public Affairs. Univ. of Texas, Austin. 26 pp.

*Kinsman, B. 1965. Wind, Waves, Their Generation and Propagation on the Ocean Surface. Prentice-Hall, Englewood Cliffs, N.J., 676 pp.

Kleinbach, M. H., and C. E. Salvagin. 1986. Energy Technologies and Conversion Systems. Prentice Hall, Englewood Cliffs, N.J. 563 pp.

Klopatek, J. M., and P. G. Risser, 1962. Energy analysis of Oklahoma rangelands and improved pastures. J. of Range Management. 35:637–643.

Knight, R. L. 1980. Energy Basis of Control in Aquatic Ecosystems. Ph.D. Dissertation, Environmental Engineering Sciences, Univ. of Florida, Gainesville. 198 pp.

Knight, R. L. 1981. A control hypothesis for ecosystems-energetics and quantification with the toxic metal cadmium. pp. 601–615 in Energy and Ecological Modeling, ed. by W. J. Mitsch, R. W. Bosserman, and J. M. Klopatek. Elsevier Scientific, Amsterdam.

Knox, G. 1983. Upper Waitemata Harbour Catchment Study, Energy Analysis. Auckland Regional Authority, Auckland, N.Z. 266 pp.

*Knox, G. 1984. The key role of krill in the ecosystem of the southern ocean with special reference to the convention on the conservation of Antarctic marine living resources. Ocean. Manage. 9:113–156.

Knox, G. A. 1986. Estuarine Ecosystems: A Systems Approach. CRC Press, Boca Raton, Fla. Vol. 1 289 pp; vol. 2. 230 pp.

*Krauskopf, K. B. 1967. Introduction to Geochemistry. McGraw-Hill, New York. 72

Krenz, J. H. 1974. Energy per dollar value of goods and services. pp. 386–388 in IEEE Transactions on Systems, Man and Cybernetics. SMC-4, 4 (July).

*Krenz, J. H. 1976. Energy Conversion and Utilization. Allyn and Bacon, Boston.

*Kruczynski, W. L., C. B. Subrashmnyam, and S. H. Drake. 1978. Studies on the plant community of a north Florida salt marsh. Bull. Mar. Sci. 28:316–334.

Kushlan, J. A., S. A. Voorhees, W. F. Loftus, and P. C. Frohring. 1986. Length, mass and calorific relationships of everglades animals. Fla. Sci. 49:66–79.

Kylstra, C. D. 1974. Energy analysis as a common basis for optimally combining man's activities and nature's. pp. 265–294 in Environmental Management, ed. by G. F. Rohrlich. Ballinger, Cambridge, Mass.

*Kylstra, C., and Ki Han. 1975. Energy analysis of the U.S. nuclear power system. pp. 138–200 in Energy Models for Environment Power and Society, ed. by H. T. Odum. Report to Energy Research and Development Administration [Contract E-(40-1)-4398]. Environmental Engineering Sciences, Univ. of Florida, Gainesville.

Lan, S. 1992. Emergy analysis of ecological-economic systems. pp. 266–286 in Advances and Trends of Modern Ecology, ed by Gao Jian. Beijing. 397 pp

*Lan, S. 1993. Chinese translation of H. T. Odum's Energy, Environment and Public Policy: A Guide to the Analysis of Systems. Nengliang Huanjing Yu Jingji, Beijing. 232 pp; includes an Appendix: EMERGY evaluation of China. See Odum et al. (1988) for English language original.

Lan, S., and H. T. Odum. 1994a. EMERGY synthesis of the environmental resource basis and economy in China. Ecol. Sci. 63–74.

Lan, S., and H. T. Odum. 1994b. EMERGY evaluation of the environment and economy of Hong Kong. J. of Environ. Sci. 6:432–439.

*Lapp, C. 1991. EMERGY Analysis of the Nuclear Power System in the United States. Research paper for M.E. degree, Environmental Engineering Sciences, Univ. of Florida, Gainesville, 64 pp.

*Larson, W. E., F. J. Pierce, and R. H. Dowdy. 1983. The threat of soil erosion to long-term crop production. Science 219:458–465.

Laughline, R. B., J. N. Guard, and H. E. Guard. 1981. Hormesis: a response to low environmental concentrations of petroleum hydrocarbons. Science 211:705–707.

Lavine, M. J., and A. H. Meyburg. 1976. Toward environmental benefit/cost analysis: measurement methodology, Center for Environmental Quality Management, Cornell University, Ithaca, N.Y. 188 pp.

Lavine, M. J., and T. J. Butler. 1982. Use of Embodied Energy Values to Price Environmental Factors: Examining the Energy/Dollar Relationship. Report to National Science Foundation (Grant # PRA-8003845) from Dept. of Environmental Engineering, Cornell Univ. Ithaca. 231 pp.

Lavine, M. J., T. J. Butler, A. H. Meyburg. 1978. Toward Environmental Benefit/cost analysis: Energy Analysis. Center for Environmental Research, Cornell Univ., Ithaca. Vol. 1, 129 pp; appendix, 219 pp.

Lavine, M. J., T. J. Butler, and A. H. Meyburg. 1979. Emergy Analysis Manual for Environmental Benefit/cost Analysis of Transportation Actions. Center for Environmental Research, Cornell University, Ithaca. vol. 1, Theories and Procedures, 129 pp. Vol II, Illustrative case histories, 257 pp.

Lawton, F. L. 1972. Tidal power in Bay of Funday and economics of tidal power. In Tidal power, ed, by Gray and Gashus, Plenum Press, N.Y.

Leach, G. 1976. Energy and food production. IPC Science and Technology Press, Guilford, Surrey, England.

Leach, G., and M. Slesser. 1973. Energy equivalents of network inputs to food producing processes. Straathclyde Univ., Glasgow. 38 pp.

*Lee, S. M., and H. T. Odum. 1994. EMERGY analysis overview of Korea. J. Korean Environ. Sci. Soc. 3(2):165–175.

Leibowitz, S. C. 1979. The embodied energy stored in the United States' soils. pp 110–122. In Energy Basis for thc United States, ed. by H. T. Odum and J. F. Alexander. Report to Dept. of Energy. Environmental Engineering Sciences, Univ. of Florida, Gainesville.

Leibowitz, S. C. 1981. A computer simulation of energy use in phosphate mining. Masters degree paper, Environmental Engineering Sciences, and pp. 119–142 in Studies on Phosphate Mining, Reclamation, and Energy, ed. by H. T. Odum. Center for Wetlands, Univ. of Florida, Gainesville.

Leibowitz, S. G. 1992. A synoptic approach to cumulative impact assessment. Environmental Protection Agency, Environmental Research Laboratory, 200 SW 35th St., Corvallis, Ore. 129 pp.

*Lem, P. N. 1973. Energy required to develop power in the United States. Ph.D. Dissertation, Environmental Engineering Sciences, Univ. of Florida, Gainesville. 213 pp.

Lem, P. N., H. T. Odum, and W. E. Bolch. 1974. Some considerations that effect the net yield from nuclear power. Paper distributed at the Health Physics Society, 19 pp.

*Leopold, L. B., F. E. Clarke, B. B. Hanshaw, and J. R. Balsley. 1971. A procedure for evaluating environmental impact. Geological Survey Circular 645:1–13.

Limburg, K. A. 1981. Energy sources and design of marine ecosystems. M.S. Thesis, Environmental Engineering Sciences, Univ. of Florida, Gainesville. 180 pp.

Limburg, K. A. 1984. Energy analysis of Sweden. Unpublished manuscript. 25 pp.

Limburg, K. E. 1985. Increasing complexity and energy flow in models of food webs. Ecol. Modelling. 29:5–25.

*Lisle, J. F. 1960. Mean daily insolation in New Zealand. N.Z. J. Sci. 9(4):992–1005.

Littlejohn, C. B. 1973. An Energetics Approach to Land Use Planning in a Coastal Community of Southwest Florida. M.S. Thesis, Environmental Engineering Sciences, Univ. of Florida, Gainesville. 187 pp.

Lockeretz, W., ed. 1977. Agriculture and Energy. Academic Press, New York. 750 pp.

*Lonergan, S. C. 1988. Theory and measurement of unequal exchange: a comparison between a Marxist and an energy theory of value. Ecol. Model. 41:127–146.

Long, T. V. 1978. Comparing methods of energy analysis in an economic framework. pp. 263–278 in Symposium Papers, Energy Modeling and Net Energy Analysis, ed. by F. S. Roberts. Institute of Gas Technology, Chicago.

*Lotka, A. J. 1922a. Contribution to the energetics of evolution. Proc. Natl. Acad. Sci. 8:147–151.

*Lotka, A. J. 1922b. Natural selection as a physical principle. Proc. Natl. Acad. Sci. 8:151–154.

*Lotka, A. J. 1925. Physical Biology. Williams and Wilkins, Baltimore. 460 pp.

Lovins, A. B. 1977. Soft Energy Paths: Towards a Durable Peace. Harper & Row, New York. 231 pp.

Lugo, A. E., and G. L. Morris. 1982. Los sistemas ecologicos y la humanidad. General secreariet of the Organization of American States, Washington, D.C. 82 pp.

*Lugo, A. E., S. C. Snedaker, S. Bayley, and H. T. Odum. 1971. Models for Planning and Research for the South Florida Environmental Study. Report on Contract 14-10-9-900-363, National Park Service, Dept. of the Interior and Center for Aquatic Sciences, Univ. of Florida, Gainesville. 141 pp.

*Lyu, W. 1986. EMERGY analysis of highway transportation in Texas. pp. 130–157 in student reports for Policy Research Project Course, L.B.J. School of Public Affairs, Univ. of Texas, Austin.

McClanahan, T. R. 1984. The effects of dispersal on ecological succession and optimal size islands. M.S. Thesis, Environmental Engineering Sciences, Univ. of Florida, Gainesville. 110 pp.

McClanahan, T. R. 1990. Hierarchical Control of Coral Reef Ecosystems. Ph.D. Dissertation, Environmental Engineering Sciences, Univ. of Florida, Gainesville. 219 pp.

McClendon, J. H. 1975. Efficiency. J. of theor. Biol. 49:213–218.

McGrane, G. 1993. An EMERGY evaluation of personal transportation alternatives. M.S. Thesis, Environmental Engineering Sciences, Univ. of Florida, Gainesville. 105 pp.

Maddox, K. P. 1978. Energy analysis and resource substitution. Unpublished Report to U.S. Department of Energy, Development Sciences, Inc., Sagamore, Mass. 14 pp.

McKellar, H. N. 1977. Metabolism and model of an estuarine bay ecosystem affected by a coastal power plant. Ecological Modelling 3:85–118.

Maltby, D. 1976. Energy and resource models of an urban ecosystem. Jacksonville, Florida. Masters Thesis, Univ. of Florida, Gainesville.

*Marchetti, C., and N. Nakicenovic. 1979. The dynamics of energy systems and the logistic substitution model. International Institute for Applied Systems Analysis, Laxenburg, Austria.

Margalef, R. 1956. Information Theory in Ecology. Mem. R. Acad. Cienc. Artes Barcelona 23:373–449. Translation in Soc. Gen. Syst. Yearb. 3:36–371.

*Martinez-Alier, J. 1987. Ecological Economics. Basil Blackwell, New York. 286 pp.

Marx, K. 1894. Das Kapital, Band III. 1962 English translation: Capital. A Critique of Politial Economy, vol. 3, The process of Capitalist Production as a Whole, ed. by F. Engels. Foreign Languages Publishing House, Moscow. 923 pp.

Mau, D., S. Bayley, and J. Zuchetto. 1978. Net energy analysis of electrical transmission lines. Can. Elec. Eng. J. 3:9–13.

Mayumi, K. 1991. A critical appraisal of two entropy theoretical approaches to resources and environmental problems, and a search for an alternative. Pp. 109–130 in Ecological physical chemistry, ed. by C. Rossi and E. Tiezzi. Elsevier, Amsterdam. 651 pp.

Meadows, D. H., D. L. Meadows, and J. Randers. 1992. Beyond the Limits. Chelsea Green Publishing, Post Mills, Vt. 300 pp.

Melcher, A. G. 1976. Net energy analysis, summary report. The Colorado Energy Research Institute, Golden, Colo. 55 pp.

Merriam, A. 1975. An Energetic Evaluation of Channel Dredging and Spoil Island Formation in a Natural Salt Marsh Ecosystem. Center for Wetlands, Univ. of Florida, Gainesville (CFW 75-13). 51 pp.

Meyers, C. D. 1977. Energetics Systems analysis with Application to Water Resources Planning and Decision-Making. U.S. Army Engineering Institute for Water Resources, Ft. Belvoir, Virginia, IWR Contract Report 77–6.

*Miller, G. A. 1966. The flux of tidal energy out of the deep oceans. J. Geophys. Res. 71:2485–2489.

Miller, M. A. 1975. Energy basis for housing systems in Collier County. M.S. Thesis, Environmental Engineering Sciences, Univ. of Florida, Gainesville. 160 pp.

Miller, M. A., and F. C. Wang. 1981. Energy analysis applied to environmental issues. pp. 781–796 in Energy and Ecological Modelling, ed. by W. A. Mitsch, R. W. Bosserman, and J. M. Kloptatek. Elsevier, Amsterdam. 839 pp.

Mirowski, P. 1989. More Heat than Light. Cambridge Univ. Press, London. 450 pp.

*Mitchell, R. 1979. The Analysis of Indian Agro-Ecosystems. Interprint, New Delhi. 180 pp.

Mitsch, W. J., and S. E. Jorgensen, eds. 1987. Ecological Engineering: An Introduction to Ecotechnology. Wiley, New York.

Mitsch, W. J., R. W. Bosserman, and J. M. Klopatek, eds. 1981. Energy and Ecological Modelling. Elsevier, Holland. 839 pp.

Mitsch, W. J., R. K. Ragade, R. W. Bosserman, and J. A. Dillon Jr., eds. 1982. Energetics and Systems. Ann Arbor Press. 132 pp.

*Monin, A. S. 1972. Weather Forecasting as a Problem in Physics. MIT Press, Cambridge, Mass.

*Moran, M. J. 1989. Availability analysis. American Society of Mechanical Engineers (ASME) Press, New York. 260 pp.

*Morrison, Philip, and Phylis Morrison. 1983. Powers of Ten. Scientific American Library, Freeman, New York. 150 pp.

*Morton, R. 1991. Emergy Analysis of Television Transmission and Image of the United States. Unpublished Class report for EES 3008. 7 pp.

*Munk, W. H., and G. F. McDonald. 1960. The Rotation of the Earth: A Geophysical Discussion. Cambridge Univ. Press, London. 323 pp.

Nagai, T., and Y. Shimazu. 1984. Science assessment of fusion power plant. J. of Earth Sciences. 32.

*National Geographic. 1981. Map including Alaska's continental shelf.

*New Zealand Dept. of Statistics. 1981. New Zealand Official Yearbook. Wellington.

Nguyen, T. H. 1984. On energy coefficients and ratios. Energy Economics 6(2):102–109.

Nilsson, P. O., S. J. Doherty, and H. T. Odum. 1992. EMERGY evaluation of forest alternatives for Sweden. Report to Swedish Power Board. Draft. 166 pp.

Noyes, G. 1977. Energy analysis of space operations. pp. 401–422 in Energy Analysis of Models of the U.S. Annual Report to Dept. of Energy (Contract EY-76-S-05-4398).

Odum, E. C., and H. T. Odum. 1980. Energy systems and environmental education. pp. 213–231 in Environmental Education: Principles, Methods and Applications, ed. by T. S. Bakshi and Z. Naveh. Plenum Press, New York.

*Odum, E. C., and H. T. Odum. 1984. System of ethanol production from sugarcane in Brazil. Ciencia e Cultura 37(11), 1849–1855.

Odum, E. C., H. T. Odum, and N. S. Peterson. 1995. Using simulation to introduce systems approach in education. Chap. 31 pp. 346–352 in Maximum Power, ed. by C. A. S. Hall. Univ. Press of Colorado, Niwot, (in press).

Odum, E. C., G. Scott, and H. T. Odum. 1982. Energy and Environment in New Zealand. Univ. of Canterbury, Christchurch, N.Z. 129 pp.

*Odum, H. T. 1967a. Biological circuits and the marine systems of Texas. pp. 99–157 in Pollution and Marine Ecology, ed. by T. A. Olson and F. J. Burgess. Wiley-Interscience, New York.

*Odum, H. T. 1967b. Energetics of world food production. pp. 55–94 in Problems of World Food Supply. President's Science Advisory Committee Report, vol. 3. White House, Washington, D.C.

*Odum, H. T. 1970. Summary: an emerging view of the ecological system at El Verde. Chap. I-10, pp. I191–I289 in A Tropical Rainforest, ed. by H. T. Odum and R. F. Pigeon. Division of Technical Information, U.S. Atomic Energy Commission (TID2470). 1660 pp.

*Odum, H. T. 1971a. An energy circuit language for ecological and social systems: its physical basis. pp. 139–211 in Systems Analysis and Simulation in Ecology, vol. 2, ed. by B. Patten. Academic Press, New York.

*Odum, H. T. 1971b. Environment, Power and Society. Wiley, New York. 331 pp.

Odum, H. T. 1972a. Chemical cycles with energy circuit models. pp. 223–257 in Changing Chemistry of the Ocean, ed. by D. Dryssen and D. Jagner. Nobel Symposium 20. Wiley, New York.

*Odum, H. T. 1972b. Use of energy diagrams for environmental impact statements. pp. 197–213 in Tools for Coastal Management. Proceedings of the conference, Feb. 14–15, 1972. Marine Technology Society, 1730 M St. NW, Washington, D.C.

Odum, H. T. 1973. Energy, ecology and economics. Royal Swedish Academy of Science. AMBIO 2(6):220–227.

Odum, H. T. 1974a. Energy cost-benefit models for evaluating thermal plumes. pp. 628–647 in Thermal Ecology, ed. by S. W. Gibbons and B. Sharitz. Proceedings of Symposium of the Savannah River Laboratory, Aiken, S.C. Division of Technical Information, Atomic Energy Commission, Oak Ridge.

Odum, H. T. 1974b. Marine ecosystems with energy circuit diagrams. pp. 127–151 in Modelling of Marine Systems, ed. by J. C. J. Nihoul. Elsevier Oceanography Series. Elsevier Scientific, New York.

*Odum, H. T. 1975. Combining energy laws and corollaries of the maximum power principle with visual system mathematics. pp. 239–263 in Ecosystems: Analysis and Prediction, ed. by Simon Levin. Proceedings of the conference on ecosystems at Alta, Utah. SIAM Institute for Mathematics and Society, Philadelphia.

Odum, H. T. 1976a. Energy analysis and net energy. pp. 90–115 in Proceedings of NSF workshop on Net Energy, Stanford, Calif. Institute for Energy Studies. Stanford Univ., Palo Alto, Calif.

*Odum, H. T. 1976b. Energy quality and carrying capacity of the earth. Trop. Ecol. 16(1):1–8.

Odum, H. T. 1976c. Net benefits to society from alternate energy investments. pp. 327–338 in Transactions of the 41st North American Wildlife and Natural Resources Conference. Wildlife Management Institute, Washington, D.C.

Odum, H. T. 1977a. Energy analysis, energy quality and environment. pp. 55–87 in Energy Analysis: A New Public Policy Tool, ed. by M. W. Gilliland. American Association for the Advancement of Science, Selected Symposium 9. Westview Press, Boulder, Colo.

Odum, H. T. 1977b. Energy analysis (letter to editor). Science 196:261.

Odum, H. T. 1977c. Energy value and money. pp. 174–196 in Ecosystem Modeling in Theory and Practice: An Introduction with Case Histories, ed. by C. A. S. Hall and J. W. Day. Wiley, New York.

Odum, H. T. 1977d. The ecosystem, energy and human values. Zygon 12(2):109–133.

Odum, H. T. 1977e. Value of wetlands as domestic ecosystems. National Wetland Protection Symposium, ed. by J. H. Montanari and J. A. Jusler. U.S. Fish and Wildlife Service, Dept. of the Interior, Reston, Va. (FWS/Obs-78/97).

*Odum, H. T. 1978a. Energy analyis, energy quality and environment. pp. 55–88 in Energy Analysis: A New Public Policy Tool, ed. by M. Gilliland. Westview Press, Boulder, Colo. 110 pp.

*Odum, H. T. 1978b. Net energy from the sun. pp. 196–211 in Sun: A Handbook for the Solar Decade, ed. by S. Lyons. Friends of the Earth, San Francisco. 364 pp.

Odum, H. T. 1979a. Energy quality control of ecosystem design. pp. 221–235 in Marsh-estuarine Systems Simulation, P. F. Dame, ed. Belle W. Baruch Library in Marine Science, Publication 8. Univ. of South Carolina Press, Columbia. 260 pp.

Odum, H. T. 1979b. Los principios del analisis de la energia neta. Energia, Universidad Argentina de La Empresa 3(21):69–80.

Odum, H. T. 1980a. Biomass and Florida's future. pp. 58–67 in A Hearing before the Subcommittee on Energy Development and Applications of the Committee on Science and Technology of the U.S. House of Representatives, 96th Congress. Government Printing Office, Washington, D.C.

Odum, H. T. 1980b. Principle of environmental energy matching for estimating potential economic value: a rebuttal. Coastal Zone Manage. J. 5(3):239–243.

Odum, H. T. 1981. Energie, economie et hierarchie de l'environment. pp. 155–163 in Etude & Recherches; Compt Rendu du Colloque de Troisiemes Assises Internationales de l'Environment, vol. 343. Society et Environment. Ministry of the Environment, Paris.

*Odum, H. T. 1982. Pulsing, power, and hierarchy. pp. 33–59 in Energetics and Systems, ed. by W. J. Mitsch, R. K. Ragade, R. W. Bosserman, and J. A. Dillon Jr. Ann Arbor Science, Ann Arbor, Mich.

Odum, H. T. 1983a. Maximum power and efficiency: a rebuttal. Ecol. Model. 20:71–82.

*Odum, H. T. 1983b. Systems Ecology. Wiley, New York. 644 pp.

*Odum, H. T. 1984a. Embodied energy, foreign trade, and welfare of nations. pp. 185–200 in Integrations of Economy and Ecology, an Outlook for the Eighties, ed. by A. M. Jansson. Asko Laboratory, Univ. of Stockholm, Sweden.

*Odum, H. T. 1984b. Energy analysis evaluation of Santa Fe Swamp. Report to Georgia Pacific Corporation. Center for Wetlands, Univ. of Florida, Gainesville. 29 pp., unpublished.

*Odum, H. T. 1984c. Energy analysis evaluation of coastal alternatives. Water Science Technology (Rotterdam, Netherlands) 16:717–734.

*Odum, H. T. 1984d. Energy analysis of the environmental role in agriculture. pp. 24–51 in Energy and Agriculture, ed. by G. Stanhill. Springer Verlag, Berlin. 192 pp.

Odum, H. T. 1985a. Water conservation and wetland values. pp 98–111 in Ecological Considerations in Wetlands Treatment of Municipal Wastewaters, ed. by P. J. Godfrey, E. R. Kaynor, S. Pelezrski, and J. Benforado. Van Nostrand Reinhold, New York. 473 pp.

Odum, H. T. 1985b. Las nuevas energia y los ecologistas: Razonar o sentir. pp. 201–222 in El Hombre, la Sociedad, El Mundo. Prospectiva del Ano 2000. Fundacion Caja de Pensiones, Barcelona. 327 pp.

*Odum, H. T. 1986. EMERGY in Ecosystems. pp. 337–369 in Ecosystem Theory and Application, ed. by N. Polunin. Wiley, New York. 446 pp.

*Odum, H. T. 1986c. Unifying Education with Environmental Systems Overviews. pp. 181–199 in Environmental Science, Teaching and Practice ed. by R. Barrass, D. J. Blair, P. H. Garnham and A. O. Moscardini. Proceedings of the 3rd International Conference on the Nature and Teaching of Environmental Studies and Sciences in Higher Education, Sunderland Polytechnic, Sunderland September 1985.

*Odum, H. T. 1987a. Living with Complexity. pp. 19–85 in Crafoord Prize in the Biosciences, 1987, Crafoord Lectures, Royal Swedish Academy of Sciences, Stockholm.

*Odum, H. T. 1987b. Models for national, international, and global systems policy. Chap. 13, pp. 203–251 in Economic-Ecological Modeling, ed. by L. C. Braat and W. F. J. Van Lierop. Elsevier Science Publishing, New York. 329 pp.

*Odum, H. T. 1987c. What is a whale worth? The Siren—News from UNEP's Oceans and Coastal Areas Programme. 33(May):31–35.

*Odum, H. T. 1988. Self organization, transformity, and information. Science 242:1132–1139.

*Odum, H. T. 1989. EMERGY and evolution. pp. 10–18 in Preprints of the 33rd Annual Meeting of the International Society for the Systems Sciences, Edinburgh, Scotland, July 2–7, 1989. Vol. 3, ed. by P. W. J. Ledington.

*Odum, H. T. 1991a. Destruction and power in general systems. In Proceedings of the 35th Annual Meeting of the International Society for Systems Science, ed. by S. C. Holmberg and K. Samuelson. Ostersund, Sweden. Vol. 1, pp. 211–218.

Odum, H. T. 1991b. EMERGY and biogeochemical cycles. In Ecological Physical Chemistry, ed. by C. Rossi and E. Tiezzi. Proceedings of an International Workshop, November 1990, Siena, Italy, Elsevier Science Publishing, Amsterdam. pp. 25–65.

Odum, H. T. 1991c. Principles of EMERGY analysis for public policy. In Ecological Economics: Its Implications for Forest Management and Research, ed. by D. P. Bradley and P. O. Nilsson. Dept. of Operational Efficiency (Research Notes 223), Swedish Univ. of Agricultural Sciences, Faculty of Forestry, Garpenberg, Sweden. 242 pp.

Odum, H. T. 1992. Simulating ecological economic parameters. pp. 1–10 in Proceedings of the International Society for Systems Science Meeting, ed. by L. Peeno. Denver, Colo.

*Odum, H. T. 1994a. Ecological and General Systems: An Introduction to Systems Ecology. Univ. Press of Colorado, Niwot. 644 pp. Revised edition of Systems Ecology, 1983, Wiley, 644 pp.

Odum, H. T. 1994. Ecological economics. pp. 159–161 in The Encyclopedia of the Environment, ed. by R. A. Eblen and W. R. Eblen. Houghton-Mifflin, New York.

*Odum, H. T. 1994b. The EMERGY of natural capital. Chap. 12, pp. 200–214 in Investing in Natural Capital, ed. by A. Jansson, M. Hammer, C. Folke, and R. Costanza. Island Press, Washington, D.C. 504 pp.

Odum, H. T. 1995a. Energy systems concepts and self organization: A rebuttal. Oecologia, (in press).

Odum, H. T. 1995b. Self organization and maximum power. Chapter 28, pp. 311–364 in Maximum Power, ed. by C. A. S. Hall. Univ. Press of Colorado, Niwot, (in press).

Odum, H. T. 1995c. Tropical forest systems and the human economy. Pp. 343–393 in Tropical Forests: Management and Ecology, ed. by A. Lugo and M. Lowe. Ecological Studies 112, Springer-Verlag, New York.

Odum, H. T. 1995d. Tropical forest systems and the human interface. Chap. 14, 117 pp., in Tropical Forests, Centennial Volume of the Institute of Tropical Forestry, Rio Piedras, P.R., A. Lugo, ed. Springer Verlag, New York.

Odum, H. T. and J. F. Alexander, eds. 1977. Energy Analysis of Models of the United States. Annual Report to the Department of Energy. Environmental Engineering Sciences, Univ. of Florida, Gainesville, (Contract EY-76-S-05-4398). 457 pp.

*Odum, H. T., and J. E. Arding. 1991. EMERGY Analysis of Shrimp Mariculture in Ecuador. Report to Coastal Studies Institute, Univ. of Rhode Island, Narragansett. Center for Wetlands Univ. of Florida, Gainesville. 87 pp.

*Odum, H. T., and M. T. Brown, eds. 1975. Carrying Capacity for Man and Nature in South Florida. Final report to National Park Service, U.S. Dept. of Interior and state of Florida, Division of State Planning. Center for Wetlands, Univ. of Florida, Gainesville. 886 pp.

Odum, H. T., and S. Brown. 1976. Energy: the dynamo word. pp. 146–151 in Born of the Sun, ed. by J. E. Gill and B. R. Read. Florida Bicentennial Commission, Worth International Communications Corp., Hollywood, Fla.

Odum, H. T., and B. J. Copeland. 1972. Functional classification of the coastal ecological system of the United States. pp. 9–28 in Environmental Framework of Coastal Plain Estuaries, ed. by B. Nelson. Geological Society of America, Boulder, Colo., (Memoir 133),

*Odum, H. T., and D. Hornbeck. 1995. EMERGY Evaluation of Florida Salt Marsh and Its Contribution to Economic Wealth. Paper presented at Salt Marsh Conference, Florida A & M Univ., Tallahassee, 1989; in Florida Salt Marsh, ed. by C. L. Coultas.

Odum, H. T., and E. C. Odum. 1978. Japanese translation of first edition of Energy Basis for Man and Nature. Japan UNI Agency.

*Odum, H. T., and E. C. Odum. 1979. Energy system of New Zealand and the use of embodied energy for evaluating benefits of international trade. pp. 106–107 in Proceedings of Energy Modelling Symposium. Technical Publication 7. N.Z. Ministry of Energy, Wellington. 247 pp.

Odum, H. T., and E. C. Odum. 1981. Hombre y Naturaleza Bases Energeticas. Transl. by J. M. Serrano Farrera and R. M. Dauder. Ediciones Omega, Barcelona, Spain. 311 pp. Spanish ed. of Energy Basis for Man and Nature, published in 1976, McGraw-Hill.

*Odum, H. T., and E. C. Odum. 1976. Energy Basis for Man and Nature. McGraw-Hill, New York. 296 pp.

Odum, H. T., and E. C. Odum. 1982. Energy Basis for Man and Nature, 2nd ed. McGraw-Hill, New York. 331 pp.

*Odum, H. T., and E. C. Odum, eds. 1983. Energy Analysis Overview of Nations, with sections by G. Bosch, L. Braat, W. Dunn, G. de R. Innes, J. R. Richardson, D. M. Scienceman, J. P. Sendzimir, D. J. Smith, and M. V. Thomas. Working Paper. International Institute of Applied Systems Analysis, Laxenburg, Austria (WP-83-82). 469 pp.

Odum, H. T., and E. C. Odum. 1993. Energy Systems in Ecology. pp. 193–197 in Concise Encyclopedia of Environmental Systems, ed. by P. C. Young. Pergamon Press, New York.

Odum, H. T., and E. C. Odum. 1994. Environmental Minimodels and Simulation Exercises. Center for Environmental Policy, Univ. of Florida, Gainesville. 273 pp.

Odum, H. T., and N. Peterson. 1995. Simulating Energy Systems Models that Evaluate EMERGY with Blocks Object-programmed for EXTEND. (Manuscript submitted for publication to Ecol. Model.)

*Odum, H. T., and R. C. Pinkerton. 1955. Time's speed regulator: the optimum efficiency for maximum power output in physical and biological systems. Am. Sci. 43:321–343.

Odum, H. T., and D. M. Scienceman. 1986. Commonalities between hierarchies of ecosystems and political institutions. Yearbook of General Systems 29:23–32.

*Odum, H. T., B. J. Copeland, and E. A. McMahan. 1969. Coastal Ecological Systems of the United States. Report to Federal Water Pollution Control Administration. Dept. of the Interior. Dept of Environmental Science and Engineering, Univ. of North Carolina, Chapel Hill, N.C. 3 vols., 1405 pp.

Odum, H. T., B. J. Copeland, and E. A. McMahan. 1974a. Coastal Ecological Systems of the United States. Conservation Foundation, 1717 Massachusetts Ave. NW, Washington, D.C. 4 vols., 1976 pp.

*Odum, H. T., M. Sell, M. Brown, J. Zucchetto, C. Swallows, J. Browder, T. Ahlstrom, and L. Peterson. 1974b. Models of herbicide, mangroves, and war in Vietnam. Part B (302 pp.) in The Effects of Herbicides in South Vietnam. Working Papers, National Academy of Sciences, Washington, D.C.

Odum, H. T., M. T. Brown and R. Costanza. 1976a. Developing a steady state for man and land: energy procedures for regional planning. pp. 343–361 in Science for Better Environment. Proc. Int. Congress of Scientists on the Human Environment, Kyofo. Science Council of Japan. Asahi Evening News, Tokyo, Japan.

Odum, H. T., C. Kylstra, J. Alexander, N. Sipe, M. Brown, S. Brown, M. Kemp and others. 1976b. Net Energy Analysis of Alternatives for the U.S. Final Report for the Committee on Energy and Power of the U.S. Congress. Center for Wetlands, Univ. of Florida, Gainesville (CFW 76-19) 85 pp.

*Odum, H. T., C. Kylstra, J. Alexander, N. Sipe, and P. Lem. 1976c. Net energy analysis of alternatives for the United States. pp. 258–302 in Middle and Long-term Energy Policies and Alternatives. Hearings of Subcommittee on Energy and Power, 94th Congress (Serial No. 94-63). U.S.Government Printing Office, Washington, D.C.

*Odum, H. T., W. Kemp, M. Sell, W. Boynton, and M. Lehman. 1977a. Energy analysis and the coupling of man and estuaries. pp. 297–315 in Environmental Management, vol. 1, no. 4. Springer-Verlag, New York.

Odum, H. T., W. Kemp, M. Sell, W. Boynton, and M. Lehman. 1977b. Energy analysis and the coupling of man and estuaries. Environ. Manage. 1:297–315.

*Odum, H. T., T. Gayle, M. T. Brown, and J. Waldman. 1978. Energy analysis of the University of Florida. Center for Wetlands, University of Florida, Gainesville. Unpublished manuscript.

Odum, H. T., J. F. Alexander, F. Wang, M. T. Brown, M. Burnett, R. Costanza, P. Kangas, D. Swaney, S. Leibowitz, and S. Lemlich. 1979. Energy Basis of the United States. Report to Dept. of Energy (Contract FY-76-S-05-4398), Environmental Engineering Sciences, Univ. of Florida, Gainesville, 444 pp.

Odum, H. T., C. Kylstra, S. Brown, J. Zucchetto, T. Ballentine, W. Kemp and N. Sipe. 1980. Energy Transformation and the Economy of the United States. Sciences Distinguished Lecture Series. Louisiana State Univ., Baton Rouge. 50 pp.

Odum, H. T., P. Kangas, G. R. Best, B. T. Rushton, S. Leibowitz, and J. R. Butner. 1981. Studies on Phosphate Mining Reclamation and Energy. Center for Wetlands, Univ. of Florida, Gainesville. 142 pp.

*Odum, H. T., M. J. Lavine, F. C. Wang, M. A. Miller, J. F. Alexander, and T. Butler. 1983a. A manual for using energy analysis for plant siting with an appendix on energy analysis of environmental values. Nuclear Regulatory Commission (NUREG/CR-2443). National Technical Information Service, Springfield, Va. 242 pp.

*Odum, H. T., M. A. Miller, B. T. Rushton, T. R. McClanahan, and G. R. Best. 1983b. Interaction of Wetlands with the Phosphate Industry. Florida Institute of Phosphate Research. Center for Wetlands, Univ. of Florida, Gainesville (FIPR Publ. 03-007-025). 164 pp.

*Odum, H. T., W. T. Brown, and R. A. Christianson. 1986. Energy Systems Overview of the Amazon Basin. Report to The Cousteau Society. Center for Wetlands (Publication 86-1). Univ. of Florida, Gainesville. 190 pp.

*Odum, H. T., E. C. Odum, and M. Blissett, eds. 1987a. Ecology and Economy: EMERGY Analysis and Public Policy in Texas. L.B.J. School of Public Affairs and Texas

Dept. of Agriculture (Policy Research Publication 78). Univ. of Texas, Austin. 178 pp.

*Odum, H. T., F. C. Wang, J. F. Alexander, Jr., M. Gilliland, M. Miller, and J. Sendzimer. 1987b. Energy Analysis of Environmental Value. Center for Wetlands, Univ. of Florida, Gainesville. 97 pp.

*Odum, H. T., C. Diamond, and M. T. Brown. 1987c. Energy Analysis Overview of the Mississippi River Basin. Report to The Cousteau Society. Center for Wetlands (Publication 87-1), Univ. of Florida, Gainesville. 107 pp.

*Odum, H. T., E. C. Odum, M. T. Brown, D. Lahart, C. Bersok, J. Sendzimir, G. B. Scott, D. Scienceman. 1987d. Environmental Systems and Public Policy. Ecological Economics Program, Phelps Laboratory, Univ. of Florida, Gainesville. 253 pp.

*Odum, H. T., E. C. Odum, M. T. Brown, D. Lahart, C. Bersok, J. Sendzimir, G. B. Scott, D. Scienceman, and N. Meith. 1976e. Environmental Systems and Public Policy, Florida Supplement. Ecological Economics Program, Phelps Lab, Univ. of Florida, Gainesville.

*Odum, H. T., E. C. Odum, M. T. Brown, D. Lahart, C. Bersok, J. Sendzimir, G. B. Scott, D. Scienceman, and N. Meith. 1988. H. T. Odum: Energy, Environment and Public Policy: A Guide to the Analysis of Systems. United Nations Environment Programme, UNEP Regional Seas Reports and Studies (No. 95), Nairobi, Kenya. For Chinese Translation: see Lan (1993).

Odum, H. T., G. R. Best, M. A. Miller, B. T. Rushton, R. Wolfe, C. Bersok, and J. Feiertag. 1990. Accelerating Natural Processes for Wetlands Restoration after Phosphate Mining. Florida Institute of Phosphate Research, Bartow (FIPR 03-041-086). 408 pp.

Odum, H. T., B. T. Rushton, M. Paulic, S. Everett, T. McClanahan, M. Munroe, and R. W. Wolfe. 1991. Evaluation of Alternatives for Restoration of Soil and Vegetation on Phosphate Clay Settling Ponds. Florida Institute of Phosphate Research (FIPR 86-03-076R), Bartow. 184 pp.

*Odum, H. T., E. C. Odum, and M. T. Brown. 1993. Environment and Society in Florida. Center for Environmental Policy, Univ. of Florida, Gainesville. 446 pp.

Odum, W. P., E. P. Odum, and H. T. Odum. 1995. Nature's pulsing paradigm. Chapter in Proceedings of the International Estuarine Research Federation, Nov. 1993, Estuaries, (in press).

Odum, H. T. and N. Peterson 1995. Simulating energy systems models that evaluate EMERGY with blocks object-programmed for EXTEND. Manuscript submitted for publication.

O'Neill, R. V., R. H. Gardner, L. W. Barnthouse, G. W. Studer, S. G. Hildebrand, and C. W. Gehrs. 1982. Ecosystem risk analysis. Environ. Toxicol. Chem. 1:167–177.

Ostwald, W. 1907. The modern theory of energetics. Monist 17:511.

Patten, B. C. 1978. Systems approach to the concept of environment. Ohio J. Sci., 78:206–222.

*Patten, B. C. 1992. Energy, EMERGY and environs. Ecol. Model. 62:29–69.

Patten, B. C. 1994. Network integration of ecological extremal principles—Exergy, EMERGY, Power, Ascendency. (Manuscript submitted for publication to Ecological Modeling.)

*Patten, B. C. 1995. Environs, EMERGY, transformity and energy value. Chapter in Maximum Power, ed. by C. A. S. Hall. Univ. Press of Colorado, Niwot. (in press).

*Patten, B. C., M. Higashi, and T. P. Burns. 1990. Trophic Dynamics in Ecosystem Networks: Significance of Cycles and Storage. Ecol. Model. 51:1–28.

Paterson, D. 1978. Methane from the bowels of the earth. New Sci. (June):896–898.

*Paynter, H. M. 1960. Analysis and Design of Engineering Systems. MIT Press, Cambridge, Mass.

Peet, J. 1992. Energy and the Ecological Economics of Sustainability. Island Press, Washington, D.C. 309 pp.

Percebois, J. 1979. Is the concept of energy intensity meaningful? Energy Economics (July) 1:148–165.

Phillipson, J. 1966. Ecological Energetics. St. Martin's Press, New York, 57 pp.

*Philomena, A. 1991. Shrimp fishery: energy modelling as a tool for management. Ecol. Model. 52:61–71.

Pillet, G. 1987. Case-study of the role of environment as an energy externality in Geneva vineyard cultivation and wine production. Environ. Conserv. 14(1):53–58.

Pillet, G. 1992. Le comptes economiques de l'environnement. Office federal de la statistique, Soceiete pour la protection de l'environment, Geneva, Switzerland, 189 pp.

Pillet, G. 1993. Economie Ecologique. Georg Editeur SA, Geneva. 223 pp.

*Pillet, G., and H. T. Odum. 1984. Energy externality and the economy of Switzerland. Schweiz Zeitschrift for Volkswirtschaft und Statistik, 120(3):409–435.

Pillet, G., and T. Murota, eds. 1987. Environmental Economics. Roland Leimgruber, Geneva. 307 pp.

Pillet, G. and H. T. Odum. 1987. Energie, Ecologie, Economie, Georg Editeur SA, Geneva, Switzerland, 257 pp.

*Pimental, D. 1980. Handbook of Energy Utilization in Agriculture. CRC Press, Boca Raton, Fla. 475 pp.

*Pimentel, D., and M. Pimentel. 1979. Food, Energy, and Society. Halstead, Wiley, New York. 268 pp.

*Pimentel, D., L. E. Hurd, A. C. Bellotti, M. J. Forster, I. N. Oka, O. D. Sholes, and R. J. Whitman. 1973. Food production and the energy crisis. Science, 182:443–449.

Pimentel, D. G., and C. H. Hall. 1984. Food and Energy Resources. Academic Press, Orlando, FL., 268 pp.

Prigogine, I., and J. M. Wiaume. 1946. Biology of thermodynamique des phenomenes irreversibles. Experientia 2:451–453.

*Pritchard, L. 1992. The ecological economics of natural wetland retention of lead. M.S. Thesis, Environmental Engineering Sciences, Univ. of Florida, Gainesville. 138 pp.

Redfield, A. C. 1934. On the proportions of organic derivatives in sea water and their relation to the composition of plankton. pp. 176–194 in James Johnstone Memorial Volume, Liverpool Univ. Press, Liverpool, England.

*Regan, E. J. Jr. 1977. The natural energy basis for soils and urban growth in Florida. M.S. Thesis, Environmental Engineering Sciences, Univ. of Florida, Gainesville. 176 pp.

*Regan, E. J. Jr. 1978. Statistical Test of Energy Models for Prediction of Urban Growth in Florida. p. 91 in Systems in Sociocultural Development, ed. by K. Balkus. Proceedings of the Society for General Systems Research, Southeastern Meeting, Tallahassee, Fla.

Repetto, R. 1992. Accounting for Environmental Assets. Sci. Am. (June):94–100.

*Richardson, J. F. 1988. Spatial Patterns and Maximum Power in Ecosystems. Ph.D. Dissertation, Environmental Engineering Sciences, Univ. of Florida, Gainesville. 254 pp.

*Richardson, J. R., and H. T. Odum. 1981. Power and a pulsing production model. pp. 641–647 in Energy and Ecological Modeling, ed. by W. J. Mitsch, R. W. Bosserman, and J. M. Klopatek. Elsevier Press, Amsterdam.

*Ricklefs, R. E. 1979. Ecology, 2nd ed. Chiron Press, New York.

Riehl, H. 1970. Climate and Weather in the Tropics. Academic Press, New York. 611 pp.

Roberts, R. 1973. Energy Sources and Conversion Techniques. Amer. Sci. 61:66–75.

Romitelli, S. 1995. Geopotential and Chemical Energy of Watersheds. Ph.D. Dissertation, Environmental Engineering Sciences, Univ. of Florida, Gainesville, in preparation.

Ronov, A. B. 1982. The earth's sedimentary shell. Int. Geol. Rev. 24:1313–1331.

Rosnay, J.de. 1975. Le Macroscope, vers Une Vision Globale. Editions du Seuil, Paris. 295 pp.

Rossi, C. and E. Tiezzi, eds. 1994. Ecological Physical Chemistry. Elsevier, Amsterdam, 651 pp.

*Rushton, B. T. 1981. Preliminary energy analysis calculations of phosphate mining. ed. by H. T. Odum. pp. 91–118 in Studies on Phosphate Mining, Reclamation, and Energy, Center for Wetlands, Univ. of Florida, Gainesville. 142 pp.

Rushton, B. T. 1988. Wetland Reclamation of Accelerating Succession. Ph.D. Dissertation, Environmental Engineering Sciences, Univ. of Florida, Gainesville. 267 pp.

Ruth, M. 1994. Integrating economics, ecology, and thermodynamics. Kluwer Academic Publishers, Boston. 251 pp.

Ruttenber, A. J. 1979. Urban Areas as Energy Flow Systems. Ph.D. Dissertation, Emory Univ., Atlanta, Ga. 354 pp.

*Ryabchikov, A. 1975. The Changing Face of the Earth. Progress Publications, Moscow.

Science. 1990. When a radical experiment goes bust—R.A.K. Science 247:1177–1178.

*Scienceman, D. M. 1987. Energy and EMERGY. pp. 257–276 in Environmental Economics, ed. by G. Pillet and T. Murota. Roland Leimgruber, Geneva. 308 pp.

*Scienceman, D. M. 1989. The emergence of emonomics. Vol. 3, pp. 62–68 in Proceedings of the 1989 meeting of the ISSS in Edinburgh, Scotland. International Society for the Systems Sciences, Louisville, Ky.

Scienceman, D. M. 1992. Emvalue and Lavalue. 11 pp. in Proceedings of the 1992 meeting of the ISSS in Denver, Colo., ed. by L. Peeno. International Society for the Systems Sciences, Louisville, Ky.

Scienceman, D. M. 1993. The system of EMERGY units. pp. 214–223 in Proceedings of the ISSS meeting in Western Sydney, Australia, ed. by R. Packham. International Society for the Systems Sciences, Louisville, Ky.

Scienceman, D. M. 1995. EMERGY Synthesis: Collected Papers by David M. Scienceman. Center for Environmental Policy, Univ. of Florida, Gainesville. Draft manuscript, 275 pp.

*Scott, H. 1953. Introduction to Technocracy. J. Day, New York.

Schipper, L. 1976. Raising the productivity of energy utilization. Ann. Rev. of Energy 1:455–517.

*Sclater, J. F., G. Taupart, and I. D. Galson. 1980. The heat flow through the oceanic and continental crust and the heat loss of the earth. Rev. Geophys. Space Phys. 18:269–311.

*Sedlik, B. R. 1978. Some theoretical considerations of net energy analysis. pp. 245–261 in Symposium Papers: Energy Modeling and Net Energy Analysis, Institute of Gas Technology, Chicago. 795 pp.

*Sell, M. G. 1977. Modeling the response of mangrove ecosystems to herbicide spraying, hurricanes, nutrient enrichment, and economic development. Ph.D. Dissertation, Environmental Engineering Sciences, Univ. of Florida, Gainesville. 389 pp.

Shabman, L. A., and S. S. Batie. 1978. Economic value of natural coastal wetlands: a critique. Coastal Zone Manage. J. 4:231–237.

Shannon, C. E., and W. Weaver. 1949. Mathematical Theory of Communication. Univ. of Illinois Press, Urbana.

Shimazu, Y. 1976. Ecological approach in national planning: conversion into nature oriented Japan. pp. 335–343 in Science for a Better Environment. Proc. Int. Congress of Scientists on the Human Environment, Kyoto. Science Council of Japan. Ashai Evening News, Tokyo, Japan, 991 pp.

*Siegel, F. R. 1974. Applied Geochemistry. Wiley, New York.

Sillen, L. G. 1961. The ocean as a chemical system. Science 156:1189–1197.

Silvert, W. 1982. The theory of power and efficiency in ecology. Ecol. Modelling 15:159–164.

Simon, A. L. 1975. Energy Resources. Pergamon Press, New York. 164 pp.

*Simon, J. I., and H. Kahn. 1984. The Resourceful Earth: A Response to Global 2000. Basic Blackwell, Oxford, England. 585 pp.

Singh, R. P., ed. 1990. Energy in Food Processing. Elsevier, Amsterdam. 375 pp.

*Sipe, N. G. 1978a. A Historic and Current Energy Analysis of Florida. M.S. Thesis, Univ. of Florida, Gainesville. 94 pp.

*Sipe, N. G. 1978b. An Overview Energy Analysis of Sarasota County. Urban and Regional Planning (Publication 12)., Univ. of Florida, Gainesville, 53 pp.

Sipe, N. G., D. P. Swaney, and J. McGinty. 1979. Energy Basis for Hillsborough County. Center for Wetlands, Univ. of Florida, Gainesville. 112 pp.

Slack, E. B. 1979. Energy cost of fishing. Catch 6:3–5.

Slesser, M. 1973. Energy subsidy as a criterion in food policy planning. J. Sci. Food Agric. 24:1193–1207.

*Slesser, M. 1978. Energy in the Economy. St. Martin's Press, New York. 162 pp.

Slesser, M., and C. Lewis. 1979. Biological Energy Resources. Wiley, New York.

Smil, V. 1976. China's Energy: Achievements, Problems, Prospects. Praeger, New York.

Smil, V. 1983a. Biomass Energies. Plenum, New York. 453 pp.

Smil, V. 1983b. Energy Analysis and Agriculture. Westview Press, Boulder, Colo., 191 pp.

*Smil, V. 1991. General Energetics, Energy in the Biosphere and Civilization. Wiley, New York. 367 pp.

*Smith, R. F. 1970. The vegetation structure of a Puerto Rican rain forest before and after short-term gamma irradiation. pp. D103–D133 in A Tropical Rain Forest, ed. by H. T. Odum and R. F. Pigeon. Atomic Energy Commission, Division of Technical Information (TID 2470), Oak Ridge, Tenn.

Smith, P. J. 1973. Topics in Geophysics. MIT Press, Cambridge, Mass.

Smith, C. C. 1976. When and how to reproduce, the tradeoff between power and efficiency. Am. Zool. 16:763–764.

Smith, G. A. 1980. The teleological view of wealth: A historical perspective. Chap. 14, pp. 215–237 in Economics, Ecology, Ethics: Essays toward a Steady-State Economy, ed. by J. E. Day, New York.

Smith, B., P. Leung, and G. Love. 1986. Intensive Grazing Management, Forage, Animals, Men Profits. Graziers Hui, Kamuela, Hawaii, 350 pp.

*Sollins, N. 1970. Comparison of species diversity at El Verde with some other ecosystems. pp. 1245–1247 in A Tropical Rain Forest, ed. by H. T. Odum and R. F. Pigeon. Atomic Energy Commission, Division of Technical Information (TID 2470), Oak Ridge, Tenn.

Soma, J. 1985. The new energy hyper-equation and its implications. Energy Engineering 92:62–71.

Soma, J. 1988a. Thermodynamic Mass Equivalence. Energy Convers. Management 28:241–243.

Soma, J. 1988b. The kernal role of the Soma number in deriving De Facto energy efficiencies and costs from casual ones. Energy Convers. Management 28:271–274.

Soma, J. 1991b. The energy ecology derived from thermodynamic mass equivalence reveals critical energy reliability faults. Energy Convers. Management 31:567–584.

Soma, J. 1993. Extensions of the natural energiomaterial Ecology. Energy Convers. Management 34:169–185.

*Spreng, D. T. 1988. Net-Energy Analysis and the Energy Requirements of Systems. Praeger, New York. 289 pp.

*Stanhill, G., ed. 1984. Energy and Agriculture. Springer-Verlag, Berlin. 192 pp.

*Steele, J. H. 1985. A comparison of terrestrial and marine ecological systems. Nature 313:355–358.

Steinhart, J. S., and C. E. Steinhart. 1974. Energy Sources: Use and Role in Human Affairs. Duxbury Press, North Scituate, Mass. 362 pp.

*Stellar, D. L. 1976. An energy evaluation of residential development alternatives in mangroves. M.S. Thesis, Environmental Engineering Sciences, Univ. of Florida, Gainesville. 124 pp.

*Stone, J. H., and S. M. S. Wong. 1991. The environment and economy of Hong Kong and Singapore via EMERGY Analysis. Final Report of HKP Project (340/420), Dept. of Civil and Structural Engineering, Hong Kong Polytechnic, Hong Kong. 168 pp.

Sundberg, U. 1984. Ed. by B. Andersson and S. Falk. The charcoal burner at Bastjarnm. pp. 16–17 in Forest Energy in Sweden. Swedish Univ. of Agricultural Sciences, Garpenberg. 111 pp.

*Sundberg, U., H. T. Odum, and S. Doherty. 1991. 18th Century charcoal production. Uppsatser och Resultat nr 212. Swedish Univ. of Agricultural Sciences, Garpenberg.

*Sundberg, U. 1992. Ecological economics of the Swedish Baltic empire–an essay on energy and power, 1560–1720. J. Ecol. Econ. 5:51–72.

*Sundberg, U., J. Lindegren, H. T. Odum, and S. Doherty. 1994a. Forest EMERGY basis for Swedish Power in the 17th Century. Scandinavian Journal of Forest Research, Suppl. pp. 1–50.

*Sundberg, U., J. Lindegren, H. T. Odum, and S. Doherty. 1994b. Skogens Änvandning Och Roll Under Det Svenska Stormaktsvaldet Perspektiv PÅ Energi Och Makt. Kingl. Skogs-Och Lantbruksakamemien. pp. 1–62.

*Sussman, M. V. 1980. Avilability (Exergy) Analysis. Milliken House, Lexington, Mass. 99 pp.

Suter, F. W., ed. 1993. Ecological Risk Assessment. Lewis Publ., Boca Raton, Fla. 538 pp.

*Swaney, D. P. 1978. Energy Analysis of Climatic Inputs to Agriculture. M.S. Thesis, Dept. of Environmental Engineering Sciences, Univ. of Florida, Gainesville. 198 pp.

*Sweden National Energy Administration. 1990. Energy in Sweden. Stockholm. 30 pp.

Templeton, G. J. 1981. A Systems Approach to Inshore Fishery Management Options. M.S. Thesis, Joint Centre for Environmental Sciences, Univ. of Canterbury and Lincoln College, Christchurch, N.Z. 87 pp.

*Tennenbaum, S. 1988. Network Energy Expenditures for Subsystem Production. M.S. Thesis, Environmental Engineering Sciences, Univ. of Florida, Gainesville. 132 pp.

*Texas Dept. of Agriculture. 1982, 1984. Field Crop Stat. Austin. 100, 103 pp.

Thayer, G. W., W. E. Schaaf, J. W. Angelovic, and M. S. Lacroix. 1973. Caloric measurements of some estuarine organisms. Fish. Bull. 71:289–296.

Tiezzi, E. N., Marchettini, and S. Ulgiati. 1991. Integrated agro-industrial ecosystems: an assessment of the sustainability of a cogenerative approach to food, energy, and chemical production by photosynthesis. pp. 459–473 in Ecological Economics, ed. by R. Costanza. Columbia Univ. Press, New York.

*Todd, D. K., ed. 1970. The Water Encyclopedia. 2nd ed., Lewis Pub., Chelsea, Mich.

*Toebes, C. 1972. Surface water resources of New Zealand. pp. in The 12th Pacific Science Congress, Wellington, N.Z.

*Ton, S. S. 1990. Natural Wetland Retention of Lead from a Hazardous Waste Site. M.S. Thesis, Environmental Engineering Sciences, Univ. of Florida, Gainesville. 89 pp.

*Ton, S. S. 1993. Lead Cycling through a Hazardous Waste-Impacted Wetland. Ph.D. Dissertation, Environmental Engineering Sciences, Univ. of Florida, Gainesville. 149 pp.

*Tribus, M. 1961. Thermostatics and Thermodynamics. Van Nostrand, New York. 641 pp.

*Tribus, M., and E. D. McIrvine. 1971. Energy and information. Sci. Am. 225:179–188.

Tseng, G. H., My Kao, and J. C. Yang. 1991. Optimal design of a pilot OTEC Power plant in Taiwan. Trans of ASME 113 (Dec.):284–299.

Turner, M. G., E. P. Odum, R. Costanza, and T. M. Springer. 1988. Market and nonmarket values of the Georgia landscape. Envir. Mgmt. 12(2):209–217.

Ulanowicz, R. 1980. An hypothesis on the development of natural communities. J. Theor. Biol. 85:223–245.

Ulanowicz, R. E. 1983. Identifying the structure of cycling in ecosystems. Math. Bioscie. 65:219–237.

*Ulanowicz, R. E. 1986. Growth and Development: Ecosystems Phenomenology. Springer-Verlag, New York. 203 pp.

Ulanowicz, R. E., and W. M. Kemp. 1979. Toward canonical trophic aggregations. Am. Nat. 114:871–883.

Ulgiati, S., H. T. Odum, and S. Bastianoni. 1992. Modeling Interaction between Environment and Human Society in Italy: An EMERGY Analysis. pp. 1121–1133 in Proceedings of the 36th Annual Meeting of ISSS in Denver, Colo., International Society for System Sciences, Louisville, Ky., ed. by L. Peeno.

*Ulgiati, S., H. T. Odum, and S. Bastianoni. 1993. EMERGY analysis of Italian agricultural system. The role of energy quality and environmental inputs. pp. 187–215 in Trends in Ecological Physical Chemistry, ed. by L. Bonati, U. Cosentino, M. Lasagni, G. Moro, D. Pitea, and A. Schiraldi. Elsevier, Amsterdam.

*Ulgiati, S., H. T. Odum, and S. Bastianoni. 1994. EMERGY use, environmental loading and sustainability. An EMERGY Analysis of Italy. Ecol. Model. 73:215–268.

Ulgiati, S., M. T. Brown, S. Bastianoni, and N. Marchettini. 1995. EMERGY based indices and ratios to evaluate the sustainable use of resources. Ecological Engineering. In Press.

United Nations 1981. Statistical Yearbook. United Nations, New York.

*United States. 1982, 1984, 1985, 1990, 1994. Statistical Abstract. Government Printing Office, Washington, D.C.

U.S. Coast and Geodetic Survey. 1956. Wave Height Tables. Washington D.C., U.S. Dept. of Commerce.

U.S. Dept. of Commerce, National Bureau of Standards. 1981. The International System of Units (SI) (NBS Special Publication 330). 48 pp.

U.S. Dept. of Commerce. 1984. Foreign Economic Trends and Their Implications for the United States: South Africa. International Trade Administration (FET 84-20), Washington, D.C.

U.S. Dept. of Interior. 1983. Mineral Facts and Problems. Washington, D.C.

*Van Moeske, P. 1981. Aluminum and Power: The Second Smelter Report. Economics Discussion Papers. Univ. of Otago, Denedin, N.Z. 45 pp.

*Veizer, J. 1988a. The earth and its life: systems perspective. Origins Life Evol. Biosphere 18:13–39.

*Veizer, J. 1988b. The evolving exogenic cycle. Chapter 5, pp. 175–219 in Chemical Cycles in the Evolution of the Earth, ed. by C. B. Gregor, R. M. Garrels, F. T. Mackenzie, and J. B. Maynard. Wiley, New York. 276 pp.

*Veizer, J., P. Laznicka, and S. L. Jansen. 1989. Mineralization through geologic time: recycling perspective. Am. J. Sci. 289:484–524.

Vernadsky, V. 1929. The Biosphere. Abridged and reprinted, 1986, Synergetic Press, London.

Vernadsky, W. I. 1944. Problems of biogeochemistry II. The fundamental matter-energy difference between the living and inert natural bodies of the Biosphere. Trans. Conn. Acad. Arts Sci. 33:483–517.

*Von der Haar, T. H., and V. E. Suomi. 1969. Satellite observations of the earth's radiation budget. Science 169:657–659.

Walker, R., and S. Bayley. 1976a. An energetic and economic simulation of the evaluation of transportation policy. Center for Wetlands (CFW-76-30), Univ. of Florida, Gainesville. 41 pp.

*Walker, R., and S. Bayley. 1976b. Quantitative assessment of natural values in benefit-cost analysis. Report to Florida Dept. of Transportation. Center for Wetlands, Univ. of Florida, Gainesville. 23 pp.

Wang, F. C. 1975. Application of Energy Flow Technique to Watershed Model. Center for Wetlands (CFW 75-23), Univ. of Florida, Gainesville. 27 pp.

Wang, F. C., and H. T. Odum. 1979. The Evaluation of Environmental Services as an Objective in Planning. Center for Wetlands (CFW 79-11), Univ. of Florida, Gainesville. 7 pp.

Wang, F. C., and Y. H. Wang. 1979. Energy Analysis and Water Value in Water Resources Planning. Abstract prepared for the 18th International Association for Hydraulic Research Congress, Calgiari, Italy. Center for Wetlands (CFW-75-37), Univ. of Florida, Gainesville. 4 pp.

Wang, F. C., H. T. Odum, and M. E. Lehman. 1977. Concepts and techniques for evaluation of energy-related water problems. Final Report to Office of Water Research and Technology (Grant 14-34-0001-6236 [C-7516]). Center for Wetlands, Univ. of Florida, Gainesville. 62 pp.

Wang, F. C., H. T. Odum, and R. Costanza. 1978. Concepts for Assessment of Energy Related Impacts of Water. American Society of Civil Engineers (Preprint 3246), Pittsburgh, PA. 27 pp.

Wang, F. C., H. T. Odum, and R. Costanza. 1979. Energy Criteria for Water Use. J. of the Water Res. Plan. and Man. Div. ASCE, vol. 105, No. WR1, Proc. Paper 15266, Mar.

*Wang, F. C., H. T. Odum, and P. C. Kangas. 1980. Energy analysis for environmental impact assessment. J. Water Resour. Plan. Manage. Div. 106:451–465.

Watt, K. 1973. Land Use, Energy Flow and Decision Making in Human Society. Univ. of California, Davis.

Watt, K. E. F. 1989. Evidence for the role of energy resources in producing long waves in the United States economy. Ecol. Econ. 1(2):181–195.

Watt, K. E.F., L. F. Molly, C. K. Varshney, D. Weeks, and S. Wirosard-jono 1977. The unsteady State, Environmental Problems, Growth and Culture. Univ Press of Hawaii, Honolulu, 287 pp.

Watt, K. E. F. 1992. Taming the Future. Contextured Web Press, Davis, Calif. 232 pp.

Weyl, P. K. 1970. Oceanography. Wiley, New York.

Whipple, S. J., and B. C. Patten. 1993. The problem of nontrophic processes in trophic ecology: toward a network unfolding solution.

*Whitfield, D. F. 1994. EMERGY Basis for Urban Land Use Patterns in Jacksonville, Florida. Master's Thesis, Dept. of Landscape Architecture, Univ. of Florida, Gainesville. 224 pp.

Williams, M. H. 1989. EMERGY Analysis of Cocaine and the Major Cocaine Producing Nations: Peru, Columbia and Boliva. Unpublished report for the class Energy and the Environment, Dept. of Environmental Engineering Sciences, Univ. of Florida, Gainesville. 12 pp.

Winarsky, I. 1979. Art, Information, and Energy. Essay for President's Award, Univ. of Florida, Gainesville. 102 pp.

Winberg, G. G. 1956. Rate of Metabolism and Food Requirements of Fish. Canada Translation Series No. 194. Fisheries Research Board,

Winteringham, F. P. W. 1992. Energy Use and the Environment. Lewis Publishers, Boca Raton, Fla. 175 pp.

*Woithe, R. 1992. EMERGY Analyses of the T/V Exxon *Valdez* Oil Spill and Alternatives for Oil Spill Prevention. Master's Thesis, Environmental Engineering Sciences, Univ. of Florida, Gainesville. 128 pp.

*Woithe, R. 1994. EMERGY Evaluation of the United States Civil War. Ph.D. Dissertation, Environmental Engineering Sciences, Univ. of Florida, Gainesville. 206 pp.

*Wulff, F., J. G. Field, and K. H. Mann, eds. 1989. Network Analysis in Marine Ecology. Coastal and Estuarine Studies, Springer-Verlag, New York. 284 pp.

*Wylie, P. J. 1971. The Dynamic Earth. Wiley, New York. 415 pp.

*Young, D., H. T. Odum, J. Day, and T. Butler. 1974. Evaluation of regional models for alternatives in management of the Atchafalaya Basin. Report to U.S. Dept. of the Interior, Bureau of Sport Fisheries and Wildlife. 31 pp.

Zucchetto, J. J. 1975a. Energy Basis for Miami, Florida, and Other Urban Systems. Ph.D. Dissertation, Univ. of Florida, Gainesville. 248 pp.

Zucchetto, J. J. 1975b. Energy-economic theory and mathematical models for combining the systems of man and nature, case study: The urban region of Miami, Florida. Ecol. Model. 1:241–268.

Zuchetto, J. 1983. Energy and the future of human settlement patterns: Theory, models and empirial considerations. Ecol. Modelling 20:85–111.

Zucchetto, J. 1981. Energy diversity of regional economies. pp. 543–548 in Energy and Ecological Modelling, ed. by W. J. Mitsch, R. W. Bosserman, and J. M. Klopatek. Elsevier, Amsterdam.

Zucchetto, J. 1984. Relationship between total energy consumption and economic activity for the Florida economy. Fla. Sci. 47:145–153.

Zuchetto, J., and S. Brown. 1976. Comparison of fossil fuel energy requirements of solar, natural gas, and electrical water heating systems. Resources, Recovery, and Conservation (Elsevier) 2:283–300.

Zucchetto, J., and A. Jansson. 1979a. A multi-level analysis of the renewable and non-renewable energy flows on Gotland, Sweden. pp. 1063–1067 in Modeling and Simulation. Proceedings of the 10th Annual Conference, Instrument Society of America, Pittsburgh.

Zucchetto, J., and A. Jansson. 1979b. Integrated regional energy analysis for the island of Gotland, Sweden. Environ. Planning A, 11:919–942.

Zucchetto, J., and A. Jansson. 1979c. Total energy analysis of Gotland's agriculture: A northern temperate zone case study. Agro-ecosystems 5:329–344.

Zuchetto, J., and G. Bickle. 1984. Energy and nutrient analyses of a dairy farm in central Pennsylvania. Energy in Agr. 3:29–47.

Zuchetto, J., and A. Jansson. 1981. System analysis of present and future energy/economic developments on the island of Gotland, Sweden. pp 487–494 in Energy and Ecological Modelling, ed. by W. J. Mitsch, R. W. Bosserman, and J. M. Klopatek. Elsevier, Amsterdam.

*Zuchetto, J., and A. Jansson. 1985. Resources and Society. Springer-Verlag, New York. 246 pp.

*Zucchetto, J., S. Bayley, L. Shapiro, D. Mau, and J. Nessel. 1980a. The direct and indirect energy costs of coal transport by alternative bulk commodity modes. Resour. Recovery Conserv. 5:161–177.

Zucchetto, J., A. Jansson, and K. Furugane. 1980b. Optimization of economic and ecological resources for regional design. Resour. Manage. Optimization 1(2):111–143.

Zwick, P. D. 1985. Energy Systems and Inertial Oscillation. Ph.D. Dissertation, Environmental Engineering Sciences, Univ. of Florida, Gainesville. 240 pp.

INDEX